ALSO BY GARY GREENBERG

The Noble Lie
The Self on the Shelf

MANUFACTURING DEPRESSION

The Secret History of a Modern Disease

Gary Greenberg

Simon & Schuster
NEW YORK · LONDON · TORONTO · SYDNEY

Simon & Schuster
1230 Avenue of the Americas
New York, NY 10020

First Simon & Schuster hardcover edition February 2010

SIMON & SCHUSTER and colophon are registered trademarks
of Simon & Schuster, Inc.

For information about special discounts for bulk purchases,
please contact Simon & Schuster Special Sales at
1-866-506-1949 or business@simonandschuster.com.

The Simon & Schuster Speakers Bureau can bring authors
to your live event. For more information or to book an event
contact the Simon & Schuster Speakers Bureau at
1-866-248-3049 or visit our website at www.simonspeakers.com.

Text designed by Paul Dippolito

Manufactured in the United States of America

1   3   5   7   9   10   8   6   4   2

Library of Congress Cataloging-in-Publication Data
Greenberg, Gary, date.
Manufacturing depression: the secret history of a modern disease / Gary Greenberg.
p.   cm.
Includes bibliographical references.
1. Depression, Mental—United States—History. 2. Antidepressants—
United States—History. I. Title.
[DNLM: 1. Depression—history—United States. 2. Antidepressive Agents—
history—United States. 3. Depressive Disorder—history—United States.
4. Psychotherapy—history—United States. WM 11 AA1 G795m 2010]
RC537.G722 2010
362.2'5—dc22   2009024310

ISBN 978-1-4165-6979-4
ISBN 978-1-4165-7008-0 (ebook)

Ring the bells that can still ring
Forget your perfect offering
There is a crack, a crack in everything
That's how the light gets in
—*Leonard Cohen, "Anthem"*

CONTENTS

MANUFACTURING
DEPRESSION

CHAPTER 1

MOLLUSKS

When Betty Twarog opens the door to her cavernous rooms at the University of Maine's Darling Marine Laboratory, you're smacked in the face with mist and the smell of brine, and the sound of water everywhere. Pumped out of Boothbay Harbor, it hisses and sprays and gurgles through pipes overhead and sluiceways underfoot, flowing through huge dark tanks full of sea urchins and starfish and other gnarly marine creatures before pouring back into the harbor. With a finger raised to her lips and a sharp shake of her head, she shushes the questions I shout over the din. At first I think she is afraid I will disturb her spat, the baby clams and scallops gestating in the bucket she's leaning over. But, she later explains, her job—to measure out precise portions of the three algae concoctions that are bubbling in tall plastic tuns in an adjacent room and feed them to her tiny charges—requires her total focus. So for a half hour, she attends to her task with silent concentration. A slightly built woman with ramrod-straight posture and long dark hair drawn back tightly from a dramatic widow's peak, she moves with the fluid grace of someone who has been doing chores like these for just over a half century. You wouldn't know it to look at her, but Betty Twarog is seventy-seven years old.

Something else you wouldn't know as she tends her mollusks is that Betty Twarog made one of the most important scientific discov-

1

eries of the twentieth century, one that changed the course of neu-roscience and medicine and set off a revolution in the way we think of ourselves. In 1952, when she was a twenty-five-year-old woman in a man's world, armed with nothing but a fresh Ph.D. and a hunch about an old scientific mystery, Twarog discovered serotonin in the brain and laid the cornerstone of the antidepressant revolution.

That's not what she had in mind. All she really wanted to do was to answer a question first posed in 1884 by Ivan Pavlov—yes, *that* Ivan Pavlov—when he took a brief excursion into the world of invertebrates. Pavlov, on a postgraduate fellowship in Leipzig, was trying to figure out the secrets of digestion. In large part, mov-ing food along the alimentary tract is a matter of smooth muscle functioning, and Pavlov decided to investigate the byssus retractor, the smooth muscle that *Mytilus edulis,* the common mussel, uses to close its shell. He was particularly interested in how it was possible for the creature to hold its shell shut against the outside world with-out expending far more energy than it could possibly take in.

His interest in this question didn't last long, and in the single paper he published on the subject before resuming the inquiries that led to his Nobel Prize (and to his eventual fascination with the salivation reflex in dogs) he offered only the merest hint of an answer. Seventy years later, Betty Twarog, for reasons she can't quite explain, found the remaining mystery irresistible. And she thought she had the answer, but it was too fantastic, too off the charts to be credible—until Abbott Pharmaceuticals just happened to mail her the means to check out her hunch.

Abbott had offered samples of a compound it had just synthe-sized to leading scientists around the country, including John Welsh, Twarog's mentor at Harvard. The molecule didn't have a name yet, or, more accurately, it had a number of them. Chemists called it 5-hydroxytryptamine after its molecular structure. Some biologists were calling it enteramine because they had found it in the guts of squid and octopi, while the biologists who had found it in blood called it serotonin. Abbott wanted the scientists to use their free samples to figure out what exactly the stuff was, what it did, and how it could be used. The company was hoping to find a way to make a drug, or a target for drugs, out of the new compound. They had no idea what they had stumbled upon.

But Twarog did, or so she believed. Pavlov, she thought, had gone much farther toward a solution than he knew. "It's perfectly beautiful," Twarog told me, "because to this day his paper summarizes the control of these muscles. He insisted that they contract under nervous stimulation and that they hold that contraction until they are signaled by relaxing nerves that turn it off." The mussel, that is, didn't clamp down its byssus retractor and then squeeze it tight like you or I would clench our fist around a quarter; instead, Twarog hypothesized, it closed the shell and threw a lock, which remained latched until a signal opened it like a key.

Twarog, unlike Pavlov, had the benefit of a discovery made in 1921 by a German scientist, Otto Loewi. Loewi wondered exactly how nerves signaled muscles—in particular, whether the process was purely electrical or somehow mediated by chemicals. He claimed that an answer came in a dream on Easter night. He sprang out of bed and rushed to his lab, where he cut the hearts out of two frogs and bathed them separately in salt water. Dissected hearts in saline will continue to beat, and Loewi had left intact the nerves that control the pulse rate—the vagus nerve, which slows it, and the accelerator nerve, which does what you think it does. He sent an electric charge from a battery into the vagus nerve; the heart slowed, just as he expected. But then he took the salt water from that bath and dripped it into the other heart's solution. When that heart slowed without any electrical stimulation, Loewi concluded that a chemical released from the vagus nerve and into the saline, and not electricity, had slowed down the heart. He repeated the experiment on the accelerator nerve, with the same result, and by 5:00 a.m. on Easter Monday had proved the principle of chemical neurotransmission.

By the time Twarog became intrigued by her mussels, Loewi's principle had been firmly established, but most scientists had settled into the belief that Loewi's chemicals—acetylcholine and epinephrine—were the only two neurotransmitters in the body. Twarog, however, was sure that there had to be another—the one that the mussel used to lock and unlock its shell—and she had a hunch that it was the chemical Abbott had sent.

In May 1952, Twarog and Welsh laid out the mussels on a lab bench. As soon as Abbott's serotonin hit them, the byssus retractors retracted. Twarog was right. Serotonin was the missing neurotransmitter.

As disturbing as the news of a new neurotransmitter might have been to scientific orthodoxy, Twarog's next idea was downright heretical. She said that serotonin would be found in the mammalian brain, which meant, of course, the human brain. At the time most biologists believed that humans were different from the rest of the animal kingdom, and the brain different from the rest of the body. In particular, they thought that electrical signals leapt around the brain like sparks, a throwback perhaps to René Descartes' idea that the pineal gland sent out ethereal messengers bearing the soul's instructions to the body.

Twarog thought this kind of reasoning was "sheer intellectual idiocy." It didn't make scientific sense—"what was the difference really between the brain and the rest of the body?" she says, still incredulous after all these years. "This is how nerves worked, no matter where they are." And, maybe more important, it didn't make philosophical sense either. "You know Tennyson's poem 'Flower in the Crannied Wall'?" She quoted from memory: "'Little flower— but if I could understand / What you are, root and all / and all in all, I should know what God and man is.' This is how it had to be."

Two years later, Twarog moved to Ohio to follow her husband to a university job. Restless, she applied for a position with Irvine Page, a Cleveland Clinic doctor who was trying to understand the role of serotonin in regulating blood pressure. On the day of her interview, it was pouring rain and, she recalls, "I looked like something the cat had dragged in." Still dripping on Page's floor, Twarog described her ideal job: a lab, an assistant, and the time to study the distribution of serotonin in the brain. He grilled her—after all, her hypothesis went against everything he'd been taught about the nervous system—but finally agreed to give her the bench space and a technician. Within a year, she had found serotonin in the brains of rats, dogs, and monkeys.

Twarog's first paper—the one about her experiment at Harvard—didn't get published until 1954. She didn't even hear back from the editor of the *Journal of Cell Physiology*—Detlev Bronk, the president of Johns Hopkins University—until John Welsh, the Harvard professor, called to inquire about the status of the article. Bronk told him that he wasn't about to ask his peers to review a speculative article by an unknown girl on such an important topic. While the paper was moldering on Bronk's desk, other scientists, much more prominent than Twarog, were arriving at a similar conclusion about serotonin. Once they had published their findings, it was safe for Bronk to let the girl have her say. Her paper with Irvine Page on cerebral serotonin also had to wait until the big boys said it first. But today no one disputes that she was the first with both discoveries.

Betty Twarog soon returned to marine biology, her first love. But many of the others went on to figure out the biology of neurotransmission, establishing within a decade that electricity really didn't fly from neuron to neuron like angels, that the brain really ran on chemicals like the rest of the body. And more than a half century later, new neurotransmitters are still turning up under the microscope, the subtleties of their metabolism still emerging.

None of these discoveries would be of much interest outside the lab were it not for some chance observations made in the early 1950s—that, for instance, an antitubercular drug that had induced an unexpected (although not unwelcome) euphoria inhibited an enzyme that breaks down serotonin, or that lysergic acid diethylamide (LSD), already famous for its profound effects on consciousness, has a chemical structure similar to serotonin. Out of these and other

findings, scientists began to cobble together a theory: that mental illness in general and depression in particular are caused by imbalances
in neurotransmitters, and especially in serotonin. This theory was of
obvious interest to pharmaceutical companies, and by 1958 drugs had
come to market designed to cure depression by fixing these supposed
imbalances. In 1988, Prozac was introduced, and by 2005, the last year
for which reliable figures are available, 27 million Americans—10 percent of the adult population—were taking antidepressants, most of
which act on serotonin, at an annual cost of more than ten billion dollars. It was a success far beyond anything Abbott could have dreamed
of when they sent their serotonin to John Welsh's lab.

This is how the best science stories start—with a chance discovery that leads to a vast change in our everyday lives. Take the brilliant insight, the dogged determination, and the sheer good fortune
behind an achievement like Betty Twarog's, throw in the poignant
contrast between her anonymity and the significance of the knowledge she uncovered, add to it some gee-whiz-interesting science,
and the next thing you know you have not only a great tale, but also
an excellent example of the way that scientists take us up toward
Parnassus—in this case the heights of happiness and health. A good
science story can make you feel even better about progress and the
prospects for humankind.

That's not the kind of story I'm going to tell in this book.

The invention of antidepressants is not the kind of achievement
that contributes unambiguously to the betterment of our species.
You probably already suspect that. Unless you've been living off the
grid for the last half century, and especially the past two decades,
you already know that serotonin has become a household word
and Prozac and its chemical cousins—known collectively as selective serotonin reuptake inhibitors, or SSRIs—have become staples
of the American medicine chest. And you know about the contro-
versies that have ensued. You've had conversations about them with

friends or family—or with yourself, when you wondered if your unhappiness or worry were signs of the disease we call depression, or when your doctor wrote a prescription for you and you hesitated to fill it, or when you took the pills, felt better, and wondered what that meant about you. And you've probably found that these discussions leave you just as confused as you were before. One thing antidepressants don't do is end confusion about antidepressants. You'd need a different drug to do that.

You'll also need a different book to do that. I'm not going to end this confusion for you. In part, that's because my subject is not the drugs so much as the condition they purport to treat, the disease of depression. But it's also because ongoing uncertainty is a hazard of reading a book by an old-fashioned psychotherapist like me, who believes that when it comes to important and complex questions, the best approach is to leave yourself in doubt for as long as possible, to live with inner conflict rather than to end it, to withstand yourself rather than to become someone different, to understand how you arrived at an important juncture rather than strike out down a road simply for the sake of getting on with life.

In this case, the crossroads that we've all arrived at is as crucial as it gets, and what I will do in this book is to show you how we arrived here, how we got to a point in our history where it is common, if not mandatory, to think of our unhappiness as a disease. And I'm going to do something else here: to try to convince you that what is at stake with antidepressants and the disease they treat isn't only the question of whether or not to take drugs for our unhappiness, or even whether or not it's really a good idea to call our unhappiness clinical depression. What's at stake is who we are, what kind of people we want to be, what we think it means to be human.

If that seems like a stretch, then you should listen to Peter Kramer.

One of the strangest things about the antidepressant revolution, and one indication that more is going on here than biochemistry, is

that the drugs that started it—the SSRIs, which first appeared in the
United States in 1988—are no more effective at treating depression
than the generation of drugs invented in the immediate aftermath
of Betty Twarog's discovery. And that's not very effective. Nearly
half the time, the drugs fail to outperform placebos in clinical trials. In real life (which generally lasts longer than a clinical trial and
allows for modifications in dosage and brand), they seem to make
a positive difference in perhaps 60 percent of the people who take
them. You would think that if depression were really biochemical in
nature and the drugs were really targeted at the culprit, they'd work
better than that. Of course, the first part of that statement remains
speculative: despite their best efforts—and notwithstanding what
doctors tell their patients when they prescribe them antidepressants—scientists have yet to find a single brain anomaly that is correlated with all cases of depression, let alone one that causes it.

There are many reasons that antidepressants took hold despite
these inconvenient truths, but one of the most important factors
in their ascent was Kramer's *Listening to Prozac,* which began to fly
off of bookstore shelves in the mid-1990s, about the same time that
Prozac prescriptions began to fly off of doctors' pads. Kramer managed to articulate something that all of us—patients, their families
and friends, doctors, and drug companies—needed: a credible justification for taking drugs whose principal effect was to make us feel
better about ourselves. *Listening to Prozac* helped make the world
safe for antidepressants.

In his book, Kramer starts out like many of us do about this subject—tentative, searching, ambivalent. As he gathers momentum,
however, his case for using the drugs—not only to treat depression,
but to "remake the self," as his subtitle put it—grows stronger, until
it turns into a restrained but unmistakable endorsement. And while
you have to wonder about that title—Eli Lilly himself couldn't have
asked for better product placement—the fact that permission came
not from an ad man but from a neutral expert, a sensitive and hon-

est and articulate eyewitness to the revolution, only strengthened the case for the drugs.

Listening to Prozac ends with a prophecy. Having spent the better part of three hundred pages worrying over the complexities of using drugs to solve our problems, Kramer speculates that questions like these may already be pointless.

> By now, asking about the virtue of Prozac . . . may seem like asking whether it was a good thing for Freud to have discovered the unconscious. Once we are aware of the unconscious, once we have witnessed the effect of Prozac, it is impossible to imagine the modern world without them. Like psychoanalysis, Prozac exerts influence not only in its interaction with individual patients, but through its effect on contemporary thought. In time, I suspect we will come to discover that modern psychopharmacology has become, like Freud in his day, a whole climate of opinion under which we conduct our different lives.

Antidepressants' most important side effect, Kramer seems to be saying, is the way they change our understanding of ourselves—altering not only our neurochemistry but our sense of its importance. And once that has happened, there's no more point in inquiring into their virtues than there is in wondering if winter ought to be so cold and snowy. It's an ironic end to a book that asks about little besides Prozac's virtue—and which did so much to usher in the climate of opinion under which we think of our unhappiness as a disease.

Kramer borrowed the phrase *climate of opinion* from W. H. Auden's elegy "In Memory of Sigmund Freud." Freud, Auden wrote, was no longer just a person:

> *he quietly surrounds all our habits of growth*
> *and extends, till the tired in even*

the remotest miserable duchy
have felt the change in their bones and are cheered

There was a time, and it wasn't very long ago, when people didn't feel in their bones that they "had depression," when the Centers for Disease Control weren't calling depression "the common cold of mental illness," when the World Health Organization wasn't claiming that depression was "the leading cause of disability . . . and the 4th leading contributor to the global burden of disease." It is possible that doctors have gotten better at recognizing depression. It's possible that contemporary life imposes demands that exceed the neurochemistry bequeathed to us by natural selection. It's even possible that global warming, widespread warfare, the worldwide economic collapse—that these seemingly irremediable conditions are making us sick with worry. Indeed, all of these explanations for the apparent depression epidemic could be true at the same time, but there is another possible explanation: every new climate of opinion about who we are has its distinctive form of lousy weather. Clinical depression—unhappiness rendered as disease—is ours.

Climates of opinion don't descend fully formed from the heavens any more than occupying governments do. If they did, if Betty Twarog's discovery had simply led to a sudden and cataclysmic change in the way we think of our unhappiness and what to do about it, then the skirmish that broke out in 1995 between David Wong and Arvid Carlsson in the pages of the journal *Life Sciences* would never have happened. Wong, the Eli Lilly scientist who first formulated Prozac, claimed in passing that his drug was the first SSRI—an assertion to which Carlsson, who won the Nobel Prize for his pioneering work in the neurochemistry of Parkinson's disease, took exception. Carlsson knew better because he had invented the first SSRI, zimelidine, which the Swedish pharmaceutical company Astra brought to market as an antidepressant named Zelmid in 1982, five years before Prozac. *Life Sciences* was forced to print a retraction and apology.

The reason that the editors of *Life Sciences* didn't catch Wong's overreaching—and that you have most likely never heard of Zelmid either—is that Astra never took its drug very seriously, at least not as a big moneymaker. Or so you must conclude from the fact that on the eve of its introduction into the United States, when it began to seem that patients taking Zelmid were prone to contracting the rare neurological disorder Guillain-Barré syndrome, Astra decided not to do the studies necessary to investigate the connection. Instead, it simply pulled the drug from its shelves. The company's executives just didn't think there was enough of a market for an antidepressant to make it worth the shareholders' while. Or to put it another way, they didn't think there were enough depressed people out there.

To judge from the industry's willingness to spend huge amounts of money to minimize their drugs' association with violence and suicide and other, less dramatic side effects, that's not a problem anymore.

The climate changes slowly and imperceptibly, and once it's settled in, it's as invisible to us as the sea is to a fish. But if you start to look for it, it's awfully hard to miss.

For instance, let's say you haven't been able to shake off a setback or a loss, and you find yourself preoccupied and worried, prone to tears, avoiding sex and other pleasures, overeating and undersleeping and just plain not enjoying life as much as you once did. And let's say you resist this idea that you have an illness, but on the other hand, you're mighty tired of feeling this way, and one sleepless night cruising the Internet, you end up at depressionisreal .org, a coalition of "seven preeminent medical, advocacy, and civic groups who have joined forces to educate the public about the true nature of depression and how people can live and thrive with this biological disease." There you can tune into a podcast of the *Down*

& Up Show, which promises to "separate fact from fiction" about depression. You can find out about depression rates in the United States. You can read about depression and women or depression and the Latino community. You can download a mood tracking calendar. You can even take a test that tells you whether or not you have depression. And if it turns out that you do, you can read about resources that you can contact tomorrow, or you can click over and get some comfort right now from Paul Greengard, a doctor who, as it happens, shared the Nobel Prize with Arvid Carlsson. Dressed in his white lab coat, Greengard gazes reassuringly from beside this message:

> Some say depression is all in your head. Well, that's right. And wrong. It's right because depression is in the head, or more precisely, the brain. In fact, we've seen how it destroys the connections between brain cells.
>
> But saying depression is all in your head is also wrong. There's nothing imaginary about depression. It's a serious medical condition that affects every aspect of a person's health.

Greengard is hardly the only doctor—or even the only Nobel laureate—to deliver this message. It has saturated American popular culture to the point that it is nearly inescapable. And it has done some good. The idea that depression is a treatable medical condition has given people permission to talk to their doctors about suffering for which they might otherwise never seek relief. It has saved lives by preventing suicide, kept families together, helped people to stay productive. And it has had enormous benefits for basic neuroscience: industry interest in finding drugs to treat depression has opened up the coffers to researchers trying to figure out how the brain works.

Nonetheless, there is indeed something, if not quite imaginary, then certainly invented, about depression. Greengard gives us only two choices: that it is *real,* which in the current fashion means that it is the result of neurochemical events, or that it is *fake,* a product of our fickle imaginations or our weak wills. He overlooks a third possibility: that it is made up not *by* us, but *for* us, that depression—or at least the version of depression that Greengard is describing—is manufactured.

Depression is surely an affliction, one that at least in some cases may well have a specific, although still undiscovered, brain pathology—a disease in the usual sense of that word. This is a powerful and compelling idea: if you are unhappy in a certain way, then you are suffering from a brain illness, no different in principle from any other illness. That idea has become part of the way we think about ourselves, part of the incessant chatter of our own minds (or is it our brains?), of the constant self-evaluation by which we mark our lives.

Am I happy enough? has been a staple of American self-reflection since Thomas Jefferson declared ours the first country on earth dedicated to the pursuit of happiness. *Am I not happy enough because I am sick?* on the other hand is a question that has just arisen in the last twenty years. This is the sense in which depression has been manufactured—not as an illness, but as an idea about our suffering, its source, and its relief, about who we are that we suffer this way and who we will be when we are cured. Without this idea, the antidepressant market is too small to bother about. With it, the antidepressant market is virtually unlimited.

My first bout of depression began in 1987, at the same time my first marriage ended. It's not that I didn't want the divorce. In fact, it was my idea, an idea I had expressed in a time-honored if ignoble way: by falling in love with another woman. I now think of this transgression as a merciful sort of wickedness, my adultery putting us out of a misery that neither my first wife nor I had the wisdom or courage to end. We were like two comets that had crashed into each other deep in interstellar space. The collision nearly consumed us and left behind nothing but cold and darkness, and, at least for me, a smoldering pile of self-reproach.

I was thirty years old, a psychotherapist by day, a psychology doctoral student by night, and you would think that at some point—I would nominate the time I found myself on the floor watching dust specks float through sunbeams for hours (because they happened to be in my line of sight, because looking at anything else or closing my eyes and staring at my own black insides would just take too much effort), racked by some unspecifiable pain, like my whole being was a phantom limb, and thinking about that lady in the Life-Fone pendant ad, the one who has fallen and can't get up—you would think that at a point like this it would have occurred to me that I was depressed. Come to think of it, that probably did occur to me. But in 1987, *depressed* didn't mean what it has come to mean in the years since. Then it was a convenient description, something to say to a friend or to myself, a shorthand that left the details to the imagination. Now it's an illness.

To be fair, depression was already an illness in 1987. It just wasn't quite so famous as it is now. In fact, it had been an official disease in more or less its current form since the 1980 release of the third edition of the American Psychiatric Association's *Diagnostic and Statistical Manual of Mental Disorders.* The DSM, as it is known in the industry, is a compendium of psychological troubles, sorted into groups (affective disorders, substance use disorders, psychotic disorders) and from there into individual diagnoses (*major depressive disorder, alcohol dependence, schizophrenia*). And it is indispensable to the business of therapy. Not only does it provide a taxonomy of mental disorders, which in turn gives us therapists a private language in which to talk to one another and a way to feel like we're part of a guild; it also assigns to each species of anguish a five-digit code. Written on a bill, that magic number unlocks the insurance treasuries, guaranteeing that because we therapists are treating a disease rather than, say, just sitting around and talking to people about what matters to them, we will get paid for our trouble. This is why the most recent edition of the DSM (we're now on the fourth, with the fifth due in 2012) sits on the shelf of virtually every therapist in the country, including me.

The DSM is an unparalleled literary achievement. It renders the varieties of our psychospiritual suffering without any comment on where it comes from, what it means, or what ought to be done about it. It reads as if its authors were standing on Mars observing our discontents through a telescope.

As we will see a bit later, that was exactly the intent of the authors of the third edition, which was a radical departure from the two previous DSMs and the model for subsequent versions. They meant to incite a revolution in psychiatry, a discipline that previously had not hesitated to comment on theoretical (some would say metaphysical) matters such as the origin, nature, and causes of mental anguish. It took a decade or so, and the introduction of a few new drugs, before the revolution was complete, at least with respect to depression. Had my troubles occurred later in that decade, I'd have been much more likely to reach for the LifeFone, to get my diagnosis and the meds to go with it, and to become part of the CDC and WHO statistics.

I wouldn't have avoided this path on principle; in fact, as I'll describe later, drugs (although not the drugs you might expect) did help me finally bring my black dog to heel. But it simply never occurred to me to think of myself as sick. I just figured I'd had a disaster in my life and my unhappiness was the consequence of it, as surely as whacking my thumb with a hammer would have left me injured and in pain and really mad at myself. I worried that I might never get over this, that I would be alone forever, that my finances would never recover, that my divorce was also my initiation into the reality of how hard life really was. I talked about this in therapy, of course, about this and many other subjects. I learned all sorts of things about myself that I didn't want to know. I marveled at the ability of mercifully long-forgotten chapters of my private history to insinuate themselves into waking life, at the bad faith I could engage in and the pain it could cause myself and others, but my therapist and I never, to my recollection, talked about me as a sick person. Whatever I had seemed like a bad spell that I had to outwait or at least get used to while I did my best to overcome it.

I did think of depression as a disease; at least I did in my professional life. But I associated *that* depression with patients like Evelyn. A young woman who was already weeping when I went to fetch her in the waiting room, she told me right off that her life was unmitigated agony. Every success was its own punishment, and her professional achievements, the love of her family, even the sun coming up on this gorgeous spring day only made her feel worse— as she put it, it was as if she were Frankenstein's monster watching through the window while the human family lived their happy lives in their warm hut. She said she had called me because she had recently accepted an invitation for a free vacation to Hawaii, and as the date approached she was beside herself with dread. "Because of the expectations to have a good time like everyone else, and the light, that relentless sunshine, which is just going to crush me," she said. "I know this is as good as it gets, and it's not good enough for me, which just makes me hate myself more." She stopped, fixed me in her gaze, and lowered her voice to a near whisper. "I hope the fucking plane crashes."

And then there was Ann, the biologist who ended her promising research career to marry a truck driver who beat her and then left her, taking their son with him. She was sure she deserved this and any other failure or indignity that visited her, and her day could be ruined if someone praised her. She was a connoisseur of anguish who had more words for her blue moods than Eskimos allegedly have for snow, who wrung her hands ceaselessly and cried rivers as she talked, but who always seemed surprised when I pointed these things out or expressed concern about them. And not only surprised—she told me frequently that the fact that I paid that kind of attention to her and still seemed to like her reflected poorly on me.

Or Barbara, who phoned me one night demanding, "You have to tell me why. Give me a reason to go through all this pain." I told her I knew that she was suffering, that I would listen to her and stand by her and get up in the middle of the night to comfort her, that I would remind her of all the other people who loved her, all the things she still wanted to do, but that beyond that I couldn't give her what she was asking for. She was dead the next morning, lying in bed next to one of those people who loved her. She had overdosed on her antidepressants.

That's the disease of depression as I saw it then: severe, disabling, and deadly, unrelated to circumstance, resistant to comfort (let alone to treatment)—and, thankfully, rare. Wounded as I was, my suffering—and that of most of my patients—wasn't even in the same ballpark as theirs, and surely not in the same diagnostic category. Which isn't to say we weren't unhappy—after all, why else would we be spending our time and money complaining to therapists about our lives?—just that it didn't seem to me (or, so far as I know, to them) that we had *depression*.

Or so I thought at the time. Now, it could be that I just didn't want to place myself in the category of the mentally ill; after all, when it comes to those diagnoses, most of us therapists are better at dishing it out than taking it. Or that because I wasn't looking for depression, I didn't see it except in the most dramatic cases. If that's so, then the last twenty years, in which it has become unthinkable for clinicians and laypeople alike *not* to consider unhappiness as a symptom, constitute a period of unparalleled triumph for public health.

But it could also be that depression has expanded like Walmart, swallowing up increasing amounts of psychic terrain, and that, also like Walmart, this rapidly replicating diagnosis, no matter how much it helps us, and no matter how economical, is its own kind of plague. It could be that the depression epidemic is not so much the discovery of a long-unrecognized disease but a reconstitution of a broad swath of human experience as illness. Depression is, in this sense, a culturally transmitted disease, the contagion carried not by some microbe or gene, but by an idea transmitted by subtle and not-so-subtle means, including clever direct-to-consumer prescription drug advertising; ruthless drug company dominance of medical education, research, and practice; those dire statistics; state laws ordering insurance companies to pay for the treatment of depression as they would for diabetes or cancer therapies; a new DSM with even more subspecies of depression; and casual conversations with diagnosed and medicated friends. Borne on these vectors and others, the notion has spread that our sorrows, our discontents, our unhappiness, and our hopelessness are the signs of a pervasive disease, until it (the idea, not the disease) has taken up residence in nearly all of us. Twenty years (and a few more bouts of intense unhappiness) after my spell on the floor, I would be very unlikely to feel as I did that day and not conclude that I was probably sick. I resist this thought, but I live under the same climate of opinion as you do, so I must confess that I still don't know whether that resistance is a mistake.

I have a couch in my therapy office. People often make nervous jokes about it before seating themselves in one of my chairs. Every once in a while, someone will lie down on it, consciously parodying the Freudian stereotype. These patients might have noticed that the office would be better off without the couch. It's not only a cliché; it's also ugly and too big for the space. But it's an outstanding place to take a nap, which is really why it is there. As long as I can remember, the hand of Morpheus has reached up from the underworld and grabbed me by the neck every afternoon at around two o'clock. He is very hard to resist, so I have never scheduled midafternoon appointments, lest I embarrass myself and infuriate my patient by nodding off in the midst of their travail.

It turns out that frequent naps—more than half an hour a day, four or more days per week—are a symptom of depression. (There's no explicit exception for countries, such as Spain, that have siesta schedules, but one imagines that therapists in those places adjust the criteria accordingly.) I did nap more after my first marriage collapsed, although I never kept track. One day during that period, I was awakened from a particularly lovely nap by a phone call from my father. I immediately forgot everything about the conversation other than the way I felt as I fought through my grogginess: anxious to the point of nausea.

"Dread," I said to my therapist, whom I happened to be seeing later that day. "Just a feeling of dread and self-loathing. Like there he was working hard, being productive, *functioning*"—he was calling me from his office, where he spent ten-hour days until he was well into his seventies—"and here I was wasting time, crashed out on the couch in the middle of the day."

"Well, what do you think this means?" she asked. I had, I knew, lobbed her a huge hanging curve—all that Oedipal drama captured in a single scene.

"Maybe nothing. I have to say, it felt, I don't know, *biological*."

"Biological? You mean, like there are little bugs swimming in your blood or something, making you feel dread?"

She said this as if it were the most preposterous idea in the world, as if anyone who believed it was either evading the truth or just plain deluded.

It's not preposterous anymore. There are many ways to distinguish various depressive states from one another. You could, for instance, listen to the stories I'm telling here and conclude that there are three sorts of depression—the temperamental kind that seems to sum up a considered view of the world as a not-so-happy place, the kind that seems always to have been there and has no particular reason behind it, and the kind that comes on after a setback. Evelyn's depression is a good example of the first, Ann's of the second, and, if I have to place myself in a category, mine belongs in the last. And then there are formal distinctions. For example, in the old days, which is to say before the DSM-III, doctors talked about manic-depressive illness, in which patients alternated between those two poles; involutional psychotic reaction, a condition of delusional guilt and self-loathing that came on in middle age; and depressive neurosis, the garden-variety unhappiness that psychoanalysts treated in the Freudian heyday. Whether these distinctions were valuable or not or based on anything other than current fashion is hard to say. But what is clear is that they no longer exist. Sometime in the twenty years since my therapist made fun of me, the "bugs" have gnawed them into so much powder.

In *Against Depression,* his sequel to *Listening to Prozac,* Peter Kramer wrote "Depression is neither more nor less than illness, but illness merely." Being depressed is not simply a response to circumstance, he argued, although it can be kindled by events in our lives. Neither is it a sign of sensitivity or intelligence or insight, nor a branch of suffering with roots in the social or political world—a despairing apprehension, say, of the world we have made. Nor is it a response to the tragedies inherent to human life—mortality, for instance, and the inevitability of loss. Indeed, he claimed, the failure to grasp the fact that depression is just another disease, just another way our bodies have of betraying us, as purposeless and meaningless as tuberculosis (which, he points out, was once seen as a mark of refinement), is itself a symptom of a widespread and longstanding, but deeply wrongheaded, view: that melancholy signals a profound grasp of the true nature of existence.

Kramer likened depression to "an occupying government," one that has apparently colonized our collective consciousness, propagandized us, as it were, into believing that it is more than illness. Under this regime, we don't understand that when you're lying on the floor of your study and it feels as if someone has turned up the gravity, you're in the throes of a disease as frank and indisputable as, say, appendicitis—and that you are just as much at risk as you would be if you ignored that pain in your lower abdomen. Kramer confessed to having fallen prey to this ideology himself—not as a practicing melancholic, but a practicing psychiatrist. He learned this, he wrote, from a patient who, once the drugs had kicked in, chided him for paying too much attention to what her depression might actually mean. But he reeducated himself, and in his book urged the rest of us—doctors and patients alike—to do the same.

We are on the brink of an epochal shift, Kramer went on—to a time when "the eradication of depression [will] seem unremarkable as a . . . social goal." Only one thing stands in the way of achieving that goal, Kramer wrote: ignorance. It takes many forms, but one of them is people like me and the other critics of the depression industry who are, according to *Against Depression,* unwittingly in thrall to that colonial power and who therefore insist on pointing out certain facts. Like, for instance, that the prevalence of depression magically skyrocketed just after the drug industry introduced the SSRIs, that the diagnostic criteria underlying this increase can't distinguish between grief and depression, and that as a result the diagnosis threatens to swallow everyday sorrows. People who continue to believe these things, as the title of his book implies, must be, wittingly or not, *for* depression.

At the risk of sounding like the man who says no when asked if he's still beating his wife, I'll tell you that I'm really not on the side of the suffering that afflicted Evelyn and Ann and killed Barbara, the kind that drives people to their knees, or their beds, for months or years at a time. In fact, I'm not in favor of suffering at all. By criticizing the idea of depression as a disease, I'm not wishing anguish upon us. (Nor do I think that we need to safeguard pain against the depression doctors' attempts to do away with it; something tells me that psychic suffering will never be in short supply.) Pain, psychological and otherwise, is just a brute fact, neither noble nor evil, neither redemption nor scourge. It may play some important evolutionary role—designed, perhaps, to alert us to the fact that something is wrong or to create the necessity for invention—but it's not hard to imagine a different mechanism fulfilling these functions, one that doesn't hurt so much.

The division of the world into forces in favor of and against depression is as false as every other Manichaean scheme. Everyone is against depression, just as everyone is against war and child abuse and global warming. The argument is really over *who* is depressed, which is to say over whose inner life gets pathologized under the new depression regime and what the depressed people are going to do about it. That's why it's important to figure out just what the depression doctors mean by the diagnosis and where that meaning came from: because there are burdens to being declared ill. Unless you are a drug company, in which case the only burden of a widespread illness for which you own the treatment is figuring out what to do with the profits.

I wish I could tell you that this very lucrative notion about unhappiness has been brought to us by the marketing departments of the big drug companies. That would make convincing you to resist it an easier job. But while I will tell you plenty of stories about shrewd and sometimes questionable corporate behavior, proving that drug companies will do what they have to do in order to sell their product is no more or less illuminating than uncovering gambling in Casablanca. It's worth noting when the usual suspects behave suspiciously—when, for instance, a website like depressionisreal.org is funded by Big Pharma, but it would be a mistake to see this as evidence that the drug companies are conspiring to change the way we think about ourselves in order to make us dependent on them for our well-being.

The captains of the pharmaceutical industry are merely doing what they get paid the big bucks to do—to sail their corporate ships expertly on the winds and currents of the times. And the times, with some help from Big Pharma, have delivered them an ideal consumer for their product: someone convinced that unhappiness is a problem for their doctors to treat.

The history of the invention and production of depression is a strange and elusive kind of secret. Most of what I'm going to expose here isn't buried in corporate files. It's as obvious as a commercial for Prozac—or, for that matter, as the fifty thousand copies of *Recognizing the Depressed Person* that Merck distributed to doctors in 1963 or *Symposium in Blues,* the compilation album of blues songs that they paid RCA to press and send out three years later with prescribing information for their latest antidepressant inserted up its sleeve. It's laid out in black and white in the scientific literature, which documents, in addition to all that breathtaking neuroscience, the poor performance of antidepressants and the failure of the serotonin imbalance theory to explain depression. It's right there in the way that over the last century or so, medicine has shaped a climate in which we feel a bone-deep conviction that disease is something biochemical, that health and illness are scientific categories, and that doctors are dispensers of magic bullets aimed at molecular bad guys. It's on the front page of the newspaper where stories about America's drug war stand as daily reminders that we are very confused about taking drugs to change our moods—a confusion that is largely circumvented when we instead take drugs to treat a disease.

These are the raw materials of depression, and they've been assembled in the clear light of day, hidden, like Poe's purloined letter, in plain sight. I'm going to show you how depression has been manufactured right before your eyes—not in order to deny that depression exists or even that it can, in some cases anyway, rightly be considered a disease that can be cured by drugs, but in order to provide you with another tool to figure out what to do if recalcitrant sadness sets in and sends you to your doctor's office. Because Peter Kramer is both right and wrong about the climate of opinion—right that psychopharmacology is a sign of a major change in the climate, wrong that it is not worth your time to "ask about the virtues" of the new climate. Once you find out how unhappiness has become an illness to be treated with drugs, and once you grasp that there is a history to your depression that has nothing to do with your biochemistry, you have another choice besides "all in your head" and "all in your brain." If the idea that depression is a disease is as much a matter of history as it is of science, if it is, in short, a story about our suffering, then you are free to look for other stories, or to tell your own. You are free to arm yourself with information that your doctor might not even know about, to seek alternatives, to resist the regime—or to choose, because it makes sense to you and not because a drug-industry-fattened doctor told you so, to subscribe to that story.

I'm not going to tell you that I don't have a dog in this hunt. I'm writing this book in part because I think that the medical industry, regardless of its intentions, has acquired far too much power over our inner lives—the power to name our pain and then sell us the cure one pill at a time. But even though I am a psychotherapist, I don't think the only alternative is what I sell in my office one hour at a time—although I will point out that it is probably the only profession built on the idea that changing the story we tell about our suffering can relieve it. And I know, through my own experience as both a therapist and, as I'll detail later, as an officially depressed person, that drugs—although not necessarily the drugs that Pharma is selling—do work. But that doesn't mean that depression—yours, mine, or anyone else's—is the disease the depression doctors say it is.

CHAPTER 2

JOB VERSUS HIS THERAPISTS

I t is customary for histories of depression to start with Hippocrates, the ancient Greek physician.* There are good reasons for this. In addition to originating the oath by which physicians pledge not to harm or kill or seduce their patients, Hippocrates set Western medicine on its current course by insisting that the doctor's job was to use his own senses to acquire the actual details of his patients' suffering. When he told his disciples to seek the truth by examining the phenomenon of the illness itself, Hippocrates was urging them to kick the gods out of the clinic; as Hippocrates said of epilepsy, known in his time as the *sacred disease,* "it appears to me to be no more divine nor more sacred than other disease, but has a cause from which it originates." This idea—that illnesses exist in nature and that it is the doctor's job to find and, if possible, to heal them—is exactly the idea behind most medicine today, including the treatment of depression.

---

* Historians of antiquity tell us that while there was a physician named Hippocrates, the body of work attached to his name (much of which is in the first person) was mostly written by his disciples. For purposes of simplicity, I use "Hippocrates" to refer to the composite character who emerges from the Hippocratic corpus, not the individual (about whom we know very little).

One of the conditions that Hippocrates took note of looks something like our depression. "Fear and sadness that is prolonged means *melancholia*," he wrote, and the melancholic patient, who suffers from an excess of black bile (which is how *melancholia* translates from the Greek) has an "aversion to food, sleeplessness, irritability, and restlessness." He is rumored to have cured the king of Macedonia's melancholia by deducing that he was secretly in love with his recently deceased father's concubine and prescribing a consummation of his desire (which makes you wonder if Hippocrates had heard about Oedipus).

It's easy to understand the depression doctors' eagerness to enlist the father of medicine in support of their contention that depression is a disease. The winners get to write history, after all, so why wouldn't they claim this patrimony? But even leaving aside for the moment the fact that so much of what Hippocrates and his followers wrote is fanciful at best—for instance, that "it is a deadly symptom . . . when the patient sleeps constantly with his mouth open" or that lying "with the hands, neck, and legs tossed about in a disorderly manner and naked . . . indicates aberration of intellect"—you have to wonder why, if depression is such a common disease and Hippocrates such a voluble commentator on a vast range of human suffering, his work on melancholia is so scant. His notes on the subject are scattered throughout his works, and he doesn't tell us much about it, not even how the problem was related to the other black bile disorders, which ranged from hypersexuality to hemorrhoids (a condition that, unlike melancholia, he devoted an entire book to).

Hippocrates' lack of attention to melancholia doesn't mean that depression isn't an illness. But to cite Hippocrates as an authority for its existence is a little like citing George Washington as an authority on wooden teeth or cherry tree removal: just because he was a great man who had some interest in the matter, we shouldn't necessarily privilege his opinion about it. There is, however, an ancient account of depression that is much more robust than Hippocrates'—and much more like our current version of the malady. It's also much older. In fact, according to one scholar, as soon as people started taking enough notice of themselves to put stylus to clay tablet—in around 5000 B.C., 4,500 years before Hippocrates, in the Mesopotamian society known as Sumer—they wrote down a story about a whopping case. (It would be a mistake to conclude from this ancient lineage that depression, like, say, the common cold, has been with us from the beginning. After all, we know virtually nothing of the inner lives of *Homo sapiens* for the 200,000 or so years prior to the advent of writing, so its appearance at the dawn of history could just mean that people became despondent as soon as they started paying enough attention to themselves to take notes.) The Sumerian version of this story is in fragments, but the Hebrews eventually incorporated it into their Bible. It has since become one of the Western world's best-known, if not best-loved, stories.

Poor Job! Pillar of Uz and patriarch of a large family, a God-fearing, evil-shunning "mark among all the people of the East," he's just minding his own business—which is considerable, according to the Bible's detailed list, including "seven thousand sheep, three thousand camels, five hundred yoke of oxen and five hundred she-donkeys, and many servants besides"—when Satan challenges Yahweh to a duel over Job's righteousness.

> Job is not God-fearing for nothing, is he? Have you not put a wall round him and his house and all his domain? You have blessed all he undertakes and his flocks throng the countryside. But stretch out your hand and lay a finger on his possessions: I warrant you, he will curse you to your face.

Rabbis and priests have long argued about what role Yahweh plays in the mayhem that follows, whether he commissions the hit like a godfather or just turns a blind eye while Satan does his mischief, but from Job's point of view this doesn't really matter. Either way Yahweh's wager spells disaster for Job. In one day, nomads swipe the oxen and donkeys and slay the servants, lightning strikes dead the sheep and shepherds, the camels are carried off by Chaldeans, and then, as Job discovers from the last in a line of bad-news messengers, his children are killed by a sudden storm.

To Satan's dismay, Job maintains his faith. But the Prince of Darkness prevails upon Yahweh's insecurity one more time, and Job is inflicted with "malignant ulcers from the sole of his foot to the top of his head." Job's wife nudges him toward the dark side. "Curse God," she says, "and die," but Job will have none of it. "If we take happiness from God's hand, must we not take sorrow too?"

The story would end with Job's impressive forbearance were it not for the arrival of three old friends—Eliphaz, Bildad, and Zophar—who ostensibly show up to console him. They weep and tear their clothes and then sit down silently with Job for seven days, turning their visit into the ritual week of mourning. And in that week, something happens to Job—perhaps the extent of his catastrophes sets in or the pain and disfigurement of his ulcers take their toll, or maybe he recognizes that in some way his comforters are sitting shiva *for* him rather than with him, that stripped of his wealth, family, and dignity he is as good as dead. Whatever the occasion for his change of heart, Job finally confesses the nearly unspeakable truth: he has lost his desire to live.

> *May the day perish when I was born,*
> *And the night that told of a boy conceived.*
> *May that day be darkness,*
> *May God on high have no thought for it,*
> *May no light shine on it.*

Not only that, he tells them, he actually longs for what he is sure awaits him in death.

> *I should now be lying in peace,*
> *Wrapped in a restful slumber,*
> *With the kings and high viziers of earth . . .*
> *Down there, bad men bustle no more,*
> *There the weary rest.*

Not all suicidal people are depressed. Sometimes they're angry or trapped in desperate circumstances, or, in the case of the terminally ill, already dying and ready to take matters into their own hands. But Job is also irritable and implacable, uninterested in food or prayer or any of the things that once brought him pleasure. And like Evelyn—and nearly every depressed person I've ever met—he's tortured by the bright light of day, sees it as a mockery of itself. "Why give light to a man of grief?" he asks his friends. "Why make this gift of light to a man who does not see his way?" It's no coincidence that the two most famous recent memoirs of depression—William Styron's *Darkness Visible* and Andrew Solomon's *The Noonday Demon* (a metaphor he borrowed from Psalm 91, which Job also cites)—invoke this image. This total dejection, the demoralization that turns light into reproach and darkness into anguish is what makes Job's wish to be dead a mark of what we call depression.

I wouldn't want to blame Job's pitiful psychological state on his therapists. It's nearly impossible, I think, to get used to being in the presence of someone whose "only food is sighs and [whose] groans pour out like water," as Job puts it, who is both looking to you for comfort and yet ready to tell you why the comfort you offer just makes things worse. You fight off your impatience and fear and search for the words that will shine through what Hawthorne called "the black veil" and into whatever corner of his psyche is not shrouded in gloom, and you hope you come up with something more helpful than what these men offer to Job in his anguish. But still you have to wonder how they think it will help Job to level accusations disguised as questions like these:

> Can you recall a guiltless man that perished,
> Or have you ever seen good men brought to nothing? . . .
> Was ever any man found blameless in the presence of God,
> or faultless in the presence of his maker?

Or this:

And now your turn has come, and you lose patience too;
Now it touches you, and you are overwhelmed,
Does not your piety give you confidence,
Your blameless life not give you hope?

With comforters like these, who needs ulcers? That's certainly what Job is wondering when he calls them "charlatans, physicians in your own estimation" and wishes that "someone would teach you to be quiet," or when he asks, "Will you never stop tormenting me and shattering me with speeches?" But then again, he doesn't order them to be quiet or to leave his house and take their sanctimony with them. Undoubtedly this is in part because they've cast doubt on his integrity, forcing him to defend himself, but we might also imagine that Job listens to them in hopes that he will hear something that will actually comfort him—if not a cure, then at least an explanation of what has happened to him, one that can restore his faith that life is fair and God is just or, by providing a reason for his suffering, assuage his grief.

Job isn't buying his self-appointed physicians' answers, however. Indeed, their attempts to console him just seem to egg him on. With growing stridency and in agonizing detail, Job argues his case: if this could happen to him, then life itself is so unfair as to be cruel and meaningless.

Is not man's life on earth nothing more than pressed service,
His time no better than hired drudgery,
Like the slave, sighing for the shade
Or the workman with no thought but his wages,
Months of delusion I have assigned to me,
Nothing for my own but nights of grief.
Lying in bed I wonder, "When will it be day?"
Risen, I think, "How slowly evening comes."

Slowly, inexorably, the personal becomes the universal, Job's wish to be dead in order to escape his pain escalating into an indictment of life's injustices:

Why do the wicked still live on,
Their power increasing with their age?
They see their posterity ensured,
And their offspring grow before their eyes.

And from there into the most profound pessimism—a rejection of the very terms of existence:

But man? He dies and lifeless he remains;
Man breathes his last, and then where is he?
The waters of the seas may disappear,
All the rivers may run dry or drain away;
But man, once in his resting place, will never rise again.

And, finally, into blasphemy: he will take his complaint all the way to the top. "I mean to remonstrate with God," says the man who started his mourning week demanding nothing of the sort.

Now you have to admire Job's chutzpah here: he is going to call God to account. If he knew what you know—that he has indeed gotten the royal shaft from the king of the universe—he'd also know that he has Him dead to rights. It's a little perverse, of course, the way Job is betting against himself, as if being correct that life is not worth the candle—and pressing this case with its creator—is consolation for how bad he feels. But depressed patients sometimes do just this— make such a compelling case for the pointlessness of their lives and of life in general that you don't know whether to agree with them or to assert that their pessimism, as William James once said of Arthur Schopenhauer's, is like "that of a dog who would rather see the world ten times worse than it is, than lose his chance of barking at it."

You don't know, in other words, whether to read bleakness as symptom, which is exactly what Eliphaz does.

Does a wise man answer with airy reasonings,
Or feed himself on an east wind?

Does he defend himself with empty talk
And ineffectual wordiness?
You do worse: you flout piety . . .
A guilty conscience prompts your words,
You adopt the language of the cunning
Your own mouth condemns you, and not I
Your own lips bear witness against you.

Or as a latter-day Eliphaz might put it, Job is in denial and cannot see that disorder in his inner world has led to his view of the outer world, that "it is man who breeds trouble for himself as surely as eagles fly to the height."

Job's inner life has been the subject of this story from the beginning, when God and Satan squared off about whether his piety was authentic or God-bought; Eliphaz is only taking this scrutiny one further step. Confronted with Job's abjection, he does exactly what a depression doctor does: first, he invokes an idea about what a human being is supposed to be—someone who can take these blows and still maintain his faith and piety—and then he claims that Job's problem is a failure to be that way. If Job were healthy, a modern Eliphaz would say, he wouldn't be so distraught but instead would be able to see how he had bred trouble for himself (if not in the catastrophes themselves, then in his response to them). He would be able to roll with the punches and then move on to the rosier future, where, according to Eliphaz,

You shall be safe from the lash of the tongue,
And see the approach of the brigand without fear.
You shall laugh at drought and frost,
And have no fear of the beasts of the earth . . .
In ripe age you shall go to the grave,
Like a wheatsheaf stacked in due season.

If only he were sufficiently pious, Job would have the ability to live through setbacks without losing heart.

With all of this anticipation of the modern world, it's tempting to say that Job was suffering from an undiagnosed case of depression, and that, had his comforters had prescription pads and an acquaintance with the techniques of cognitive-behavioral therapy, they could have relieved his suffering. But this misses a much more important way in which Job anticipates our modern understanding of depression: at the dawn of history, when confronted by the bewilderments of loss and by the human capacity for deep despair, self-appointed physicians sought to diagnose pessimism as a pathology within the suffering person. By the time the story leaves the comforters behind, when Yahweh thunders his answer "from the heart of the tempest," informing Job of his insignificance and impertinence, of the absolute irrelevance of his puny notion of justice to the raw majesty of creation, the fix is in: the fault lies not with creation, not with a god who would sell out his best customer or, for that matter, who would send a man with a sense of justice into a world in which the gods roll us like dice, but with the man himself. Job's desolation is an affront to the convictions of his therapists, and they desolation back with a diagnosis.

"In its commonest form," Peter Kramer wrote in *Against Depression*, "depression is a disorder of emotional assessment of experience." People, struck by the inevitable misfortune of life, lose the ability to bounce back, which Kramer calls "resilience." The result is what he describes as "a fixed tragic view of the human condition" that prevents people from moving "toward assertiveness and optimism." This is Job's problem from a psychiatrist's point of view: he fails to see not only that he is mired in negativity, but also that his attitude itself is the problem. He mistakes his pathological view of things for an apprehension of the truth about his pain and the world in which he suffers.

This misunderstanding, Kramer says, is a hidden wellspring of Western civilization, a culture that he says valorizes depression because it has been shaped by depressives. "Our aesthetic and intellectual preferences have been set by those who suffer . . . deeply," he wrote. "If the unacknowledged legislators of mankind . . . are depressives, then we might want to examine the source of our value judgments when it comes to pessimistic views of the human condition." Those judgments are flawed, Kramer believes; affronted by their victory, by the history they have written, he strikes back with a diagnosis.

Not all the depression doctors are as articulate and forthcoming as Kramer, but I think he has spoken for them here, or at least accounted for one of the reasons that so much of our psychic suffering has come under the rubric of their disease. Your sadness doesn't become depression until it has settled in for a while—officially, according to the DSM, for two weeks. So what happens on that fifteenth day? The depression doctors will tell you that the threshold is derived from statistics, but like so much about depression, it's based on circular logic: the number was derived from the experience of people the doctors already considered depressed. So it's left to us to figure out that what's at stake is persistence. After two weeks, it seems, your dejection is at risk of becoming a fixed and tragic view that is not only unpleasant but also nearly taboo in a society dedicated to the pursuit of happiness—and that was, for different reasons, taboo in the land of Uz. Your sadness becomes depression, in other words, when it turns into pessimism.

The arbitrary nature of fortune, the near certainty that unbidden catastrophe will visit each of our lives, the inevitability of mortality, a nature that is more generous with pain than with pleasure, in short, all the stacked-deck calculus of human existence—these are challenges to optimism if not outright invitations to pessimism, and that's before we even consider what a hash we've made of both civilization and nature. But I don't wish to mount a broadside against optimism or, Kramer forbid, more legislation for pessimism. Instead, I want to point out that the depression doctors have done exactly what Eliphaz and company did. Psychology may have replaced theology, but pathology is still the point: for Kramer no less than for Eliphaz, pessimism is evidence of interior disturbance.

"On this medication, I am myself at last": this is what Kramer tells us his patients say when their depression lifts. This may indeed mean that they have become healthy, that to be able to "laugh at drought and frost" is to feel the way nature intends them to feel. But it may also be that the self they have become can, thanks to the drug, give up the fixed and tragic view and live comfortably in an unchanged world. The usual—and, as we will see later, justified—rap against medicalized depression is that it doesn't really distinguish between ordinary sorrow and pathology. But it may be, indeed I think it is the case, that the diagnosis also can't distinguish adequately between disease and demoralization any more than the cure can distinguish between making people well and making them feel better about their lives. The depression doctors, in other words, may not be able to avoid the errors of Eliphaz.

I don't want to overstate this. I'm not worried that antidepressants will turn us into mind-numbed, smiley-faced zombies. The drugs aren't that effective, at least not yet. But I do think we need to pay attention to our feelings of demoralization. Pessimism can be an ally at a time of crisis, and I think we're living in one right now. Regardless of whether or not the drugs work, to call pessimism the symptom of an illness and then to turn our discontents over to the medical industry is to surrender perhaps the most important portion of our autonomy: the ability to look around and say, as Job might have said, "This is outrageous. Something must be done."

For religious people—in Job's time as well as in ours—the solution to the problem he represents is to relinquish the expectation that human sensibilities can grasp the sense of life and to replace it with a conviction that there is a divine, if inscrutable, plan behind our suffering. Job's pessimism and outrage, in this view, dissolve when he gives up that expectation. His suffering over the unfairness of his life is transformed into faith in a God whose justice surpasses understanding and whose mercy can soothe his grief. (Although the restoration of his wealth that Yahweh finally grants to Job, in an ending the rabbis tacked onto a book otherwise considered too bleak, seems to miss this point; for why would Job think that it will not be taken away again? And what about his children?)

For those of us who look to science for revelation, however, suffering has a very different fate and its cure rests on a different transformation. We place our faith in doctors and their science. Founded on the idea that knowledge moves us forward, that ignorance is all that stands between us and the best of all possible worlds, scientific medicine embodies the faith that we can figure our way out of our troubles. This belief rests on some optimistic assumptions: not only that the world will yield its secrets, but also that it has secrets to yield, that life is lawful in a way that will make sense to us. It's no wonder then that depression has fallen into the hands of the doctors: science is the natural enemy of pessimism.

To say that a particular form of suffering is a disease is always to go beyond the observation that the suffering exists. It is also to say—as Kramer does when he looks forward to the elimination of depression—that the suffering doesn't belong in our world, that we would live better lives without it, and that we ought to do so. When doctors turn suffering into symptom, symptom into disease, and disease into a condition to be cured, they are acting not only as scientists, but also as moral philosophers. To claim that an affliction ought to be eradicated is also to claim that it is inimical to the life we ought to be leading.

With some diseases, it barely matters that there is a philosophical dimension to a diagnosis. It is hard to imagine a world in which cancer and diabetes are not best understood as illnesses. But when the pathology is an attitude, a "fixed tragic view of the human condition," and when the treatment is touted as the restoration of the true human selfhood, then we really should consider whether that attitude is best understood as an illness to be eliminated. We should wonder whether doctors who urge us to come out against depression aren't, wittingly or otherwise, also urging us to adjust ourselves to a world that our pessimism shows to be deeply flawed.

And above all, we should recognize that to talk about why we suffer and what we should do about it is also to talk about how things ought to be. To say that a young man is sick when he is lying on the floor of his office, reeling from personal disaster, is to make a moral statement, and then to cloak it in the language of science, which plays the same role in our world that religion did in Job's. And just as Eliphaz and his colleagues overstepped with Job, so too the depression doctors, and their drug company sponsors, have overstepped with us. They don't know any better than you and I what life is for or how we are supposed to feel about it.

CHAPTER 3

MAUVE MEASLES

I became an officially depressed person in 2006. I received my diagnosis from a highly respected Harvard psychiatrist working at the Mood Disorders Unit at Massachusetts General Hospital who said I had major depressive disorder, recurrent, mild, with melancholic features. I wasn't entirely surprised that I turned out to be mentally ill. I had shown up at his office to enroll in a clinical trial of an antidepressant medication, and a diagnosis is a requirement for entry.

But I was expecting a different diagnosis—minor depressive disorder. This isn't an official psychiatric disease, at least not yet. It is listed in the DSM-IV, but in an appendix of "Diagnoses in Need of Further Study." Much depends on this further study, notably whether the diagnosis will get the official four-or-five-digit code that compels insurance companies to pay for its treatment, and whether the Food and Drug Administration will then be able to give a drug company an *indication*—that is, the right to claim that its drug treats the new disease.

Not all research diagnoses turn out to be winners. Take premenstrual dysphoric disorder (PMDD) for instance. The idea that PMDD is a psychiatric illness must have looked like a good idea to someone. My money is on Eli Lilly, which wanted to squeeze a few more dollars out of Prozac, or Sarafem, as the company had relabeled it for treatment of PMDD. But despite (or perhaps because of) its corporate sponsorship, the diagnosis ran into stiff opposition from feminists who objected to the way that it pathologized what they considered to be a normal variant of human behavior. PMDD has turned out to be a bust. It's still languishing in the back of the book and may even disappear completely when the DSM-V comes out in 2012.

Minor depressive disorder, on the other hand, seems like a shoo-in for advancement. To qualify, you only have to report three of the nine depression criteria, one of which has to be either a sad mood or loss of interest or pleasure in all, or almost all, activities. (Major depression requires five.) You could have trouble, for instance, eating and concentrating and, so long as you were also unhappy most of the time for most of the days in a two-week period in the last six months, you would qualify. By the time I got diagnosed, about twenty years after my bout on the floor of my study, I'd had enough disappointment and setback, not to mention nearly six years of the Bush administration, to ensure plenty of periods like that—none quite so bad as the first, but all of them unpleasant enough. I hadn't exactly come to value this experience, but I'd learned to tolerate it the same way that you tolerate a difficult friend or watch a disturbing movie, and for the same reason: that you get something out of the bargain, some insight into the world, some glimpse of the way things are.

I was, in other words, at risk of exactly what Peter Kramer, in *Against Depression,* warns about: mistaking mental illness for clarity. Of course, I had good company. Not only Job, with his conviction that a life in which everything you've lived for can be taken away so easily is not what it's cracked up to be, but a whole pantheon of unacknowledged legislators—like William James, who, in his *Varieties of Religious Experience* put it this way:

> The normal process of life contains moments as bad as any
> of those which insane melancholy is filled with, moments
> in which radical evil gets its innings and takes its solid turn.

The lunatic's visions of horror are all drawn from the material of daily fact.

Psychiatrists like Kramer cut through the philosophical question
of when the "normal process of life" becomes insanity with their
assertion that two weeks of those moments is quite enough. As
unsatisfying as this answer is—give me Yahweh in a thunderstorm
any day; at least then the fact that the line is arbitrarily drawn by
those with the power to do so is obvious—I thought it could work
in my favor. When I showed up at Mass General, veteran of the dis-
satisfactions of a middle-class, middle-aged American life—nothing
spectacular, but enough to keep me up nights and make me blue
for a couple of weeks at a time on a regular basis—I figured that a
diagnosis of minor depression was a sure thing.

I'll tell you all about how wrong I was, and why, and what happened, in due time. First, though, let me get the credibility problem out of the way. Yes, I went to Mass General with an agenda.
I figured my temperament and the American Psychiatric Association's zeal to create a new disease were made for each other—and
added up to a good opportunity to write about a little-explored
region of the medical-industrial complex. Clinical trials are the
pivot of the depression industry, the venue in which the drug companies get the government to guarantee the public that an antidepressant works and won't hurt you. But in the bargain, they also
confer legitimacy on the disease that the drug purports to treat.
Every approval of an antidepressant also ratifies the claim that the
disease it treats really exists.

Many diseases don't need this kind of advertising. Whatever *disease* means—and this question is far from settled—no one will deny
that cancer or malaria deserve the label. But sometimes the public
has to be convinced. That's why GlaxoSmithKline (GSK) once paid
a doctor to say that an "uncontrollable urge to move [the] legs, or
'creepy-crawly' sensations in the legs . . . that often leads to sleep
disruption" was actually a disease called restless legs syndrome.

RLS, said GSK, causes insomnia, marital discord, and poor job performance. This campaign was a transparent attempt to persuade people to think of their suffering as a disease, and the linchpin of the argument (and, of course, the reason GSK was bothering to make it) was that the drug Requip, which as a treatment for Parkinson's disease had reaped disappointing profits, relieved RLS. If a medicine makes a problem better, this logic goes, then the problem must have been a disease to begin with, and its sufferers are entitled to all the benefits we bestow upon the sick: sympathy, research money, insurance coverage, and so on.

So a clinical trial of a treatment for a research diagnosis like minor depression is also a trial of the diagnosis itself. That's why I went to Mass General: to see if I could catch a glimpse of the depression machinery as it cranked up to turn out a new model.

As motives go, mine was less straightforward than, say, wanting to benefit humankind or get myself cured. But neither was my visit to Mass General a repeat of one of the greatest pranks ever perpetrated on psychiatry: a study by David Rosenhan, a sociologist who in 1972 sent a cadre of his graduate students into various emergency rooms complaining, dishonestly, that they were hearing the word "thud" in their heads. The students were hospitalized, most of them for schizophrenia. Once there, they behaved normally, or what passes for normally among graduate sociology students. They read, asked questions, and took extensive notes, all of which was duly noted in their charts as more symptoms. When they were released, it was with the diagnosis of paranoid schizophrenia, in remission.

Rosenhan called the 1973 paper he published in *Science* about his caper "On Being Sane in Insane Places." You can imagine how embarrassed psychiatrists were when the story hit the press that the *One Flew over the Cuckoo's Nest* nightmare was true. The profession was already under siege, thanks to a society that for many reasons was beginning to suspect that mental illnesses weren't real, but merely ways of pathologizing nonconformity. A cottage industry sprang up to rebut, denounce, and generally scream at Rosenhan. But no one took issue with his finding that it's easy to get diagnosed with a mental illness and then much, much harder, if not impossible, to get undiagnosed. That part, as opposed to his allegedly shoddy ethics and research methods, was unassailable.

This was the best part of my going to Mass General: I didn't have to lie. Not that I hadn't thought about it. I do know the DSM-IV pretty well, certainly well enough to fake just about any psychiatric illness. But I couldn't get myself to do it just for the sake of my writing career. So when I found out there was a study I thought I could get into by telling the truth, I jumped at the opportunity.

It wasn't all ambition, however. The possibility that the trial drug, which was Celexa, might make me feel better—well, I can't deny this was intriguing. I've got nothing against better living through chemistry. I've practiced my own amateur version of it for many years, in fact. And I've spent a couple of decades listening to patients (and friends) sing the praises of antidepressants and seeing the results up close and personal. It was enough to make me curious, and sometimes envious, especially when I was depressed. Sometimes I'd wonder if it wasn't just stubbornness that stopped me from visiting a psychiatrist, some point of pride or a fear that I'd be sucked into the Prozac cult or forced to abandon some of my deepest convictions if the drug worked—reluctance that, whatever the explanation, felt insuperable, and, at times, depressing. The clinical trial gave me perfect cover from myself, a way to check out the drugs while maintaining that I was only doing research. Call it the Kinsey approach.

So when I got diagnosed with major depression instead of minor depression, I suppose I was only getting my comeuppance for trying to exploit the system, which was in turn glad to bestow a disease upon me, but not necessarily the one I wanted.

I didn't get kicked out of Mass General. To the contrary, my doctor immediately gave me five major depression studies to choose from. But it was impossible to ignore the fact that after listening to my answers to his questions, a capable and compassionate doctor told me I had a serious mental illness—something wrong with my brain that was causing the trouble in my mind—and a much worse one than I had thought to begin with. I was the last person in the world that I would have expected to believe this. But as I'll describe, the idea that my difficulties were an illness caused by biochemical imbalances grew on me during the trial—especially the part about the possibility that I could be cured of what I had long ago come to think of as myself.

But even before that happened, even as I walked to his office for the first time, it had dawned on me that this whole vast apparatus with its towers and pavilions arrayed like the castles of the Magic Kingdom, its maze of bustling streets—the doctors checking their watches, the patients, some wheeling IV stands down the sidewalk (one of them even sneaking a smoke as the liquid dripped into his veins), the family sitting crying on a bench—was a monument to one brilliant and magnificent idea: that our suffering is caused by diseases that can be cured by medicine. Well, actually, those are two ideas—that diseases exist in nature and that we can improve nature by finding the culprit and getting rid of it—and they seem, like all common sense, to be unassailable and timeless. They may even seem not to be ideas, but simple facts.

But they are ideas, invented by people rather than discovered in nature, and much newer ones than you might think. Indeed, the belief that we can turn what ails us into the target for a drug first appeared about 150 years ago and was not widely accepted until the early part of the twentieth century. Neither have illness and cure always been related in the way we usually think they are: that we identify diseases and then look for the remedy. Drug-driven diagnoses are not original to the depression industry, or for that matter to the restless legs syndrome industry. In fact, when it comes to the modern understanding of disease, the drugs have often come first.

Betty Twarog's mussels weren't the first mollusks to figure in the history of depression. *Murex trunculus* and *Murex bandaris,* two species of sea snails that litter the coasts of Italy and Asia Minor, beat her critters to the punch by a good two thousand years. Both varieties are four or five inches long, with pastel-colored bands spiraling up their shells. *M. trunculus* looks like an elephant's head, its spiny shaft like a trunk, while *M. bandaris* features spikes that stick up like the points of a child's jack. You might stoop to pick up a few of these specimens on a beachcombing walk, but they're a little too subtle to be your prize find. Unless, that is, you happen to be the unnamed dog belonging to Heracles that, according to legend, took a mouthful of snails while on a walk with his master. The legend doesn't say why Heracles was at the beach in Tyre (I suppose that even Greek half-god heroes need a vacation occasionally), but it does tell us that after the dog spit them out, its mouth had turned an extraordinary shade of purple.

It was left to the Phoenicians to figure out how to exploit this discovery by crushing, salting, and boiling the snails until they had extracted a dye that, according to Pliny the Elder, was "exactly the colour of clotted blood, and . . . of a blackish hue to the sight, but of a shining appearance when held up to the light." It was laborious to produce—the mucus of thousands of snails was needed to color a single robe—but so glorious that it eventually fetched its weight in silver at ancient markets. Tyrian purple (also known as royal purple), became the color of kings and generals and nabobs.

And, in the late 1850s, of the ladies of Paris. Inspired, some would say inflamed, by the Empress Eugénie, wife of Napoleon III, whose haute couture graced the pages of the fashion magazines just then coming into production, Parisians couldn't get enough of the scarlet-purple hue known to them as mauve. A little less red than the Phoenician original, mauve was nonetheless a gorgeous color and demand was high. "Mauve Measles," as *Punch* called it, spread quickly across the Channel, leaving Englishwomen with a "measly rash of ribbons."

The Parisian mauve came not from snail mucus but from bat guano and from certain lichens that also could stain fabric purple.

These sources were plentiful and easier to refine than the snails, but supplies still had to be found and secured, harvested and processed. Variations in sunlight and soil conditions and other vagaries of nature could affect hue and quantity. None of this would necessarily have been a problem worth solving—after all, wasn't this kind of inconsistency the way of the natural world?—if it weren't for the Industrial Revolution, which was imposing a new expectation: that commodities, particularly consumer commodities, should be uniform and easily available and certainly not made out of bat poop or snail snot if it could at all be avoided. It was left to a kid to figure out how to meet this emerging market.

By the time William Perkin entered the City of London School in 1851, at the age of thirteen, he had already considered a number of careers: carpenter (his father's trade), engineer, painter, musician. At his new school he took a shine to science and sought entry to Michael Faraday's Saturday lectures about electricity, a request that was granted by the great man himself. But nothing caught his fancy like the twice-weekly chemistry lectures taught by Thomas Hall, his writing master. Soon young Perkin prevailed upon his father to allow him to set up a lab at home where he could explore the principles he was learning from Hall.

Chemistry barely existed as a scientific discipline in mid-nineteenth-century England, where it was associated with apothecaries and other charlatans. But Faraday, along with Prince Albert and other prominent Britons, saw the advances being made by chemists on the Continent, especially in Germany, and rounded up the money for the Royal College of Chemistry, which opened in 1845 with twenty-six students. Hall had attended the first classes there, and in 1853 he urged Perkin to enroll.

Perkin's mentor at the Royal College was August Wilhelm von Hofmann, who had been recruited from Germany by Prince Albert himself. Hofmann had an interest that was perfect for demonstrating to a skeptical educational establishment the practical value of studying chemistry. He thought that natural substances could be synthesized in the lab so long as you started with materials that contained the elements that chemists were just identifying as the building blocks of the natural world: carbon, oxygen, hydrogen, sulfur, and nitrogen. Nature, according to Hofmann, assembled these atoms into molecules, and then into the substances of daily life, in much the same way that industrialists were assembling raw materials into finished products. Figure out how nature did this, and *you* could conceivably make anything it could make—and without all the bother of, say, gathering and boiling snails.

Hofmann was lucky to have this insight at a time when a rich supply of those basic elements was just coming available—and on the cheap, as an unwanted by-product of industrialization: coal tar, the stinky residue of the process by which coal was refined into the gas that fueled London's lamps. An enterprising Scotsman, Charles Macintosh, figured out how to smear coal tar on textiles to make a rubbery waterproof cloth, and soon people were wearing macintoshes in the Glasgow rain. But no one knew what else to do with the stuff, so mostly it got dumped into streams, where it killed fish and made washday a nightmare. Hofmann, however, thought he could extract value from the hydrocarbons, and he set Perkin to work on one of his pet projects: making quinine.

Malaria was not only a scourge of people living in the swamps and fens of England and the rest of Europe, it was also a problem for the armies of imperialism, which found the warmer climates overrun with the disease—a bad thing not only for the natives but, more important, for the men who would contract malaria in the course of conquering them. Cinchona bark had long been used by indigenous people as a remedy for fevers, and at the end of the seventeenth century, a British physician, in one of the earliest controlled studies of a drug, proved that its effect was unique to what was then known as tertian fever.

In the nineteenth century, a pair of Frenchmen isolated quinine as the active ingredient in the bark and developed a way to extract it. Soon, the medicine was in wide use. The trees still grew mostly in South America, however, so quinine was expensive: the East India Company's annual budget for it at midcentury was one hundred thousand pounds—enough to hinder the business of empire and to delay the invention of the preferred way to end a long day of bearing the white man's burden: the gin and tonic.

Hofmann thought he had a solution to the problem. Thanks to advances in microscopy, he knew that naphthalidine, a derivative of coal tar, differed from quinine by only a couple of hydrogen and oxygen molecules—water, in other words. It wouldn't be as simple as adding water to the naphthalidine, he said, but with enough time at the bench, he was sure he could make medicine out of coal tar. Hofmann set Perkin to the "happy experiment" of figuring out how to make oil and water mix.

Alas, the experiment was not so happy. Perkin soon determined that all the chemistry in the world wasn't going to convince naphthalidine to turn into quinine. One of his failures was intriguing, however: it had yielded a powder with a reddish tint. Perkin decided to see what color would come from another coal tar component, aniline. "Perfectly black," he described it, and when he dried it and added alcohol (wine spirits, in this case), it turned a beautiful shade of purple. Perkin, with the encouragement of some older entrepreneurs, soon proved that the new color, mauveine, was fast in wool and silk. He quickly forgot about malaria. He had achieved something stupendous—an alchemy that actually worked. He could transmute the dross of the Industrial Revolution into gold—or mauve, as the case may be.

Perkin was eighteen years old when he got his patent on mauveine, and he was soon fabulously wealthy. Dressmakers, textile manufacturers, fashionable (and now, due to the lower price, not-so-rich) women and the men who liked to look at them all benefited. Unless you count the factory workers and neighbors (and occasional customers) who suffered from the poisonous by-products of aniline dye manufacturing and the producers of natural colors, whose madder and indigo and other vegetable-based dyes plummeted in price, mauve made winners out of everyone.

But the biggest beneficiaries of all were a group of German companies whose names are still familiar: Bayer, Hoechst, Geigy, BASF (originally the Baden Aniline and Soda Factory), and Agfa. Perkin's invention proved that natural substances could be chemically synthesized out of cheap chemicals, and these companies exploited the obvious opportunity. Uncovering the structure and properties of hydrocarbons—the field we know today as organic chemistry—they eventually figured out how to manufacture plastics, textiles, pesticides, and all the other synthetic products that we've become accustomed to and dependent upon. Not least among those products were drugs, and the German companies that turned out synthetic dyes eventually became the backbone of the pharmaceutical industry.

The journey from Perkin's mauve to Prozac is not as long or winding as you might think. From the time that Hofmann tried to synthesize quinine, the medical industry and the dye industry have been thick as thieves, and not only because some dyes, as you'll see in a moment, turned out to cure diseases. The linking of haberdashery and therapeutics also helped to establish an idea that revolutionized medicine, the idea that lay underneath the clinical trial where I got my diagnosis: that drugs can be magic bullets, aimed directly at the chemical causes of our suffering.

Paul Ehrlich, the German doctor who came up with this theory, had just been born when Perkin invented mauveine, but by the time he was a teenager, he had been captivated by—some said obsessed with—synthetic dyes of all colors. "When I felt . . . miserable and forsaken," Ehrlich once told his secretary, "I often stood before the cupboard in which my collection of dyes was stored and said to myself, 'These are my friends which will not desert me.'"

Ehrlich had been introduced to his friends by his cousin, a biologist who was exploiting dyes in a way that Perkin didn't foresee: as staining agents for microscope slides. The synthetic dyes, far superior to vegetable-based colors, illuminated cellular structures as never before. Ehrlich was enchanted by his cousin's slides, but it wasn't this new glimpse into the invisible world of cells that caught his interest. Rather, he wondered, why did the same dye make different parts of the same cell show up in different shades? How did dyes perform their revelatory magic in the first place?

Ehrlich's biologist friends were uninterested in this question. "They cared so little for the *theory* of it," he complained to his secretary. But Ehrlich thought he knew where the answer lay: in chemistry. Like Perkin, he had been a basement chemist, and he claimed to have unique abilities—not only to make concoctions, but also to understand what he was doing. "I can see the structural formula . . . with my mind's eye," he once wrote with a characteristic lack of modesty, "and my chemical imagination has developed so rapidly that sometimes I have been able to foresee things that were recognized only much later by the disciples of systematic chemistry." What Ehrlich saw under the microscope with his mind's eye was not nuclei and mitochondria lit up in color, but the whirling dance of molecules that was spinning out the colors in the first place. "Substances act only when they are linked," he proclaimed. When a dye latched onto a cell, the linkage created a new molecule that had the color as one of its properties. Stained tissue samples thus opened a window into not one but two unseen realms: the biology of the cell and the chemistry of the interaction between living matter and hydrocarbons.

Even among his mad-scientist peers at medical school in the 1870s, Ehrlich was known as a "brilliant eccentric . . . wandering around the laboratory with hands that looked as though they had been thrust into innumerable paint pots up to the wrists." Upon graduation in 1878, he continued to hang out with his dyes but also took a job as a physician at Berlin's largest hospital, the Charité. Hospitals, even good hospitals, were grim places at the time, with little but comfort to offer patients with typhus, tuberculosis, syphilis, cholera, and all the other diseases (not to mention the opportunistic infections that could turn any wound into a death sentence) contributing to the forty-year life expectancy of the average Johann. But scientists had just begun to explore a new and controversial theory about illness: that tiny organisms—*germs*—were responsible for most of the scourges that brought people to the hospital or the family deathbed in the first place.

This wasn't a new idea. As far back as the first century B.C., a Roman doctor, Marcus Varro, was warning his countrymen to avoid marshland lest they encounter the "minute creatures that live [there] and that cannot be discerned with the eye and . . . enter the body through the mouth and nostrils and cause serious diseases." More recently, in the 1790s Edward Jenner had discovered that smallpox could be prevented by inoculation with cowpox, and in 1854 John Snow had demonstrated that cholera was transmitted through the water supplies.

You would think that doctors would have connected the dots by the time Ehrlich was trying to puzzle out how to use the knowledge he was generating with dyes in his medical practice. But although scientists had by then cracked open enough bodies, living and dead, to gain at least a rudimentary understanding of how blood flowed, how some of our organs worked, how we were put together, and even, in a very small way, how the brain gave us speech and thought and behavior, doctors remained firmly rooted in their Hippocratic past. Like those early physicians, they gathered their knowledge empirically. They tried something and if it seemed to work they tried it again on a similar sickness without spending much time on the whys of the result. That's why the remedies at their disposal—techniques like bloodletting and trepanning, and the herbal potions in the pharmacopoeia—were largely those of their ancient masters.

Medicine in 1850 was largely as it had been in 350 B.C.—guided by whim, by accident, and by tradition.

Some of these traditions were harmless enough. Theriac, for instance, a concoction dating back at least to the second century A.D. healer Galen that remained in the pharmacopoeia until the mid–twentieth century, probably never hurt anyone. In fact, since it contained—in addition to pulverized viper flesh, which is to say snake oil—generous quantities of opium, it probably made nearly everyone feel better. But other traditional cures probably hastened many deaths, including that of George Washington, whose fatal ague was treated with mercury, another popular ancient remedy and the same metal that doctors today tell you to avoid at all costs. To the extent that physicians were successful they relied not on knowledge of how their remedies acted biochemically to cure a disease but on luck, on trial and error, and, perhaps above all else, on the placebo effect. At least one prominent doctor of the nineteenth century—Oliver Wendell Holmes—knew this. "If the whole *materia medica* . . . could be sunk to the bottom of the sea," he told his brethren in the Massachusetts Medical Society in 1860, "it would be all the better for mankind and all the worse for the fishes."

Naturally, most doctors didn't see it that way. They even had a theory for why their medicines should work. Disease, they thought, was the result of an imbalance among the four humors—blood, phlegm, and yellow and black bile—that coursed through the body. This idea had been in force since the Hippocratic era and updated over the centuries, although the relationship between, say, blood the humor and blood the substance had never been fully clarified. Nor had the reason that a miasma—bad air—could throw the humors out of balance and thus be the source of a contagious illness. So even by the time Holmes was worrying about the fishes, the treatment of disease remained a matter of using time-proven, if poorly understood, remedies to restore the proper ratio of the humors.

To accept the germ theory, doctors would have to abandon not only all this tradition, but a worldview about balance and harmony that was more than two thousand years old. As one doctor put it, "If we were once to admit that the pox was produced by little animals swimming in the blood, then we would have as much reason to think likewise not only of the plague . . . but also of smallpox, hydrophobia, scabies, sores . . . and this would overturn the whole of medical theory."

But some scientists were beginning to doubt that what had worked for Plato and Aristotle was really the best medicine. Among them was Louis Pasteur, who in 1862 showed that heating a liquid would kill the microorganisms that fermented it, and who strongly suspected that doing so would make the milk supply safe. But it wasn't the great Frenchman who convinced the reluctant medical establishment of the truth of the germ theory. That honor went to a German country doctor, Robert Koch.

Koch's patients were farmers, and their herds were succumbing to a disease that turned their blood black and thus was known as anthrax, from the Greek for "coal." Koch had heard that a couple of scientists, using the new stains and improved microscopes, had spotted some rod-shaped structures in the blood of animals that had been killed by anthrax. He bought himself a state-of-the-art microscope and used it to elucidate the life cycle of the rods, which he called "bacilli." He then conducted a series of experiments in which he injected mice, rabbits, and frogs with the bacillus. The animals did him the favor of contracting the disease and dying, and in 1876, Koch announced that he had discovered the cause of anthrax and offered the undeniable proof under his microscope.

Paul Ehrlich and Robert Koch met two years later, but they were only distant acquaintances until 1882. That was the year that Robert Koch announced that using the same techniques he had used with anthrax, he had found the tuberculosis bacillus and killed four guinea pigs with it. He lamented, however, that he had not yet found a dye that would make the bacillus easy to spot. Within the year, Ehrlich had found that methylene violet would light it up unmistakably, and the two men began a long collaboration.

But identifying a pathogen and doing something about it were two different matters. Pasteur came to focus on vaccination, while Koch thought that sterilization and other public health measures were the best way to kill the bugs. Ehrlich, however, had a different idea. Why not exploit the chemical affinities between synthetics and organisms for purposes beyond diagnosis? He explained his idea this way:

> It should be possible to find artificial substances which are really and specifically curative for certain diseases, not merely palliatives acting favorably on one or another symptom . . . Such curative substances a priori must directly destroy the microbes provoking the disease; not by an "action from distance" but only when the chemical compound is fixed by the parasites.

"Magic bullets," as Ehrlich called these substances, would fly "straight onward, without deviation, upon the parasites." Once scientists "learn how to take aim, in a chemical sense," they could have a "marvelous effect": they would be able to draw a bead on disease and kill it.

Ehrlich initially conceived of magic bullets as the body's own weapons, and of drugs as a means to unleash them, a process he called "chemotherapy." His investigations into immunology earned him the Nobel Prize in medicine in 1908, but by then his efforts had turned to a more direct use of chemicals: as the bullets themselves.

Ehrlich had also by then gained a corporate patron: Hoechst, to which the magic-bullet model was greatly appealing and which bankrolled Ehrlich starting in 1906. Ehrlich's a priori approach promised not only huge reward, but also lower research costs. It took some of the guesswork out of drug development by using knowledge about the molecular structure of both drug and pathogen to narrow the field of candidates. "There must be a planned chemical synthesis," Ehrlich said, and he had an idea of where to start.

The ongoing attempt to synthesize quinine had led researchers to look at other tropical diseases besides malaria, including sleeping sickness, the coma-inducing result of a tsetse fly bite. In 1903, one of Ehrlich's dyes had proven toxic to sleep-sick animals. Trypan red—named after the trypanosomes, the spiral-shaped organisms that caused the disease—proved disappointing as a remedy; whatever it was doing to cure mice was not working in humans. In the course of his work, however, Ehrlich had discovered that trypan red was much more effective in the mice when it was mixed with a form of arsenic. He also heard that a couple of British doctors had used Atoxyl, a compound derived from arsenic, to treat animals infected with sleeping sickness. Atoxyl turned out not to work in humans; even worse, it destroyed their optic nerves, making them blind before they slept themselves to death.

But Ehrlich was intrigued by arsenic, which, like mercury, was an ancient remedy. And he had discovered something significant about Atoxyl and other arsenic derivatives: at the end of their chemical chain was a reactive chemical group, an open link of nitrogen and hydrogen that could easily be joined to other molecules. Ehrlich concluded that Atoxyl was effective in animals for exactly this reason: its reactive chemical group latched onto the "parasites" in the same way that dye latched onto tissues. It was perfect, in other words, for the planned syntheses that could both prove out his magic-bullet method and lead to drugs that would make Hoechst happy. And he determined that the test animals were doing a little synthesizing themselves: their metabolism converted the arsenic in Atoxyl into a more useful and less toxic compound. Ehrlich decided to follow suit, using the reactive chemical group to make a version of the arsenic metabolite. He combined this molecule with other chemicals, searching for a drug that would work in people. The first three years and 417 syntheses were failures, but the 418th attempt found its target: *arsenophenylglycin*, as the compound came to be known, reliably killed trypanosomes in animals and humans but otherwise left the host's body alone.

Taken alone, this discovery would probably not have changed the course of medicine. Sleeping sickness, after all, is a rare disease, and one that mostly afflicts people in impoverished parts of the world. But in 1905, a colleague of Ehrlich's at the Charité had discovered another spiral-shaped pathogen, a spirochete, and this one was responsible for a plague that at the turn of the century had infected as much as 10 percent of Europe's population, ravaging the lives of common folk and kings and artists alike: syphilis.

Known to the French as the disease of Naples, to the Italians as the French disease, to the Russians as the Polish disease, to the Japanese as the Chinese disease, to the English as the Spanish disease, and to all concerned as a scourge, syphilis, or the great pox, first appeared in the late fifteenth century, not long after Christopher Columbus returned from the New World. When Europeans weren't blaming one another for the disease, they speculated that its true origins were the savages that Columbus and his men met (and then some) on his voyage.

John Hunter, the eighteenth-century British physician, was one of the blame-America-first crowd. He also believed, against the prevailing wisdom of the time, that syphilis was caused by a "putrid liquid" contained in the pus that exuded from a chancre. In 1767, he injected some of that liquid, obtained from a patient's penis, into his own. His intent was to determine whether gonorrhea and syphilis were variants of one illness. He didn't consider the possibility that his patient could have both illnesses, so when he developed symptoms of each, he concluded that "gonorrhea and the chancre are the effects of the same poison." Even if he had this wrong, however, he did manage to report accurately on the natural course of venereal diseases—although when his aorta burst twenty-six years later, he probably didn't know that his experiment had killed him.

Hunter treated himself by cauterizing and applying mercury to his chancres and to the sores that erupted elsewhere in the disease's later stages. Mercury had been the treatment of choice for the pox from the beginning—probably because physicians since Galen had used it with some success to heal skin conditions. Doctors rubbed mercury on patients' genitals, injected it into their veins, vaporized it so they could breathe in the vapors, served it up as a chewing gum, dissolved it into alcohol to be imbibed, infused it into their rectums, and even, at least in Italy, coated their underpants with it. (The treatment gave rise to a rueful adage: "A night with Venus is followed by a year with Mercury.") Mercury did seem to have some effect, and physicians learned to recognize the symptoms of mercury poisoning, like salivation, before it killed their patients, but doctors' ability to publicize themselves as healers of the pox, which they did in slick pamphlets promising effectiveness and discretion, generally far surpassed their ability to actually do so.

With the success of the smallpox inoculation, some nineteenth-century doctors thought that the great pox would also be controlled by vaccine, but they soon found that infected people could get reinfected, making the disease unsuitable for that treatment. Public health measures also foundered—on both the distaste for talking openly about sex and on the near impossibility of controlling people's sexual behavior. By midcentury, at least one doctor saw syphilis as making a mockery of progress.

> Nineteenth century man has managed to do away with long distances, tunnel through mountains, harness the power of fire, and yet he has not thus far managed to preserve himself against disease. As a conqueror of matter, a new Icarus, he sets out boldly heavenwards; as the threadbare king of creation he . . . falls prey to a disease which the simplest precautions would enable him to avoid.

And in 1875, a French scientist gave even more reason for concern: syphilis, he said, was even worse than already thought, for it not only caused the genital chancres of its primary stages, and the blisters and fevers of its secondary stages, but a third, more horrifying stage, characterized by tabes dorsalis, a condition that caused people to lose muscle control and balance, and general paresis, a form of insanity. By one count, nearly half of the patients in Europe's mental hospitals were suffering from tertiary syphilis.

A pale spiral-shaped organism had been spotted in chancres as early as 1837. But it was hard to make out in the microscopes of the time, and germ theory was still a twinkle in Pasteur's eye, so there was little interest in isolating and identifying the bug until Ehrlich's colleagues found it in blood and tissues as well as syphilitic sores. Because it vaguely resembled the spiraling trypanosomes, it made sense to see whether compound 418, which had killed those germs so well, would also attack the syphilis spirochete. The answer, unfortunately, was no, but Ehrlich and Hoechst were convinced that they were on the right track, and eventually Ehrlich's 606th compound did the trick, killing the syphilis bug in infected rabbits.

Even before Ehrlich had satisfied himself that his results were not a fluke, word about 606 got out—and so, thanks to Hoechst, did samples of the drug. Doctors in Italy and Russia reported good results with early stage syphilitics, a Swiss physician reported that in addition to successful treatment of syphilitics he'd used 606 to cure a case of leukemia, and Alexander Fleming, a British doctor not yet famous for stumbling onto penicillin, reported that the drug had a "remarkable effect" on syphilis.

In April 1910, Ehrlich officially announced his success at a Wiesbaden medical conference. Almost immediately, he was besieged with requests for 606. Patients showed up at his research offices, hoping for a shot. By September, more than ten thousand patients had been treated, a number that had tripled by November. By year's end Hoechst had distributed 65,500 free doses to doctors all over the world—virtually everywhere but the United States, where doctors warned that having a cure for the disease might encourage promiscuity. Headlines soon trumpeted the success of Salvarsan, the household-friendly name Hoechst settled on instead of 606 or the even less mellifluous *dihydroxydiaminoarsenobenzene*. By year's end the company had put more than 375,000 doses into the pharmacies and was the proud owner of the world's first scientifically proven wonder drug—a magic bullet aimed directly at the heart of one of the worst diseases known to humankind.

Shortly after Salvarsan's release, a German magazine ran an article that included this clever encomium to the drug:

> *There's hardly a child who believes in the gods,*
> *They've vanished without a trace;*
> *Even those who serve at Venus's court*
> *Now laugh in Mercury's face.*

This enthusiasm was misplaced, as wonder drug enthusiasm often is. Salvarsan promised to make the world safe for adultery and fornication, but even in successful cases, treatment could drag on for months or years of repeated, painful injections and debilitating side effects. Sometimes it made people sick or even killed them—it was, after all, made from arsenic. Some doctors stopped laughing at Mercury and started augmenting Salvarsan with old-fashioned quicksilver ointments. The treatment ultimately came to be seen, as one doctor put it, as "a long, slow, painful, and expensive grind," and the world's enthusiasm, at least when it came to syphilis treatments, eventually moved on to the next wonder drug, Fleming's penicillin.

But the quatrain's optimism was, in another sense, spot-on. Salvarsan finished the job that Hippocrates started. After two millennia of stumbling around in the humoral darkness, doctors and drug companies were ready to displace the gods entirely from the clinic by taking direct aim at disease and killing it at its source. And even if the drug itself was only a qualified success, the idea behind it was a blockbuster. Soon enough, Ehrlich's magic-bullet promises came true in ways that exceeded imagination.

To cite just three familiar examples: scientists identified the lack of insulin as the culprit in diabetes in 1921; in 1922 a team in Toronto successfully treated a fourteen-year-old diabetic boy; and in that same year Eli Lilly and Company devised a method for mass-producing human-ready insulin from the pancreas of a cow. In 1942, chemists at Charles Pfizer and Company, which made its first fortune by producing citric acid, figured out a way to grow penicillin in fermenting corn liquor, allowing a drug previously great in promise but short in supply to be mass manufactured in time to treat the wounds (and syphilis) of World War II soldiers. In the late 1950s, researchers at Merck discovered that chlorothiazide, another benzene compound, could change blood chemistry enough to lower blood pressure, and Diuril was born. By the turn of the twenty-first century, Oliver Wendell Holmes's prophecy had come true, but in a way he couldn't have expected: the fishes were indeed suffering from the modern *materia medica,* but only because millions of people living better through chemistry were flushing the metabolized remains (and unused pills) down the toilet and out into the sea.

It is nearly impossible to overstate the impact of Ehrlich's idea. It has turned the suppliers of magic bullets into wealthy and powerful corporations, and doctors into dead-aim gunslingers possessed of an authority that Hippocrates could only dream of. It has also turned diseases into afflictions with specific causes that can be located in our biochemistry. By revolutionizing our view of sickness and health, in short, it has ushered in a new climate of opinion about suffering and its remedy, and even more about who we are and why we suffer: not the descendants of Job, awaiting the next inexplicable misery, but people with a biochemical essence that can be known and, when it goes wrong, corrected.

And that's not all. If scientists can figure out how to make a beautiful color the first time, every time—and out of industrial waste no less, no mucking about with snails or bats—and use that knowledge not only to make the ladies of Paris happy but also to cure the great pox, then the prospects for humankind are suddenly and dramatically enhanced. Petitioning Yahweh for an account of the wherefores of suffering—and falling into resignation or despair when no answers are forthcoming—is unnecessary when doctors can peer into the recesses of the body, find the answers in its molecules, and send the chemicals in to the rescue. The promise of a boundless future that originated with the Enlightenment and began to come to fruit in the Industrial Revolution has perhaps no better expression than in the birth of scientific medicine.

But that promise also created a temptation, one that eventually would prove irresistible. To the manufacturers of drugs, diseases are markets. The continued growth and success of the pharmaceutical industry depends on a proliferation of those markets. It was only a matter of time before doctors and drug companies started to improve upon nature in yet another way: by creating the diseases for which their potions are the cures.

Indeed, even as Ehrlich was working in Berlin, in another corner of the same hospital, another doctor was beginning to do just that. He wouldn't have described his work that way, of course. As he mapped the landscape of psychic suffering, he thought he was discovering diseases, not inventing them, and he had no intention of curing his patients. Still, simply by insisting that there were mental illnesses in nature and that he knew how to find them, Emil Kraepelin set the machinery of depression in motion.

CHAPTER 4

THE DANGERS
OF EMPATHY

When my doctor at Mass General went to determine whether or not I was minorly depressed, he did what nearly any researcher in any clinical trial for a psychiatric condition does. He sat down across his desk from me and opened up a big loose-leaf binder. In it was the script for the Structured Clinical Interview for DSM-IV (SCID), a test derived from the DSM's diagnostic criteria. The procedure is very simple. To find out if you satisfy the two-weeks-of-sadness requirement, the doctor asks you if you have been sad for two weeks. To find out if you have lost interest in the activities that usually bring you pleasure, he asks if you have lost interest in the activities that usually bring you pleasure. This goes on for forty-five minutes or so, the questions shunting you from one slot to another, like a coin in a sorter, until you drop into the drawer with all the other pennies.

What the doctor doesn't do as he scores your SCID is pay much attention to how you are actually behaving, the words you use to express yourself, or the way you come across in person—in short, the qualities that we usually think make us who we are. In fact, my doctor didn't even have my name right. He kept calling me Greg. I would have corrected him, but I didn't want to embarrass him.

In the old days—which is to say back when psychiatrists paid attention to your own account of your interior life—the fact that I didn't want to embarrass my doctor, had it somehow come up in the interview, would have mattered. A clinician might have seen it as a reflection of some aspect of my personality—a fear of conflict, perhaps, or disguised hostility, or even some compulsive need to take care of others. My suffering would have been seen as the outgrowth of that fear or need, the question of how I came to feel that way would have been central, and the diagnosis would have depended in part on the answer to that question. The symptoms alone, in other words, would not have been enough to render a diagnosis. The doctor would have needed to understand the context and meaning of my symptoms, and my illness would have been seen as at least partly a matter of biography.

It's not hard to understand why diagnosis doesn't work that way anymore. Reaching that kind of conclusion requires open-ended conversation and liberal interpretation, which would be very hard to map onto a troubleshooting chart. That's an inefficient process, and it would yield an unscientific result. The difficulties raised by this approach to diagnosis reached a crisis point in the early 1970s. In addition to the Rosenhan study, psychiatrists were confronted with research that showed that they often disagreed about what mental illness a given person had. Diagnostic trends varied from country to country, from city to city, even from hospital to hospital, and diagnoses began to seem more like folk stories than medical categories. Even worse for the industry's credibility, in 1973, after years of subjecting homosexuals to all manner of "treatment," the American Psychiatric Association voted homosexuality out of the DSM. Developments like these seemed to indicate that psychiatrists didn't know how to define mental illness to begin with. That kind of confusion could have been very bad for business.

So just a few years before Prozac came along, psychiatrists turned to what they called a descriptive nosology. In a development I'll describe in detail later on, they came out with an entirely revamped

DSM, one that focused not on personalities or causes of mental illnesses but on lists of symptoms like the one that my doctor was using to diagnose me. These lists featured more or less objective criteria—duration of unhappiness, changes in weight, length of sleep. They were designed to meet statistical standards like interrater reliability, which made them much more friendly to the quantitative tests and measures that we equate with science. And they worked. It turns out that if you standardize the questions you ask, you will come up with standardized answers. Or, to put this another way, if you go into the interview looking for what you already know, then you are very likely to see it.

The trick with the descriptive approach to diagnosis is to keep your eye on the loose-leaf notebook and not on the patient. That's why it didn't really matter whether my doctor knew my name or noticed that I was cracking jokes, engaging him in relatively sophisticated conversation about neurochemistry, talking about sad things but not being sad—or, for that matter, that I had driven eighty miles, shown up almost on time (the subway was a little slow), was dressed and groomed, and so on. Details like these would have been inconvenient, to say the least. Clinical trials are hard to fill. Even more important, the mental health industry's commitment to the DSM and the SCID is its best hope for maintaining its sometimes tenuous place among the disciplines of scientific medicine. If the SCID spits out a diagnosis that just doesn't fit the patient, then what would that mean about the psychiatric enterprise?

The problem here is that all those descriptors, in all their detail and specificity, don't necessarily add up to a disease. A good doctor would never conclude that a person with a sore throat and fever necessarily has a streptococcal infection, and a good scientist would not say that the disease of strep throat is constituted solely by a sore throat and fever. Both would insist that a bacteria must be present to complete the diagnosis. This is the great advance in diagnostics brought on by magic-bullet medicine: the symptoms of a disease are only the signs of the disease, not the disease itself.

Except in psychiatry, where the symptoms constitute the disease and the disease comprises the symptoms. William James had this tautology in mind when, remarking on another disease that doctors no longer believe to exist, he wrote, "The name hysteria, it must be remembered, is not an explanation of anything, but merely the title of a new set of problems." To say that a person who suffers from sadness and lethargy and sleeplessness and the loss of appetite and interest in pleasure is depressed is merely to give his suffering a new title—at least so long as depression is no more or less than the condition in which a person suffers in this fashion.

Psychiatrists would no doubt love to be able to skip even the SCID's superficial questions in order to diagnose depression. They would simply look into a microscope, at which point it wouldn't matter what the patient said, or what the psychiatrist thought about what he said any more than it would matter what a cancer patient said. The industry is working hard to eliminate the human element from psychiatry, but for now the best it can do is to circle the answers in notebooks and train practitioners to ignore what's in front of their eyes.

If this approach seems a little unsophisticated, a little primitive, and a little inhumane, there's a reason for that. When the APA turned to a descriptive nomenclature, they weren't exactly making an innovation. In fact, they were turning back nearly a century, to a nearly forgotten diagnostic system developed by Emil Kraepelin, a German doctor who was much more interested in weeding out the mentally ill than in curing them. Resurrecting Emil Kraepelin's system, psychiatrists also dusted off his solution to the problem that William James had noted: act as if there is science behind your nosology, and eventually the name of the disease will seem to be an explanation of everything.

Emil Kraepelin embarked on his career as a psychiatrist with an interest in his patients' inner lives. His doctoral dissertation, published in 1879, was "The Place of Psychology in Psychiatry," and early on he rebelled against his mentor, who believed that only the microscope could reveal anything important about mental illness. But within a few years, he was convinced that psychiatry, and perhaps the world in general, was much better off without psychology.

The problem, as Kraepelin saw it, was that the only source of psychological information about insanity was the patient, and the patient was, well, insane. So, he concluded, "we cannot afford to pay much attention to the patient's account of his experiences." Neither did he think that it was a good idea to indulge in "poetic interpretation of the patient's mental process. This we call empathy," he said, warning that a science-minded doctor employed it only at his own peril.

> Trying to understand another human being's emotional life is fraught with potential error. This is true in healthy people and much more so in sick ones. "Intuition" is indispensable in the fields of human relations and poetic creativity, but it can lead to gross self-deception in research.

Kraepelin felt that even his own interior was unworthy of exploration. When he published an analysis of his dreams, it was not to ferret out their meaning but to illustrate the way that the language of dreams resembled the language of insanity.

Kraepelin eventually landed a job in an asylum in Estonia, whose natives spoke a language that he didn't understand. Now that it was impossible to listen to his patients' stories, he was free to focus on the subject that he thought could put psychiatry on an equal footing with the rest of medicine: mental illnesses themselves, uncontaminated by unreliable psychology and fickle empathy. By the time he returned to Germany from the Baltic hinterlands, Kraepelin had given the modern world a conception of depression that fit the newly emerging medical model: it was, he said, a disease like any other. Indeed, he told his students, "from the medical point of view, it is disturbances in the physical foundations of mental life which should occupy most of our attention."

With this proposal, Kraepelin was entering treacherous territory. Although scientists had begun to identify brain structures that appeared to underlie specific faculties of mind—notably language—both this knowledge and the technology used to obtain it were rudimentary. Even more important, some doctors had already suggested that the mind could be reduced to a function of the brain, and they had landed in hot water as a result. The Parisian doctor Julien Offray de La Mettrie, for instance, had proposed in 1745 that the brain was a machine that produced consciousness, and that it was in this respect no different from the rest of the body: "it possesses muscles for thinking as the legs do for walking." For suggesting that the mind was nothing more than the output of a "machine that winds itself up, a living picture of perpetual motion," and the soul merely a "vain term," and for prophesying that doctors would soon understand the clockworks so well that "everything can be explained, even the surprising effects of the disease of the imagination," La Mettrie was exiled to Holland, from which he fled for Prussia, where he died in 1751.

Intolerance of materialism wasn't limited to the French and the Dutch. Fifty or so years later, Franz Josef Gall had run afoul of the Viennese authorities by proposing that the brain was divided into twenty-seven organs, each of which produced a different aspect of the mind, and all of which Gall claimed to have mapped. What's more, he said, a careful observation of the skull—measuring it, feeling its protuberances, taking a practiced look at its shape—would reveal the brain that lay beneath and with it the character and intellect of its owner. Gall's *organology* resembled in some ways the *physiognomy* already being practiced across Europe, but he was the first to claim that cranial features were expressive not of the immaterial soul, but of the flesh-and-blood brain, which in turn was the seat of the soul. His reward was a letter from the Emperor Franz II. "This doctrine concerning the head, which is talked about with enthusiasm, will perhaps cause a few to lose their heads," he wrote. Because Gall's materialism threatened "the first principles of morality and religion"—not to mention the delicate sensibilities of the women of Vienna—Franz II banned him from delivering further lectures.

The ban was good for Gall's business. The ensuing sensation made his European tour, on which he dissected brains, read skulls, and displayed his collection of skulls of the rich and famous, a huge hit. "There is always unending applause at all the public demonstrations," Gall wrote to a friend back home. But it was terrible for science, especially when Johann Spurzheim, one of his lab assistants, broke away. Gall's cartography, Spurzheim announced, was inaccurate, especially insofar as it included too many territories of evil—there was a region, for instance, for murderousness—and no room for redrawing the boundaries or to rectify deficiencies. Spurzheim claimed that people were not so bad as Gall had said, and that to the extent that their brains were weak, he could show people how to cultivate and improve them. He gave his version of organology a new name—*phrenology*—and took his show on the road, playing to even fuller houses than Gall had, not as a scientist, however, but as an entertainer.

So by the time Kraepelin proposed that doctors look to the brain to understand the mind's troubles, this idea was associated more with scandal and spectacle than with science. Kraepelin had done his part to make psychiatry more like the medicine that was emerging in the wake of the magic-bullet revolution. He had, for instance, determined that nearly half the mental patients at Berlin's Charité hospital were suffering from syphilis. But such discoveries were hard to come by, especially for Kraepelin, whose eyes were too weak for microscope work, and he was doubtful that anyone else was going to parse the newly apparent complexities of neuroanatomy anytime soon. It was one thing to find a bacillus in a blood cell and quite another to figure out what was going on in that tangle of neurons and fibers.

Kraepelin saw the problem: psychiatry risked being left in the dust as the rest of medicine galloped along on the back of science.

This disadvantage was nowhere more evident than in the chaotic state of psychiatric diagnostics. When Kraepelin moved to Estonia in 1886, psychiatry was a professional Babel. No one really knew if Dr. A's case of "masturbatory insanity" was the same as Dr. B's, or perhaps closer to Dr. C's patient with "wedding night psychosis." Without a reliable nosology—a systematic way to name the varieties of insanity—doctors could neither communicate with one another or, more important, demonstrate to a patient, his family, and the general public that they knew what they were talking about when they rendered a diagnosis. "Pathological anatomy"—the kind of findings that allowed physicians to proclaim that a sore resulted from smallpox and not syphilis—may have offered "the safest foundation for a classification," but Kraepelin thought there was another way to get at the diseases that lurked just behind patients' unreliable and idiosyncratic accounts.

Kraepelin's elegant solution rested on his insight that the advent of scientific medicine granted doctors the power not only to treat disease—which was still a mostly unrealized promise—but also, and perhaps more importantly, to give accurate names to human suffering, to render diagnoses. To the extent that doctors relied on scientific instruments to determine those names, they could claim that their diagnoses were based on empirical observation rather than whim or superstition or tradition, and that they therefore got to the truth of our suffering. Modern diagnostics eliminated the metaphysics of the ancient doctors by disclosing a world of pathogens behind the world of symptoms, a verifiable reality behind the appearance of disease. To use these findings to carve up the landscape of suffering into its diagnostic regions was thus to map the natural order of illness. So when a doctor pronounced the name of his patient's suffering, he was invoking something that existed in that hidden world: a disease that could be seen and touched, if only by specially equipped scientists. By virtue of his special tools and advanced training, the doctor could claim to know something about his patients' suffering that they themselves—or, for that matter, laymen in general—could not.

Kraepelin also realized that doctors didn't have to look at a stained specimen—or show one to their patients—to exercise their diagnostic power. They only had to know the disease's signature symptoms, which were the arrows pointing to those pathogens. The names themselves carried the authority of the microscope. And, he thought, there was no reason that psychiatrists couldn't have this kind of certainty. All they needed was the kind of reliable and comprehensive list of diseases they would undoubtedly derive from a pathological anatomy if one were available. They could, in other words, have the form of science, if not its content. He may have only had appearances to work with, but Kraepelin believed that by observing them carefully enough, he could discern the natural order that symptoms pointed to.

There was powerful precedent for this approach. In the 1730s, the Swedish botanist Carl Linnaeus had published his *Systema Naturae,* dividing the natural world into three kingdoms, each of which had its own phyla, classes, orders, families, genera, and species. *"Deus creavit, Linnaeus disposuit"* ("God created, Linnaeus organized"), he wrote, and in case the point wasn't clear, he depicted himself on the cover of his book naming the creatures in Eden, as if he were improving on Adam's first attempt. His grandiosity may have been justified, however, because in naming and sorting the natural world, Linnaeus created the common language that made plant science possible. For 250 years, until DNA testing came along, Linnaeus's classification of plants reigned supreme in botany.

Linnaeus faced a problem similar to Kraepelin's: he had nothing to work with except appearances and his senses. His job was to use what he saw and felt and smelled to establish the links and discontinuities in the botanical world. Many plants have thorns, for instance. But only some also have a thick stem with shiny green leaves. And only some of those blossom into a fragrant, voluptuous flower that turns into a hard, sour fruit. This cluster of appearances and, even more important, the course of a plant's life, is its unique signature, the set of data that separates a rose from a cactus or crown of thorns. Once you establish these associations, said Linnaeus, you can not only pronounce the name of a plant and its relation to other plants with authority, you can also know the ultimate fate of the tiniest sprout.

Kraepelin knew that most insane people had delusions. But for some, the delusions came on a few years after a bout of syphilis and led inexorably to dementia, paralysis, and death. For others, they began as hallucinations in adolescence, remitted occasionally, and rarely affected general health. Still other insane patients were driven to bed by their delusions, then flung into a frenzy of activity, and later back into a stupor. In each case, however, the disease had a particular course and outcome that could be observed over time. By looking at what happened to the patient a doctor could judge with certainty the variety of madness that had been manifest in the patient's condition; the patient's fate would tell the doctor what disease he had in the first place. And by the same reasoning that Linnaeus used, Kraepelin argued that once you know enough about the appearance and progress of a particular form of insanity, you can determine the likely fate of a patient from the first delusion. You don't have to know the biology of the process to claim that you know what will happen next. With enough observation and corroboration, you can, as Linnaeus might have put it, accurately separate the fruits from the nuts.

The important biography, then, was not of the patient—the particulars of which, between the unreliability of the insane person's account and the dangers of empathy, merely confused the issue—but of the disease. And Kraepelin set out to write those biographies by making careful observations of what the diseases did to patients over the long haul. By 1890, when he landed a job in Heidelberg, he had developed a notion of how, by matching symptoms to one another and then to outcomes, he would, as he put it, "cut nature at its joints" and identify the discrete diseases of the mind.

His method was straightforward. For every patient who entered the asylum, he started a note card. On it, he wrote down the patient's history and condition and rendered a diagnosis. The card went into the "diagnosis box" with all the other cards, and the information went on a list. During the patient's stay, Kraepelin and his staff would revisit the cards, revise the diagnoses, and make the appropriate entries, tracking the progress of symptoms and diagnoses until discharge. Eventually, he built up what we would call a database, which he took home on weekends and away with him on vacation, sorting and resorting until he felt certain that he had laid out the symptoms and course of a particular disease. In 1893, he began to present his results in his *Lehrbuch der Psychiatrie,* an often-revised textbook intended to allow physicians to diagnose their patients reliably and to give families of the afflicted what, at least according to Kraepelin, they wanted most: a prognosis.

Of course, that's probably not precisely what was desired by the parents of the young man, diagnosed by Kraepelin as hopelessly ill with *dementia praecox,* who saw a raven at his window waiting to eat his flesh, or by the wife of the farmer, so ridden with guilt about having "practiced uncleanness" with himself (and, the farmer said, with a cow) that he couldn't get out of bed, whom Kraepelin diagnosed with *involution psychosis.* They probably wanted a cure. Kraepelin, on the other hand, didn't think this was a reasonable goal because mental illnesses, like roses and pine trees, had to run their natural course. Some were caused by toxins like alcohol or syphilis and, at least in the pre-Salvarsan era, nothing could be done about them except avoiding the poison. The rest were inherited rather than acquired, and you couldn't make a constitutionally insane person sane any more than you could get a rose to bloom from a pine-cone. On the other hand, knowing what a pinecone will become tells you something important: what kind of soil it needs, whether you should plant it in full sun or shade, how you should feed it. Likewise, the point of a taxonomy of insanity—beyond the satisfaction of naming God's critters—was to figure out what to do *with* the patient, not *for* the patient.

Kraepelin's concern about how to dispose of patients grew

urgent as the end of the century approached. His audit of German asylums indicated that the numbers of the insane were swelling, in proportion as well as in absolute terms, and, as he told a group of doctors in 1899, the potential consequences to society were dire. "All the insane are dangerous. Mental derangement is the cause of . . . sexual crimes and arson, and, to a lesser extent, dangerous assaults, thefts, and impostures," he warned. "Numberless families are ruined by their afflicted members." Perhaps most regrettable, Kraepelin added, was the fact that "only a certain number of those who do not recover succumb at once . . . [and] the effects strike deeply into our national life." That left doctors with important responsibilities:

> to prevent the marriage of the insane . . . to secure a proper education and choice of occupation for children predisposed to disease . . . to recognize dangerous symptoms in time, and, by their prompt action, to prevent suicides and accidents and obviate the short-sighted procrastination which so often keeps patients from coming under the care of an expert alienist.

To Kraepelin, the point of a reliable classification scheme was to give a doctor a way to determine whom to send to the asylum so they could neither commit mayhem nor, even worse, breed. Accurate diagnosis, he told his listeners, was thus the best, if not the final, solution to the problem of "the growing degeneration of our race in the future."

To judge from subsequent events, Kraepelin didn't succeed at allaying Germans' fears about the future of their race, although he did leave a clue as to what they would eventually do about their worries. But his intertwined motivations—identification and segregation of the mentally ill and removal of their genetic stock from the race—

were quickly lost to history. So was his therapeutic nihilism, his conviction that nosology was enough, that when it came to mental illness, there was nothing to be done for the patient. In fact, Kraepelin in general was soon forgotten, eclipsed by Freud and his notions that mental illnesses begged to be understood (and cured) through language and empathy. And when, in the 1970s, American doctors returned to Kraepelin's taxonomic approach in order to save their profession from a gathering storm, they conveniently forgot about the nihilism that gave birth to it in the first place.

That's the great thing about aligning your cause with the forces of science. You can claim that you are just describing the natural order, the way things are for all time, and that as a result history doesn't matter.

Unless, that is, it favors you.

There isn't a historical account of depression that doesn't hearken back to the sixth edition of Kraepelin's *Lehrbuch,* published in 1899, as a crucial step in proving that depression is indeed a disease. By the time the book came out, Kraepelinian nosology was all the rage in psychiatric circles; a standardized language had been just what the doctor ordered, and psychiatrists everywhere awaited every new edition with great anticipation. But the master of classification was not infallible, and Kraepelin was forced to issue a recantation in the sixth edition. Previously, he had distinguished various forms of melancholia, the malady that first appeared in the Hippocratic corpus, and separated those from what he called "mania" and "circular insanity." But now the cards revealed that he had been in error. "In the course of years," he wrote in the *Lehrbuch,* "I have become more and more convinced that all the pictures mentioned are merely forms of one single disease process . . . Certain fundamental traits recur in the same shape, notwithstanding manifold superficial differences."

The disease that encompassed all these different appearances was now to be called manic-depressive insanity. Doctors might find their patient in a state of "psychomotor excitement . . . distractibil-

ity, and happy though unstable attitude" or "psychomotor retardation, absence of spontaneous activity, dearth of ideas, and depressed
emotional attitude," but either way they were seeing the same insanity, and they could be certain that sooner or later the patient would
swing to the opposite emotional pole. Manic-depressive insanity was
a main branch on the tree of madness, and most depressions, according to Kraepelin, were simply leaves.

It's not that Kraepelin was the first to propose that depression
was a disease. But he was the first to try to bring it into line with
the new idea of what a disease was: not an indirect and idiosyncratic
result of a humoral imbalance or a punishment from the gods or
a reaction to a poison, but the direct effect of a natural process,
something gone wrong with the body. Job and his comforters, in
this view, had been saddled with a false dichotomy as they argued
over whether his suffering was the result of a flaw in his soul or a
hostile external order. Impersonal nature, which operated according to its own laws, also lived in us and could wreak its havoc from
within. The suffering that resulted, in us but not of us, was on balance something that we would all be better off without. We could
eliminate it without changing our essence, if only the means of
extermination could be found. The bullets wouldn't be available for
another sixty years or so (and he was himself not terribly interested
in cures), but Kraepelin was already defining depression as a target.

American psychiatrists didn't forget only Kraepelin's nihilism when
they resurrected him. They also forgot that when he said "insanity"
he really meant it. Kraepelin never offered a definition of the word,
but he knew it when he saw it, and he described it in detail to his
students:

Gentlemen, the patient you see before you today is a mer-
chant, forty-three years old, who has been in our hospital
uninterruptedly for about five years. He is strongly built,

badly nourished, and has a pale complexion. He comes in with short, wearied steps, sits down slowly, and remains sitting in a rather bent position, staring in front of him almost without moving . . . [S]peaking gives him a great deal of trouble, his lips moving for a little while before the sound comes out . . . That the answers come so slowly . . . shows that in this patient we have not to deal with a fear of expressing himself, but with some general obstacle to utterance in speech. Indeed, not only speech, but *all action of the will is extremely difficult to him*. For three years he has been incapable of getting up from bed, dressing, and occupying himself, and since that time has lain in bed almost without moving . . .

Here is a case of a woman twenty-three years old . . . who bore her second child six weeks ago. Seventeen days later she got a great fright from a fire in her room, and she then became apprehensive and restless, saw flames, black birds and dogs, heard whistling and singing, began to pray, screamed out of the window, lamented her sins, promised to be good, and could not sleep. She sits almost motionless, with her eyes cast down, staring in front of her, and moving her lips slightly now and then . . . She nods when I ask her if she is unhappy and mutters to herself: "There are always so many carriages coming; a great number drive about outside." Now and then she uses isolated, broken expressions, in a tone of lamentation, often repeating them one after the other: "I want to go home, to get out. Alas! Alas! only let me go away. I will not let myself be done to death. I cannot stay here. Good heavens! There is poison in the food!"

There were milder forms of insanity, of course—"numberless . . . cases of maniacal-depressive insanity which never come into an asylum and, indeed, are never recognized as morbid states at all." These patients, whose insanity might manifest as indecisiveness rather than a total loss of will "often passed off without any treatment" or were treated at "different asylums and watering-places, or ordered to travel"—prescriptions that patients, unaware of "how deeply maniacal-depressive insanity is rooted in [their] intrinsic disposition," wrongly think have cured them when their disease remits. For most of the people so disposed, the mildness is only temporary: it's just a matter of time before they find themselves completely unable to exercise their will. The rest are lucky to have a variation of the disease less severe than the others.

Kraepelin's notion of mildness, however, was different from yours and mine—and from current psychiatric practice:

> The mildest form of the depressive states is . . . *simple retardation* . . . [M]ental processes become retarded, thought is difficult, and patients find difficulty in coming to a decision, in forming sentences, and in finding words with which to express themselves. It is hard for them to follow the thought in reading or ordinary conversation . . . Customary actions, such as walking, dressing, and eating, are performed very slowly, as if under constraint. When started for a walk, they halt at the doorway or at the first turning point, undecided which way to go . . . Sometimes they become bedridden.

If a person showed up at my office with this "mildest form" of depression, I would be very likely to consider hospitalization. When my doctor at Mass General determined that I was mildly (although still majorly) depressed, I showed (and reported) none of these symptoms. Indeed, today's "major depression, mild" is one of the diagnoses that cause people to wonder if the diagnosis isn't too freely handed out.

Still, even in Kraepelin's time psychiatrists worried about whether or not the new diagnostic regime encouraged doctors to diagnose too many people as mentally ill. In 1907, just a few years after the sixth edition of the *Lehrbuch* appeared, Georges Dreyfus, a former pupil of Kraepelin, published the results of a study he'd undertaken at Heidelberg Clinic. He had observed Kraepelin's patients—the same patients whose cases had led Kraepelin to conclude that he'd been mistaken about melancholia—and determined that, judging by outcomes, the master had *not* been mistaken. He had simply stopped observing his patients too early. As a result, he'd misdiagnosed the very cases that had led to the realignment.

On the basis of this faulty assessment, Kraepelin had claimed that these patients suffered from the one form of melancholia that should not be folded into the manic-depressive diagnosis. He called it "involution melancholia"—involution being the stage of life that, according to the medicine of his time, sets in soon after forty, when the body starts its long wind-down to death, or, as Kraepelin called it, "the early senile period."

> It includes all the morbidly anxious states not represented in other forms of insanity, and is characterized by *uniform despondency with fear, various delusions of self-accusation, of persecution, and of a hypochondriacal nature . . . leading in the greater number of cases, after a prolonged course, to moderate mental deterioration* [emphasis in original].

A diagnosis of involution melancholia, with its prognosis of a course straight downhill to death, was in some ways worse than manic-depressive insanity. At least manic-depressives had the benefit of remissions, times in which their cycles crossed through relatively benign emotional territory. And that's exactly what Dreyfus found: many of Kraepelin's melancholics actually got better, or at least cycled between states of sanity and insanity. The outcome of their cases, he said, proved that they hadn't had involution melancholia in the first place. Indeed, he doubted that the disease existed in its own right, that it was anything other than yet another subspecies of manic-depressive insanity.

Dreyfus had hoisted Kraepelin on his own petard, and in 1913, when he published the eighth edition of the *Lehrbuch,* Kraepelin wrote involution melancholia out of the official nomenclature. Outside Germany, this news caused some consternation. August Hoch, director of the New York State Hospital Psychiatric Institute, objected to the banishment of "one of the most frequent forms of mental disease, and . . . one of the oldest in psychiatry" from the kingdom of insanity. He and a colleague, John MacCurdy, reviewed Dreyfus's review of Kraepelin's cases and reported that Dreyfus's "zeal outran his judgment. In a number of cases he ferreted out a history of depressions so mild as to seem to be . . . merely more or less normal mood swings." And here was a problem. Bad enough that patients had been misdiagnosed and a psychiatric illness declared nonexistent as a result. Even worse, all this confusion meant that Kraepelin's brainchild—the idea that insanity could be medicalized by means of an accurate and reliable diagnostic scheme—was in danger.

> Variations of the emotional status are of great theoretic, psychologic importance, but they should not be called "psychoses" as long as their manifestations remain within certain limits. Otherwise, nearly the whole world is, or has been, insane.

Prescient as they were in this worry, Hoch and MacCurdy's concerns were purely parochial: that "individual taste"—rather than scientific knowledge—"is likely to determine the classification adopted by psychiatrists for many years." They didn't seem to grasp that the problem they had uncovered was not that Dreyfus had misused Kraepelinian nosology, but that Kraepelin had failed at what he had set out to do. *Manic-depressive insanity, involution melancholia, dementia praecox*—this was indeed a sophisticated language, and it certainly sounded medical. But mental diseases still consisted only of lists of symptoms, and the symptoms were only symptoms because they belonged to the disease. The logic was circular, the language tethered only to itself, not to something as solid as a spi-rochete or a bacillus. That's why the controversy could arise in the first place, why it could only be fought with language, and why it could not be settled. Without an ultimate referent for the sign, as a philosopher might put it, it was impossible to say with certainty whether Dreyfus or Kraepelin was correct, which meant that it was impossible to tell who was sane and who was not.

A diagnosis that renders the whole world insane is a scandal for a psychiatry that claims to have cut nature at its joints. But it's an enormous market opportunity for an industry that would aim its magic bullets at insanity. It would take eighty years and much good fortune, not to mention some very clever advertising, but American ingenuity would eventually figure out how to put the authority of science behind the immensely profitable claim that the whole world is insane—or at least the large portion of it that meets the criteria for depression, a much greater population than Kraepelin at the height of his racialist paranoia ever imagined.

CHAPTER 5

MAKING DEPRESSION
SAFE FOR DEMOCRACY

I was in the 7-Eleven one day, waiting in line to pay for my coffee. The clerk was talking to a friend, who had just asked her how she was doing.

CLERK (early twenties, long permed ringlet curls): I don't know. I'm still achy and weak. And I've just been *so* tired. I just want to sleep all the time. I feel, I don't know, you know, *blah.*

FRIEND (same age, bigger hair, belly shirt revealing four-color wraparound tattoo): What does your doctor say?

CLERK: He doesn't know. I mean, it's not like I have a fever anymore or anything.

FRIEND: Have you been depressed?

CLERK (surprised): Well, I was on antidepressants a while ago, but I stopped them.

FRIEND (voice deepening a bit): It's depression. You got a case.

CLERK (looking a little sheepish): Oh, I don't know . . .

FRIEND (insistent now): No, really. Depression can make you sick. That's how it can kill you, you know. You ought to go back to your doctor and tell him you have depression. Get him to put you back on the medicine.

It's too bad the marketing folks from Lilly or Pfizer weren't there with a camera crew. They would have gained incontrovertible evidence to show their bosses and shareholders just how deep into America their message about depression has penetrated—and they'd have a free ad in the can to boot.

But you have to wonder what Emil Kraepelin would say about the ease with which these women talked, about their familiarity with these medical terms, about the way that his language had escaped the asylum, taken on a life of its own, and turned up in what—despite the dozen coffee selections, three cup sizes with two lid choices, and phalanx of syrups and creamers and sweeteners— would have qualified as one of Auden's miserable duchies. Frightening as their numbers were to Kraepelin, the ranks of the insane in turn-of-the-century Germany were still a very small portion of the population. And he certainly expected doctors, and not the patients, to render the diagnoses. He would likely have been shocked to discover that depression had become a subject fit for conversation among the *volk* at the 7-Eleven. I think he would have found this entire display vulgar and a mockery of his science.

Given what he wanted to do with the insane, Kraepelin was unlikely to see the value of expanding the boundaries of mental illness. Indeed, his system was a sort of reverse elitism, reserving the status of insanity for only a select few doomed souls. He may well have been appalled not only at the way my doctors at Mass General used his method to diagnose me with major depression, but also with the setting where they did it: not an asylum but an unlocked (except for the restrooms) modern office building shared with dermatologists and ophthalmologists and biotech startups, with a waiting room populated by a cross section of America—old and young, affluent and poor, white and black and brown, crazy and not so crazy—and presided over by a chipper receptionist who talked on the phone about depression as if it were the common cold, who made happy banter with the patients as she validated their parking tickets. Perhaps Kraepelin would have seen this all as proof that the race had indeed degenerated.

On the other hand, the doctors at Mass General, like the 7-Eleven clerk's friend, have a distinctly non-Kraepelinian idea about depression: that something can be done about it. It's not a death sentence anymore. The possibility of cure has made this expansion possible, made mental illness safe for the masses, as common as a convenience store.

It is fitting then that the initial conversion of Kraepelin's terrible and incurable disease into a mundane problem we can solve with a widely available product took place in the United States as the first wave of mass consumer culture washed over the country. The transformation in psychiatry was wrought largely by an ambitious immigrant, ready to take old European ideas and translate them into New World successes—the kind of man who was rapidly becoming a fixture in entrepreneurial capitalism. And by the time he was finished revolutionizing American psychiatry, that man, Adolf Meyer, had focused attention on biography as the source of our suffering.

When he left his native Switzerland in 1892, at the age of twenty-six, Adolf Meyer's mother fell into a deep depression. His former professor in Zurich, Auguste Forel, wrote to him in Chicago to tell him that his mother was unable to shake the certainty, against all evidence, that her son was dead. "She had been one of the sanest persons in my experience," Meyer later said. But now, at least according to Forel, a dyed-in-the-wool Kraepelinian, she was hopelessly and forever insane with melancholia.

News of his mother's suffering was only one of the difficulties that Meyer encountered in his newly adopted country, to which he'd gone only as a last resort. After medical school, he had wanted to stay on in Forel's lab, but he had already shown an independent streak—secretly expanding his thesis research beyond the task Forel had assigned him, setting up a lab in his home to explore a novel staining technique that the master had resisted, and refusing to become a "militant total abstainer" like his teetotaling professor. Meyer had studied with leading doctors in Paris and London, even met Jean Charcot and Thomas Huxley, but nothing came of his efforts to develop these contacts into the cutting-edge medical career he wanted. His remaining option, to join his physician uncle in his Swiss country practice, was an untenable choice for a brilliant and restless young doctor—especially when across the ocean was a country that, at least by reputation, welcomed the pioneering spirit.

Before he left Europe for America, he wangled an introduction to the great doctor William Osler, visiting England from Johns Hopkins, where he was president. But Osler wouldn't have him—because, as a chagrined Osler later explained, the man who had introduced him to Meyer was an "old humbug." Then H. H. Donaldson, a Clark University professor, told Meyer that he and the rest of his department were decamping to the University of Chicago, leaving a raft of openings behind. But when Meyer applied, psychologist G. Stanley Hall, Clark's president and the man who later brought Freud to America for the first and only time, lied and told him no jobs were available. Finally, Meyer decided to follow Donaldson to Chicago and see what happened.

At first, Meyer found only an unpaid fellowship, and he was soon forced into what he had fled Switzerland to avoid: clinical practice. He wasn't about to give up his laboratory dreams, however. His office was upstairs from a shoe store, and his few patients found themselves in a suite so stuffed with preserved brains and vials of chemicals that it looked more like a mad scientist's lab than a doctor's office. For a year, he hovered on the periphery of his profession, showing up at scientific meetings and submitting papers about his experiments, until finally Ludvig Hektoen, a professor at Chicago, told him that the institution he had just left—the Illinois Eastern Hospital for the Insane, in Kankakee—was looking for a pathologist. Meyer sent off an application, complete with references from famous doctors in Europe and America, while Hektoen greased the skids with a letter to Richard Dewey, the hospital's superintendent. Meyer thought his gamble on America was about to pay off.

Which it did, but not before something else Meyer hadn't counted on intervened: American politics, in particular the regime change in Washington that swept Grover Cleveland into office for the second time, which trickled down to Springfield, Illinois, where the first Democratic governor in twenty-five years immediately purged Republicans from state government, including the people overseeing the asylums. Dewey lost his job during the same week that Hektoen sent his letter. Back in Chicago, Meyer, unaware of this upheaval, heard nothing. After nearly a month, he decided to take matters into his own hands, got on a train to Kankakee, and knocked on the door of the new superintendent, who hired him on the spot. Meyer was proving himself an excellent fit for America's can-do economy.

Meyer was eager to work at the Kankakee asylum because it had an ample supply of exactly what he needed: diseased brains and insane people. Back in Europe, Kraepelin was still looking for the pathological anatomy that would anchor his diagnostic scheme. Correlating brain pathology to insanity seemed to be the key to understanding both normal and abnormal mental functioning, and Kankakee, with its busy wards and autopsy labs, offered a perfect opportunity to make those connections, or so Meyer thought.

But he soon developed doubts about the soundness of his enterprise. Part of the problem was purely administrative: a hospital staff "hopelessly sunk into routine and perfectly satisfied with it" was not interested in exploring the physiological basis of mental illness. Indeed, they had despaired of even agreeing on their patients' diagnoses and as a result had developed an "unwillingness to really collect facts needed for diagnostic decisions." Ten years later, he still complained that

> their reasoning for diagnosis followed rather general impressions than definite and precise statements of fact, and failed

to make a sufficiently clear distinction between what was actually found and what was merely supposed to exist.

What the staff really wanted to do was dissect brains, which led Meyer to believe that "the existence of a pathologist in a hospital for the insane was a poor remedy for that which was actually needed."

At a coroner's inquest, Meyer made clear what he had come to think was a better remedy. He had just presented his analysis of the brain of an inmate (who had died from a heart attack) when a juror said, "Now, doctor, show us what you find in the mind." Meyer responded, "There [are] more mental findings in the history than the brain." Even if, as one American psychiatrist had put it in 1870, "mind cannot be diseased, only body," it was a mistake to ignore mind; its ravings may have been meaningless in themselves, but, as Kraepelin said, they revealed the nature of the disease that caused them. Doctors should dissect the disease as carefully as the tissue, Meyer thought; especially given the state of knowledge about the brain and the difficulty of neuroanatomical research, this was the more fruitful avenue. Meyer, in other words, meant to spread the Kraepelinian gospel—not only to Kankakee but to all of American psychiatry.

As Meyer taught his colleagues the art of taking patient histories, making close observations of symptoms, course, and outcome, and rendering accurate diagnoses, he ascended the professional ranks and was soon fulfilling his ambition—teaching at the University of Chicago, delivering papers around the country, hobnobbing with politicians responsible for funding his asylums. But he was still deeply troubled by his mother's depression. He thought of his meteoric ascent and the changes he was trying to make as a response to his "lasting wish that I might pay back and give some compensation during her life, something more than gratitude." The career that he devoted to his mother was more than illustrious. Inspired in part by her recovery, Adolf Meyer changed American psychiatry in ways that he never would have imagined.

★   ★   ★

While Adolf Meyer was bringing European nosology to the New World, America was working its charms on him. Kraepelin's therapeutic nihilism, his sense that there was nothing to be done besides diagnosis and segregation, didn't square well with American sensibilities. John Winthrop's vision of America as a City upon a Hill, built through individual efforts at self-improvement, had, by the end of the nineteenth century, turned into the dynamic, bustling free-for-all of entrepreneurial capitalism. The pursuit of happiness—in the original, economic sense of that phrase—was turning out to be not the privilege of only a few, but the opportunity and perhaps even the imperative for everyone. The idea that mental illness was forever, that a class of people, by dint of constitution alone, couldn't improve themselves and would thus be excluded from the hunt for freedom and riches, didn't square with Thomas Jefferson's egalitarian promise. It was only a matter of time before therapies evolved to resolve that conflict.

You never know exactly why a person gets the ideas he does, but it is significant that within a couple of years of arriving in Chicago, Meyer met Jane Addams. In 1894, he spent a week as her guest at Hull House, where Addams was helping poor people improve themselves. She was also campaigning for the protection of children from exploitative employers, violent parents, and all the other ravages of poverty. A year later, Meyer published a paper in which he raised the question of whether it was really true that parents had to "accept the disposition of the child as a gift that must be taken without grumbling"; perhaps, he said, there was a "period of plasticity" during which character was formed. If so, then "early prevention of danger," the phrase he used to title his paper, was possible and even crucial.

The implications of this line of thought went well beyond child-rearing practices. American psychiatry, no less than European, was bound by the conviction that mental illnesses were either endogenous, an immutable property of the individual's constitution, or exogenous, the result of damage done by injuries or toxins like alcohol or syphilis. In either case, the disease was the result of impersonal, biological factors. Patients in the grip of insanity were no more responsible for contracting it or for how it made them behave than doctors were for curing it. But if, as Meyer was beginning to think, parents could influence the outcome of their children's mental lives, then by extension perhaps life events in general could play a role in the development and healing of mental illness.

Meyer was ready to answer the nature/nurture question with a resounding "yes." "The human organism can never exist without its setting in the world. All we are and do is of the world and in the world," he said. And he wanted to leave such hoary philosophical questions—something Meyer associated with his native continent—at that. "Steering clear of useless puzzles liberates a new mass of energy," he said. "The question why is mind mind, and just what it is, can be as little answered as what gold is and why it is, and why it should be so." In his newfound pragmatism Meyer revealed the influence of another Chicagoan, whom he met shortly after Jane Addams: John Dewey, who in turn introduced him to his friends Charles Peirce and William James—whose *Principles of Psychology*, which also rejected nature and nurture as the only possible explanations for human behavior, appeared in 1890. Influenced by this trinity of American pragmatism, Meyer declared that the job of psychiatrists was to understand people in their natural setting and help them adapt to it. Exogenous/endogenous, nature/nurture, mind/body—none of these tiresome questions was going to tell us much about life as we lived it, which was as an active force trying to grapple with all the complexities of everyday life.

Meyer's head was filled with these new ideas when he returned to Europe in 1896. There he found his mother recovered from her depression, which now appeared in a new light. Perhaps it had not been the result of a constitutional or anatomical defect, but a reaction—not only to his departure, but to loss. Perhaps her "delusion" that Meyer was dead was an accurate reflection of her experience, one that made sense of her depression, that provided it with meaning. After all, within a few years, her husband and daughter had died, her son Hermann had moved to the French side of the country, and her remaining son, Adolf, had left for America. Maybe it *felt* to her like Adolf was dead—or maybe, as Sigmund Freud would suggest twenty years later, her depression was her only way of expressing her anger toward him for abandoning her. And perhaps her reactions to her "natural setting" had changed. Perhaps she had, as Meyer would later say, adjusted.

Meyer spared his mother's psyche too much exposure, so we can't know these specifics. But we do know that after his visit home, he went to Heidelberg to spend six weeks with Emil Kraepelin, who had just published the fifth edition of his *Lehrbuch*. His sojourn at Heidelberg did nothing to restore his faith in Kraepelinian nosology, and by the time he returned to America he was ready to renounce it. The idea of finding patterns of symptoms was a good one, he said. But all too often, he thought, "the supposed disease back of it all is a myth and merely a self-protective term for an insufficient knowledge of the conditions of reaction and inadequacy of our present remedial skill." Indeed, Kraepelin's vaunted diagnostic system was filled with what Meyer would come to call "neurologizing tautologies."

Meyer had an idea about how to correct Kraepelin's error, and it was much more in keeping with life in his adopted home. Instead of relying on a specious classification of diseases, he asked,

> can we not use general principles and valuable deductions without pulling them into the service of a vicious attitude of mind, the attitude of that medical conceit which delights in surrounding the diagnosing and prescribing with a mystic halo so much adored by the patients trained to see wonders in the wise terms? Why not regard the "diagnosis" as merely a convenient term for the actually ascertained facts which . . . tell a clear and plain story?

In other words, instead of anchoring diagnosis in a yet-to-be-discov-ered neuroanatomy, why not simply tie it to life as it is lived? Why not move beyond the "appeal to cell-biology and correlation of sci-ences" and toward the "plain facts of history and the reactions of the patient"?

Meyer wasn't merely modifying Kraepelin or reinterpreting his statistics. He was repudiating the German master, reversing his dic-tum to ignore the patient and eschew empathy in favor of a psychia-try that listened, and listened carefully, to the actual experience of his patient. "There is no advantage," he told his fellow doctors, in merely looking for "'symptoms' of set 'disease entities' that would allow us to dump all the facts of each case under *one term or heading*" [emphasis original]. Searching for pathology, a doctor "surrenders his commonsense attitude" and fails

> to view the abnormal mental trend as a genuine but faulty attempt to meet situations, an attempt worthy of being ana-lyzed as we would analyze the blundering of a distracted pupil, or the panic of a frightened person, or the bumbling of one who reacts poorly in trying to meet an unusual situation.

The cure for mental illness was not to breed insanity out of the human race. Rather, psychic suffering should be seen as a sign that a person was having difficulty doing what we all have to do: adapt to a demanding environment. It wasn't long before Meyer concluded that people, even psychologically troubled people, could reinvent themselves, and psychiatrists, in Meyer's view, could and should help them to do it.

It took Meyer only a few years from the time he arrived in America to figure out something important about his adopted country. "The public here believe in drugs," he wrote to the governor of Illinois in 1895, "and consider prescription as the aim and end of medical skill." Americans, that is, wanted their doctors to *do something* for them. That was the last thing that psychiatrists, with their life-sentence diagnoses, could offer.

Just before he made his report to the governor, Meyer visited the Battle Creek Sanitarium in Michigan. Battle Creek was one of the biggest spas that sprang up in the last half of the nineteenth century in the United States and Europe, where doctors, but not psychiatrists, offered multifarious treatment for nervous disorders. There Meyer saw doctors doing all kinds of something for (and to) their patients: enforced bed rest, cold baths, tonics, enemas, electric therapy, pelvic massage, and, of course, lots and lots of corn flakes, which the Kellogg brothers, who ran Battle Creek, invented.

Convinced by all that prescribing, Meyer formed an Association of Assistant Physicians of Hospitals for the Insane, which he envisioned as a forum to "give us a clue for progress" toward actually helping patients. (This was in some ways a rearguard action; there was already an Association of Superintendents of Hospitals for the Insane—which would eventually become the American Psychiatric Association—but Meyer thought it was too much concerned with administrative rather than clinical matters.) But with his nascent ideas about mental illness as a maladaptive reaction to the world, he couldn't have failed to notice that this was exactly how Battle Creek's doctors thought of their patients' problems—as the result not of constitutional weaknesses or infections like syphilis, but of the difficulties of everyday life. And he must also have noticed that these doctors, who were more than willing to minister to their patients' psychic suffering, were not psychiatrists but members of other specialties, especially neurology.

In Europe, neurologists had already cornered the market for treating life's problems. Even as Kraepelin was tweaking his categories and Ehrlich was concocting his potions, neurologists like Sigmund Freud and his French mentor Jean Charcot were treating respectable, educated, and well-heeled people whose suffering stopped well short of the kinds of madness that landed patients in the asylum. These neurologists were glad to reassure their patients that they were not insane, but merely *nervenkranken* ("nervous patients"), as the Germans called them, and suffering from illnesses like *l'erithisme nerveux* ("nervous weakness")—a malady characterized by irritability, avoidance, and depression—and to treat them in *Nervenkliniks,* or in private offices, rather than asylums.

Successful as they were, however, the Europeans paled in comparison to the Kelloggs and their colleagues in the United States, who by the turn of the century had built a thriving industry on treating nerves, and especially a single nervous disorder: neurasthenia, or, as George M. Beard, neurologist and inventor of the diagnosis called it, "American nervousness." Beard's book by that title came out in 1881. It was the *Listening to Prozac* of its day, a runaway bestseller in which a doctor gave voice to common, if not yet articulated, worries about emotional life and what doctors proposed to do about it.

The ranks of the afflicted were legion. William James, his brother Henry, and their sister Alice, Theodore Roosevelt, Edith Wharton, W. E. B. DuBois, Frederic Remington, Mary Baker Eddy, Jacob Riis, Emma Goldman, Samuel Clemens—"the list," says one scholar, "could go on until it included the majority of well-known cultural producers of the time"—not to mention the regular people, most of them affluent, whose doctors told them they had neurasthenia. And no wonder so many of them received the diagnosis! Beard's list of symptoms takes up two pages of *American Nervousness,* from "Insomnia, flushing, drowsiness, bad dreams" through "ticklishness, vague pains and flying neuralgias" to "exhaustion after defecation and urination," and, finally, just in case he missed something, "etc."

The cause of all this trouble, Beard said, was a failure of the nervous system to keep up with the demands of "modern civilization," which he listed as: "steam power, the periodical press, the telegraph, the sciences, [and] the mental activity of women. When civilization, plus these five factors, invades any nation," he wrote,

"it must carry nervousness and nervous diseases along with it."
The main disease vector, apparently, was one young man, Thomas
Edison, whose "experiments, inventions, and discoveries . . . are
making constant and exhausting draughts on the nervous forces
of America . . . [and keeping] millions in capital and thousands of
capitalists in suspense and distress." But even without Edison, rapid
innovation and industry had become a whirlwind that was leaving Americans dizzy. Put Edison together with democracy and the
rise of "agnostic philosophy" inspired by the rise of Darwinism—
which, Beard said, had led to an expectation that everyone would
become "an expert in politics and theology"—throw in "the liberty
allowed . . . to Americans to rise out of the position in which they
were born," and cap it off with the possibility of knowing instanta-
neously what is going on everywhere and anywhere in the world,
and the next thing you know, you have an epidemic of severely over-
worked nervous systems on your hands.

Not that Beard thought there was anything wrong with modernity. His catalog of stresses was not a jeremiad, but a celebration
of progress. He prophesied not social and moral collapse but
exhaustion, which was, with the help of doctors, eminently treatable. Neurasthenia was not a sign of degeneracy but the mark of
the elect—the "brain workers" whose refined nature both qualified them to manage the new world and made them susceptible to
its difficulties. It was the new white man's burden, the stigma of
the elite. "Of our fifty millions," Beard estimated, "but a few millions have reached that elevation where they are likely to be nervous." Our natural allotment of nervous energy may have been
sufficient for "the lower orders," but for "the very highest classes"
these demands were like a new set of lamps interposed in an electrical circuit:

Sooner or later . . . the amount of force is insufficient to
keep all the lamps actively burning; those that are weakest
go out entirely, or, as more frequently happens, burn faint

and feebly—they do not expire, but give an insufficient and unstable light.

It was up to the doctors to "bulk up the blood," as Beard put it, in order to increase the output of our dynamos.

Charlotte Perkins Gilman made the nerve doctors infamous with her chilling story "The Yellow Wallpaper," in which she chronicled the descent of a neurasthenic woman through her regime of "tonics, and journeys, and air, and exercise"—and an absolute prohibition of work—into a psychotic obsession with her sickroom's wallpaper. But her doctors had help.

> If a physician of high standing, and one's own husband, assures friends and relatives that there is really nothing the matter with one but temporary nervous depression—a slight hysterical tendency—what is one to do?

The entire society had heard Beard's gospel—that there was a form of psychological suffering, of *nervous depression*, that had nothing to do with insanity. The patients weren't the Tom O'Bedlams of the world but the leaders of society, who declared their neurasthenia much as celebrities today confess their depressions. Neurasthenia was as well known and accepted as Teddy Roosevelt, with his rough-riding military career, exaltation of open spaces, and his energy-conserving soft speaking/big stick philosophy. And the illness was simply to be accepted and dealt with, no different from the electric light bulb and the automobile. Even William James, in the midst of writing his *Principles of Psychology,* couldn't resist the diagnosis. He took not only the rest cure at various spas but also injections of the extracts of goat's lymph and bull testicles—all this despite the fact that "I have no confidence in three-quarters of what the doctors tell me," as James wrote to a friend. But, he added, "I have so little independent hold on the situation that I am hypnotized by the remaining quarter, and by the prestige of their authority."

James doesn't even mention neurasthenia—his own or anyone else's—in his *Principles.* That's because no one thought that the affliction that had Americans submitting en masse to the nerve doctors was psychological in nature. It was a *physical* problem, a reaction of the body to an environment that demanded more than it could supply. If the mind was involved, it was only as a by-product of these economics. There was nothing wrong with the inner selves of neurasthenics; they were merely victims of their put-upon nervous systems. All they had to do was to let themselves be hypnotized (sometimes literally) by the neurologists and all would be well.

It's not hard to see why neurasthenia was such a hit and the neurologists who purveyed the cures so successful. The diagnosis gave a name to anxiety about the dizzying pace of change even as it reassured patients that as soon as their nervous system caught up, the disease would remit and all would be well—not to mention that their illness was a sign of their superior refinement. But no psychiatrist, least of all an ambitious psychiatrist like Adolf Meyer, could fail to see that the disease benefited one group of healers at the expense of his own. While neurologists, despite all their quackery, had become the go-to guys for otherwise reasonable people like William James, psychiatry was languishing. By World War I, according to historian Edward Shorter, it had "become marginal to the mainstream of medicine." It was left to Adolf Meyer to reclaim the everyday psychological suffering of Americans for his profession, and he did it in part by making depression less like insanity—and more like neurasthenia.

In 1932, nearly four decades after he returned from Europe, Meyer recounted to the membership of the American Psychiatric Association (of which he was then president) about the time when he first came to America—"days when real science in medicine was identified with the deadhouse and the use of microscopes" and psychiatrists wanted nothing more than to adopt that identity. Meyer told his audience that he was proud that psychiatry had made the shift from the deadhouse to the study of "plain facts."

But, he went on, something still troubled him. When he was a country doctor in Switzerland, he said, he had gone on rounds with his uncle at a "small cotton mill in which about a hundred . . . demented women were employed and cared for." He ministered to their coughs and complaints, discussed their diseases with their keepers, but "nobody asked or told me anything about their personalities." Meyer was incredulous at his own former cluelessness. "How was it that a practical vision of psychiatry took shape with me so slowly?" He answered his own question:

> Psychiatry became real to me only when I had to handle patients whom I also had known without the mental disorder and who were viewed not as mere derelicts but as persons to be readjusted . . . giving me many an opportunity to incorporate well-known human facts in my more strictly medical thought of the time.

Psychiatry had been led astray by Kraepelin. In his effort to hitch his profession to science's star, he had lost sight of humanity. "The great mistake of an over-ambitious science has been the desire to study man altogether as a mere sum of parts . . . as a machine, detached, by itself." This, Meyer told his assembled colleagues, was the error that had led him to ignore the personalities of his Swiss patients when he was a callow youth. It was also the error that had pushed psychiatry to the margins.

In place of "strictly medical thought," Meyer proposed what he called "commonsense psychiatry." It was an approach that fit in well with his adopted home by placing people on an equal footing with their environment, transforming external circumstance into a force to be met with the mind—aided, when necessary, by psychiatrists—just as the physical landscape had been transformed from primeval forest to civilization over a couple of hundred years. Freed from Kraepelinian constraint, from its racialist agenda and its therapeutic nihilism, psychiatrists could help patients understand what was wrong with their reactions to the world and then help them to change. They didn't have to wait for people to become insane, nor did they have to resort to outlandish theories about nervous weakness. Taking the optimistic view that the mind was up to the challenge of curing itself, psychiatry could become a force for self-improvement in everyday life.

The commonsense approach meant that a diagnosis of a mental illness could be as benign as the diagnosis of neurasthenia, requiring nothing more or less than a prescription. You could be sick without being ruined, Meyer said, but only if you weren't tossed into Kraepelin's bins. This was especially true of depression.

> Kraepelin's manic-depressive insanity and dementia praecox . . . do not exhaust the material that presents itself to us . . . There, are, for instance, many *depressions* which command our attention . . . without their belonging to the above groups.

Many people, Meyer went on, suffer from *essential depressions* that have nothing to do with manic-depressive insanity.

Some of these people suffered from what Meyer called "constitutional depression," "a pessimistic temperament that is inclined to see the dark side of everything and is led to gloominess and despondency upon slight provocation." Others, the simple melancholics, experienced "an excessive or altogether unjustified depression . . . a susceptibility for the unpleasant and wearing aspect of things only." Still others tended to complain of a strange malaise, which they sometimes attributed to stray electricity or nocturnal rape or other problems of "an absurd character." There were also cases of postpartum depression, depressions that accompanied other diseases, presenile depressions—depressions indeed for every stage and style of life. Meyer wasn't entirely clear about whether these were all diseases in their own right or just symptoms of other diseases, but he also said that this didn't really matter. The important diagnostic distinction was about the role of the patient's mental life in his suffering.

> There are conditions in which disorders of function of special organs are the essential explanation of a mental disorder . . . and in these, the *mental* facts are the *incidental* facts . . . But there are cases in which . . . we must use terms of psychology—not of mysterious events, but *actions* and *reactions* . . . a truly dynamic psychology.

Meyer never clarified how to distinguish the two kinds of conditions, but his proposal that at least some depressions were psychogenetic opened up an entirely new possibility in psychiatry: that people who were not hopelessly insane could still be psychiatric patients. They fell into some middle ground, into a region that Kraepelin had overlooked when he mapped the landscape of mental illness. Indeed, they were only suffering an exaggerated version of a universal experience, different from "normal depression," as one of Meyer's disciples put it, only "in its greater fixity, depth, and . . . disproportion to its causative factors." Where the insane were delusional, the pathologically depressed were merely irrational. They were more like Job in the view of his comforters—unhappy beyond what the conditions of their lives warranted—than like Kraepelin's psychotic patients. But unlike Eliphaz, commonsense psychiatrists could offer real comfort: the news that something indeed had gone wrong with the depressed patient, but because it was in his mind rather than in his constitution or his brain, he could be cured through self-knowledge.

In this respect, the psychogenetically depressed were the lucky ones. They were spared the life sentence, their families dodged the stigma of degeneracy, and they could be cured—but only if the psychiatrist did exactly what Kraepelin warned against: listen with empathy, interpret, pay attention to the patient's experience. Psychiatrists, in other words, should offer patients exactly what Freud and Charcot and some other European neurologists had recently begun to offer: psychotherapy. By placing the patient's own story at the center of treatment, psychiatrists could help "the *person himself* transform the faulty and blundering attempt of nature" that had brought him to unhappiness in the first place. All the doctors had to do was to replace their fixation on disease entities and their presumed biochemical causes with a recognition that diagnosis is a storytelling device more than a medical category. If they renounced their "mystic halo," if they made themselves less strictly scientific, or at least less dependent on what Meyer called "brain mythology," psychiatrists could actually help patients—at least those with psychogenetic depressions—figure themselves out and get better. Psychiatrists could do what neurologists had been doing—minister to everyday suffering—without resorting to "neurologizing tautologies," and, more importantly, without eliding the life of the mind.

You've probably never heard of Adolf Meyer before. But as of 1941, according to the *History of Medical Psychology,* written by two psychiatrists and published that year, he had been "the dominant figure in American psychiatry" for fifty years. His fall into obscurity isn't all that mysterious. Meyer's own taxonomy of mental illnesses, the one he proposed to replace Kraepelin's, was anything but commonsensical—a self-contained mythical world populated by reaction types with quasi-Greek names like *thymergasia* and *kakergasia*. His ideas about therapy, on the other hand, were the kind of bromides that give common sense a bad name.

> The physician can offer a patient with a depression a sense of security by communicating understanding based on his personal knowledge of him and of the situation. He must be

able to maintain the patient's contact with a well measured regime . . . to avoid inducing any antagonistic attitude which would interfere . . . with the rapport that may keep him reasonably in touch with the condition of the patient.

Shorter sums up history's verdict: Meyer, he says, was "a second-rate thinker and a verbose writer" whose conclusion—"that everything is very complex"—turned out to be "poisonous to the advance of scientific discipline."

But if that's so, then what are we to make of those first fifty years? Why would a second-rate thinker be judged a giant in his own time? The answer has little to do with nosology or therapeutics and everything to do with marketing.

Meyer's attempt to free psychiatrists from their deadhouse and asylum ghettoes depended on his affirmation of what Hoch and MacCurdy had denounced: the possibility that the whole world could be insane, or at least maladjusted enough to need psychiatric services. Everyday people, he insisted, could be depressed without being crazy, and their difficulties could be understood and treated by expanding psychiatric reasoning beyond "strictly medical thought." By focusing on "life problems," psychiatry, as Shorter put it, was "acquiring a Main Street beachhead" and transforming itself into the mental health industry.

Meyer's efforts intersected with another development in early twentieth century America. Men like John Watson—a prominent psychologist who took his talents to the J. Walter Thompson Company after he was forced out of Johns Hopkins in 1920 for his affair with a graduate student—and Sigmund Freud's nephew Edward Bernays were teaching manufacturers how to use the mass media to sell their products. Their efforts were informed by psychological knowledge. Watson was famous for his extravagant tabula rasa claim:

Give me a dozen healthy infants, well-formed, and my own specified world to bring them up in and I'll guarantee to take

any one at random and train him to become any type of specialist I might select—doctor, lawyer, artist, merchant-chief
and, yes, even beggar-man and thief, regardless of his talents, penchants, tendencies, abilities, vocations, and race of
his ancestors.

And Bernays, known as the "father of public relations" advised his
clients to "make customers" by understanding the "structure, personality, the prejudices of a potentially universal public."

The ad men also capitalized on what psychology, thanks to
efforts like Meyer's, was creating as it ministered more and more
to everyday concerns: what the cultural historian Jackson Lears has
called the "therapeutic ethos," in which soap is sold not on its ability
to get you clean, but on its advertised "promise of psychic security
and fulfillment." With the careful guidance of men like Bernays and
Watson, manufacturers could turn people's psychological suffering into a market for their products by convincing them that their
troubles were a particular kind of problem—the kind for which the
company just happened to have a solution. The ad men discovered,
in other words, that if you could name people's pain, you could sell
them a cure.

Which is exactly what Meyer accomplished. Lowering the bar
for entry into the psychiatrist's office, he gave his profession unique
and privileged access to the average citizen, the one whose life
wasn't as happy or productive or fulfilled as he thought it should
be. Meyer claimed that cure could be found in the one resource
that everyone, especially every American, had: a life story. This
democratization of mental suffering was enhanced by other developments in American life, notably the mental hygiene movement,
spearheaded by activist (and former asylum patient) Clifford Beers,
that made "mental health" a subject of polite conversation. People
could now talk about their "life problems" without fear that they
would be carted away to the loony bin. They could be depressed
without being insane and they could be cured.

In this sense, Meyer was the victim of his own success. For all the time that he was making mental illness safe for democracy, Sigmund Freud and his followers were busy in Europe crafting a different kind of antidote to suffering. Like Meyer's, it steered a course between Job and Eliphaz, between the conviction that life is hopelessly rigged against us and the certainty that only a hopelessly flawed soul suffers, by claiming that our suffering was inflicted by history, which meant that redemption could come through narrative. The kind of life story the Freudians encouraged their patients to tell was anything but commonsensical, but it was largely thanks to Meyer that when psychoanalysis came across the Atlantic, America—psychiatrist and patient alike—was ready to think that their psychic suffering, and the hope for its relief, lay in their biographies.

CHAPTER 6

WHAT YOUR MUM AND DAD WILL DO TO YOU

I had two doctors during my clinical trial—George Papakostas and Christina Dording. They were both at least a decade younger than I and accomplished enough in their field to have been accepted into the clinical faculty at Harvard Medical School. Papakostas, who did the first four interviews, was soft-spoken and affable, with a boyish face framed by wire-rimmed glasses, and seemed to enjoy nothing more than a good conversation about serotonin transporters. He was unfailingly kind, greeted me at each visit with a smile and handshake. And he managed to inject genuine concern into the questions about sleep and appetite, headaches and constipation, and worry and guilt that he asked each week, in exactly the same order, using exactly the same words.

Dording was not quite so likable. But then again we met under less than ideal circumstances. She came out to fetch me on my fifth visit, announcing that she was my doctor as if it were the most normal thing in the world to hand patients from one psychiatrist to another without any notice. Which, she explained to me, it was. What was strange, in fact, was that we hadn't met before. "By this point in the study, I've usually had you," she explained. Somehow that sounded ominous to me.

Still, I had to wonder whether it was something I said, whether my repeated attempts to get Papakostas to talk about what we were actually doing here, about the enterprise of turning my inner life into algorithm-friendly numbers, about the complexities of neuroscience and its still tenuous relationship to the consciousness it was supposed to explain, had led him to want shut of me. I took Dording at her word when she explained that the switch was just the standard operating procedure. But when we passed by Papakostas's door and he was at his desk, peering at his computer screen, I felt a little pang. After all, we depressives are rejection-sensitive.

You would think they'd factor that into your treatment, at least when it comes to abruptly changing your doctor. But this wasn't really treatment, at least not in the usual sense. They surely wanted me to feel better, but the real focus wasn't me. I was just the guy they had to go through to get at depression, the carrier pigeon whom they had to treat well while they carefully unwrapped the message from his leg—or at least well enough to return home for the next dispatch. It didn't really matter which pigeon flew in the window or which soldier debriefed him—or, for that matter, what the bird's name was.

This detachment makes sense, especially if you are a Kraepelinian. Disease is disease, after all. If your oncologist or your ophthalmologist is tied up, indisposed, or just plain tired of you, then why shouldn't he send in an associate to see you? Your tumor or your glaucoma don't care who treats it, so why should your depression? And if you complain to your psychiatrist about this, or betray your injured feelings—let alone if you want to talk about what you think makes you depressed—then your psychiatrist should remember what Emil Kraepelin told his colleagues: "Trying to understand another human being's emotional life is fraught with potential error."

The depression industry is dedicated to not making this mistake. The major instrument of its vigilance is the Hamilton Depression Rating Scale. The HAM-D, as it is known to the depression doctors, was developed in the late 1950s by British psychiatrist Max Hamilton. At the beginning of the antidepressant era, he saw the need for a way to measure depression and its absence so that the benefits of the drugs could be assessed. This was back in the days when depression was an incapacitating illness that required hospitalization, so Hamilton was able to observe many patients closely and over a long time. No one seems to know whether he was familiar with Kraepelin's method (it would still be a couple of decades before descriptive psychiatry was rehabilitated by the American Psychiatric Association), but Hamilton's approach was similar: he proceeded empirically, teasing out the symptoms in order to break down the illness into its constituent parts.

Hamilton determined that depression had seventeen hallmark features. Some of them were psychological—feelings of sadness and guilt, for instance—but the majority were physical: quality and quantity of sleep, various physical complaints, weight gain or loss. He gave each of these symptoms its own item and assigned point values based on the severity of each item. The absence of guilt was a zero, hearing voices of denunciation earned the maximum of four; the physical symptoms of anxiety like stomach cramps or palpitations were worth four points if they were incapacitating and zero points if they were absent. Hamilton gave doctors one gimme— an item that awarded two points toward depression for a patient who "denies being ill at all"—and proposed that a patient's improvement (although not his original diagnosis; the test was standardized on people whom Hamilton had already decided were depressed) could be measured by repeatedly assessing these symptoms, adding up the scores, and charting the trends. Which is exactly what Drs. Dording and Papakostas were doing on my visits to Mass General.

Our conversations—"In the past two weeks, Greg, have you been feeling especially self-critical?" "Um, I don't know, maybe, yeah, I guess so"— were strangely, nearly perversely, at odds with the business at hand. We were talking about my sadness, my disappointment, my self-regard—matters that were not only intimate, but that seemed to exist primarily in language, in the way I talked to myself about myself—in an almost unimaginably circumscribed manner.

This, much more than being shunted from doctor to doctor, I found hard not to take personally. It was even a little depressing: didn't they care about me? They didn't have to get my name right, but didn't they at least want to hear the story behind the answers, the reasons I felt guilty or self-critical, that I woke up at four in the morning in cold sweats or collapsed into sleep at two in the afternoon? It was a story I'd been assembling over twenty years, that I inevitably thought about on my eighty-mile journeys to Mass General, and that, as these stories go, seemed pretty good to me, certainly deserving of more than a question or two about my guilt or my insomnia, more than a number circled on a page in a loose-leaf binder. It even had an ironic twist at the end of the most recent chapter, the psychiatric version of an O. Henry tale.

I was all dressed up with nowhere to go, and when I wasn't feeling indignant about this, I found it disheartening. I felt like a fool, and a narcissistic one at that, for thinking that I mattered.

That, by the way, is a classic example of depressive thinking. Something bad happens to you, maybe just a minor insult. A store clerk is unkind. You fight with your wife. A day at work goes badly. You're looking out your window at a gorgeous spring day, the first strong sun of the season, and you realize that the projects that you abandoned last fall are reemerging from under the receding snow, rebuking you for your disorganization. Or you just feel a twinge in your body, a queasy little ripple, and you would like to ignore it but, as with that funny noise you just heard in your car, you know you're not making it up and you know it's not going to just disappear.

Whatever the occasion, it is as if a trap door somewhere inside you opens up and you fall through. You start to think about all your failures, all the mistakes, all the effort and opportunity squandered, the love lost and withheld, and the time, the irreplaceable, irretrievable time that you have misspent, that you are right now, as you slog through the tar pits of your unhappiness, misspending on self-excoriation and self-reproach and other forms of unpleasant self-absorption, your depression a serpent eating its own tail. (You might even, if you happen to have once gotten a doctorate in psychology or just read the health section of your local paper, know about the studies that show that depressed people have shorter lives. All those stress hormones that you are constantly bombarding yourself with—even if you're not suicidal, you are surely killing yourself, you jerk.)

The story I would have told Papakostas and Dording was in its broad outlines probably nothing they hadn't heard before, even if they'd never conducted a therapy session in their lives (and you can get to be a psychiatrist these days without a moment of nodding and stroking your beard). Psychiatrists are, no less than you and I, soaking in the blood of therapeutic confession, veins opened in memoirs and on couches—if not theirs, then Oprah's. They've heard these stories—mine, yours, and anyone else's on whom they care to risk your empathy—and they all follow the same outlines that Philip Larkin sketches so succinctly in his poem "This Be the Verse":

> They fuck you up, your mum and dad.
> They may not mean to, but they do.
> They fill you with the faults they had
> And add some extra, just for you.
>
> But they were fucked up in their turn
> By fools in old-style hats and coats,
> Who half the time were soppy-stern
> And half at one another's throats.
>
> Man hands on misery to man.
> It deepens like a coastal shelf.

Get out as early as you can,
And don't have any kids yourself.

My version went like this: Mum and Dad were nineteen and twenty-two respectively when they got married. It was no shotgun wedding, just the kids doing what kids in the early fifties did: falling in love, courting, adding it up on paper—two bright Jewish kids, Wellesley and Yale, their ambitions meshing nicely with postwar prosperity, each appearing to be the balm to the other's loneliness, pulled toward marriage like rocks rolling down a hill. They discovered quickly that they'd made a big mistake.

It took them a couple of decades before they found a way out. When I was young, and they were busy trying to destroy each other while pretending to have a happy family, I would stare at their wedding album, as if I could soak up whatever mutual joy they were feeling then and cast it on our roiling misery.

But I couldn't, and we careened along from eruption to eruption, and the molten anger poured over us children, burning us in different ways. My sister hid in the shade of powerful men until she found her voice and became proud and ambitious and successful. My brother learned to lie low, to stand at the edge of the fire while he watched and slowly, patiently, inexorably amassed the evidence of who was to blame. I, on the other hand, pursued trouble like a heat-seeking missile. From the earliest age, I learned to enjoy the conflagrations that ensued.

We all became journalists. My sister, a part-timer like me, writes about torture at places like Abu Ghraib and Guantánamo. My brother worked his way up from the bottom to the top of the television news industry, where he produces investigative reports in which he plays the Porfiry Petrovich to all manner of Raskolnikovs. I broke into the business by writing about my long and disturbing correspondence with Ted Kaczynski, using the Unabomber to blast a career opening for me. My sister says this proclivity toward witnessing and revealing violence and injustice is not a coincidence.

As Larkin says, misery of the unhappy family variety deepens invisibly underneath you, as treacherous as a drop-off in the sea. So when I met a beautiful and witty woman in the early 1980s, when I was in my midtwenties, it stood to reason that I'd fail to see certain things. Like that I was in a state of mourning, having just lost the place where I was living—a cabin I'd built in the woods and in which I'd eked out my own Kaczynski-like subsistence (although without the letter bombs)—and that it probably wasn't such a good time to decide to get married. Or that my intended was like my mother to an extent that was later embarrassing to admit—high-strung, sure that the world owed her a living, prone to rage when it didn't come through, and most likely to hurt the people closest to her in the process. Or that submerged in my desire to please her, to soothe her, to make her love me, was white-hot rage, hostility bordering on hatred for forcing me to work so hard in the first place.

Which I was pleased to let her act out through all the thousand or so days of our marriage. She broke windows, heaved coffee cups, once even picked up a piano bench and smashed it on the floor. I don't think she liked being that way, and I'm sure she hated the ways I provoked her to it. That's probably why it only took her about five seconds to decide what to do when she discovered the Other Woman. And when she left, she took away my opportunity to duck my own anger, my self-loathing, my sadness, all the counterparts inside me to her tempestuousness.

If I'd known what else to do with those feelings, I probably wouldn't have had to marry her in the first place. Or, after we split, to pummel myself to the floor of my study. Nearly twenty years later, when I turned up at Mass General, I'd managed to marry a grown-up woman, to grow up a little myself, and to limit my exposure to that kind of craziness to my therapy practice, where my ability to withstand the violent emotions conjured by love and loss allowed me to take the risks of empathy. But my misery had a life of its own, and as I got older and my love life settled down, it seemed most likely to appear as I approached success. Some people thwart their ambitions by getting in their own way, but that wasn't exactly my problem. Indeed, I was having improbable success, fulfilling ambitions that I didn't know I had, reaping benefits that I was sure I didn't deserve, raising expectations that I was sure I could never fulfill.

Like the contract for the book before this one, in which I was going to write a chapter about my visit to a clinical trial, which I expected would yield a mother lode of evidence about the fissures underlying the antidepressant revolution. Signing it set off a barrage of self-criticism and worry and doubt well out of proportion to the difficulty of actually writing the book. I responded in the depressive's native fashion: I procrastinated. Until I was stuck on the couch, and on an oppressive June day, when I began to feel like I was tumbling down my personal oubliette, I finally reached for the phone and called Mass General.

I'd once written a magazine story about the placebo effect in antidepressant trials. The woman who was the star of that piece had responded so well to her pill that both she and her clinician were shocked to discover that she had been taking a placebo all along. (They wasted no time ignoring that fact; her payment for participation was a year's worth of Effexor.) She told me that immediately after her initial call to the researchers, she felt better. And when I hung up the phone and right on schedule I felt my mood lift, it was as if I had stumbled into a bad novel, or a parody of a bad novel. The absurdity of my situation—that I had created both the grounds and the opportunity for cure of my depression by endeavoring to write about it—made me laugh, which itself made me feel even better.

Call me self-involved (and after one thousand words of this tale, how could you not?) but I thought this episode made a good story, one that captured the self-consuming experience of my depression, that provided a natural springboard into its history, that commented on my professional interest in the subject and my skepticism about what the doctors were up to. That's the one I was prepared to tell my Mass General doctors, had they asked.

In 1971, when Philip Larkin wrote "This Be the Verse," psychiatrists assuredly would have asked about these particulars and would at least have feigned interest. We lived under a different climate of opinion then, the one that Adolf Meyer, with his focus on personal biography as the source of everyday suffering, did so much to usher in. Larkin, incidentally, stole his title from Robert Louis Stevenson's "Requiem," written in 1879, a poem that invokes its own climate of opinion:

> Under the wide and starry sky,
> Dig the grave and let me lie.
> Glad did I live and gladly die,
> And I laid me down with a will.
> This be the verse you grave for me:
> Here he lies where he longed to be;
> Home is the sailor, home from sea,
> And the hunter home from the hill.

Stevenson got his wish; those are exactly the lines engraved on the stone underneath which Larkin no doubt meant to set him spinning. In a century, Larkin was claiming, human life had changed from the celebration of a journey that ends in peace to a calamity that can't end soon enough.

The novelist Stanley Elkin once wrote, "Life is either mostly adventure or it's mostly psychology. If you have enough of the one then you don't need a lot of the other." You can see the truth of this zero-sum equation in the vast difference between Stevenson's wide-eyed world traveler and Larkin's grumbling explorer of the interior world, between the man who lives and dies glad to have been given the chance and sees even in death the evidence of benevolence in the universe, and the man who has no end of complaint about his lousy childhood and, by extension, about the raw deal of human existence.

Whether or not this is the change in human character that Virginia Woolf famously claimed occurred on or about December 1910, it is clear that even as Adolf Meyer was democratizing mental illness, a transformation was under way. And while Woolf and her Bloomsbury circle, with their avant-garde literature and art, may have fancied themselves to be at the cutting edge of the change, it was a Viennese Jew who led the Victorians from adventure into psychology by pointing out that our lives were conducted over the deepening shelf of ocean that he called the unconscious.

Without Sigmund Freud, who would think to tell the kind of story I just did about the immolated family? Who would think to locate the source of his misery in the intimate details of his own private past? Like the idea of magic-bullet medicine, the notion that our personal history makes us who we are, that our troubles come from a mum and dad whose troubles were handed to them by theirs, is so much in our bones that it is easy to forget that it once didn't exist.

Take the idea that depression is anger turned inward, for instance. You've probably heard that truism before—maybe from a friend or a therapist, or in a novel or movie—but without knowing that it originates in Freud's essay "Mourning and Melancholia." Those two states, Freud says, are clearly related to each other, but in their differences we can see just what the problem is in depression. Mourning, with its "painful mood, the loss of interest in the outside world . . . turning away from any task that is not related to the memory of the deceased" is an extreme state, to be sure. But it also seems in some way fitting—after all, what would it say about love if we did not grieve over the loss of the beloved?—and temporary. It may not be rational in itself, but mourning makes enough sense for doctors to "consider interfering with it to be pointless, or even damaging."

Not so with melancholia, whose cause is hard to discern and which overstays its welcome until interfering seems like a good idea. For Freud, however, the true mark of melancholia, what distinguished it from its domesticated cousin, was not so much its irrationality or its persistence as the melancholic's loss of self-esteem. "In mourning, the world has become poor and empty," he wrote, "in melancholia it is the ego that has become so." It may be hard to see why a person is mourning, or why he mourns for so long, but it isn't until mourning turns to self-loathing that you can say for certain that the patient has crossed the border into pathology.

But the fact that the melancholic is ill and that his illness takes the form of being sick of himself should not lead the doctor to conclude that the patient is wrong in his assessment.

> It would be fruitless . . . to contradict the patient who levels such reproaches against his ego in this way. In all likelihood he must in some way be right . . . He seems only to be grasping the truth more keenly than others who are not melancholic . . . [If] he describes himself as a petty, egoistic, insincere and dependent person, who has only ever striven to conceal the weakness of his nature . . . he may as far as we know come quite close to self-knowledge and we can only wonder why one must become ill in order to have access to such truth.

Some melancholics may be mistaken, Freud argued, but the validity of their self-evaluations is not germane to the question of whether they are suffering from melancholia. The true mark of illness is the melancholic's failure to maintain the sense that he is not petty, egoistic, etc., *even if he is.*

And rest assured he is. Your mum and dad may have fucked you up, but they had plenty of help from you. Like every other child, you loved your parents when they gratified you and hated them when they didn't, and you started doing that as soon as your mother's breast was offered and withdrawn (or perhaps as soon as you were ejected from your timeless, painless, intrauterine life into a world of hunger and need). If you have a strong constitution and parents who know how to do their jobs, you eventually learn to control your love and your hate, to grow an ego that can find strategies to make life less confusing and chaotic, that spares you (and those you love) from your titanic feelings. And one of the first thing that the ego does is fool itself into thinking that it is better, more substantial, and less in thrall to our darkest impulses than it really is.

In "Beyond the Pleasure Principle," Freud describes how the ego gets this idea in the first place—and how important it is that it does so. He tells us about the time he observed his eighteen-month-old nephew playing a game with a wooden reel designed to be pulled along the floor by a string. The nephew never used it in this fashion; instead, he repeatedly threw it out of his crib and retrieved it, saying *"fort"* ("gone") and *"da"* ("there") as he did. This game, Freud said, was "related to the child's great cultural achievement—the instinctual renunciation (that is, the renunciation of instinctual satisfaction) which he had made in allowing his mother to go away without protesting." By finding a way to control his rage at being abandoned by his mother, the boy had renounced the pleasure of vengeance and taken a decisive step toward reining in his destructive impulses. This, Freud said, was the basic building block not only of maturity but of civilization itself.

Freud also noticed that his nephew seemed much more interested in the throwing than the retrieving, which he took to mean that mastering loss was more compelling than seeking the pleasure of return, that getting hold of himself "carried along with it a yield of pleasure of another sort but none the less a direct one." It's the pleasure of self-control, of efficacy and self-reliance, and of living in a world where effect follows upon cause, where what happens next can be predicted and controlled. The boy's delight in his game was the precursor of a grownup's comfort in the civilized world.

No matter how well a child accomplishes this task, however, the hatred and violent impulses remain, held at bay by individual effort—particularly the conscience—and by the collective force of civilization. Without this achievement, life would be nasty, brutish, and short; with it, however, the nastiness is merely pushed back to the distant reaches of our minds, into the unconscious. The unconscious never forgets, and trouble like melancholia is kindled when the original hurt and the violence it conjured reemerge to consciousness—a catastrophe that doesn't require anything as dire as death, that could be as simple and banal as the demise of a bad marriage or any other disappointment through which an opposition of love and hate can be introduced to a relationship or an ambivalence already present can be intensified.

> Whatever insult set off the hatred in the first place can't be avenged—maybe the loved one is dead or just not available or doesn't see what he has done—and whatever history gives the ambivalence its peculiar shape is long past. Not only that, but unconscious longings are deep and intense, their satisfaction forbidden.

For some unlucky people—those with parents who repeat those insults and disappointments, who thus provide a feast to feed our ambivalence—melancholia is the only available currency with which to buy off their hatred.

> Patients manage to avenge themselves on the original objects along the detour of self-punishment, and to torment their loved ones by means of being ill, having taken to illness in order to avoid showing their hostility directly.

This is why, Freud says, "Prince Hamlet has ready for himself and everyone else" a catalog of his own shortcomings, and why so many other melancholics do the same: "Because everything disparaging that they express about themselves is basically being said about someone else"—the person who died or left or didn't come through, and all the other people who did that, and so on back to the original objects of ambivalence: mum and dad. Whom you cannot hate or kill because, incompetent or mean or neglectful as they may be, they are all that you have.

This is also why self-reproach is the identifying mark of melancholia, whether the occasion is the loss of love or the rise of ambition and success: because it signals that the crucial fiction—that we are wholly on the side of our own better angels, that we don't also hate those whom we love or want to destroy the people who have hurt us—has failed. "The loss of the love-object is an excellent opportunity for the ambivalence of love relationships to come to the fore," Freud wrote, which means that it is a likely time for us to glimpse the bottomless, destructive desires that haunt our conscious lives. The melancholic is the person whose mum and dad made it impossible for him to maintain the illusion that normally keeps this awareness at bay.

The biggest weakness in Freudian theory—and perhaps the major factor in its fall from grace—is that it is, as philosophers of science like Karl Popper would put it, nonfalsifiable and therefore not subject to scientific test. Psychoanalysis is a self-contained system, its basic tenets impossible to verify. How do you test for the presence of the unconscious, which exists largely as an absence? The answer is that you don't. You accept it on faith and go from there. In this respect, psychoanalysis is a throwback to Hippocratic medicine, to a time when wise men postulated forces that no one could see but that must be lurking behind, and causing, what was visible.

But every so often, one of Freud's theories is inadvertently supported by science. The central idea of "Mourning and Melancholia"—that the disease consists of the loss of an illusion—is one instance. In the modern version of melancholia, the story that I would have told my doctors—the one in which I concluded that I had no business writing books, that my success was at least as much a fluke as the just reward for my effort—is not a clearheaded assessment, but the sign of pathology. Indeed, according to a theory developed in the 1960s, depressives make themselves sick by persistently and pervasively overestimating just how bad things are. This cognitive distortion is actually the pathogen. Something has gone haywire in a patient's thinking—and, in later versions of this theory, in a patient's brain—and caused him to become unduly negative. Cognitive-behavioral therapy, in which a therapist helps a patient correct this bleak outlook, is the cure.

This theory, it turned out, could be tested. You could, for instance, break the unspoken rat-experimenter pact, the one that says you reward and punish a rat depending on what behavior you want to reinforce, and instead administer electrical shocks at random. And when you find that the rat eventually just curls up in a ball and stops eating, you can call that *learned helplessness* and extrapolate that this is what happens in depressed people—they get the idea that they can't make things better and give up. Then you can offer to help them by showing them that they aren't helpless, that they can improve their circumstances, that their lot is not as bad as they think. And you can turn to your Freudian friends and say that things just aren't that complicated and dark. You don't even need a human mind to get depressed, just an expectation that no matter what you do you are going to get hurt.

But a funny thing happened to learned-helplessness theory. Cognitivists predicted that depressed people would be significantly more likely than non-depressed people to blame themselves when things go wrong. In 1979, a couple of psychologists, Lauren Alloy and Lyn Abramson, decided to check out this hypothesis. They set up a series of studies revolving around a green light and a button. In the first experiment, subjects were told to push the button and decide whether or not it made the green light come on, a condition that was controlled by the experimenter. Over and over again, the depressed people were better than their normal peers at assessing their role in the light's status.

Then Alloy and Abramson introduced money into the equation. They gave some subjects five dollars and told them that they'd lose money every time the green light failed to light. They gave other subjects no money but told them that they'd get money if the light came on. What they didn't tell them was that the button was completely irrelevant and that everyone who started with money was going home broke, while everyone who started with nothing was going to win five bucks. Then they asked them to estimate the extent to which they were responsible for their fortunes—a task at which depressed people excelled. And when the experimenters started to give subjects control over the light, the nondepressed people turned out to think that they deserved to win but not to lose regardless of the actual facts. Depressed people, in the meantime, continued to be superior at figuring out their role in events. The experimenters concluded that "depressed people are 'sadder but wiser' . . . Nondepressed people succumb to cognitive illusions that enable them to see both themselves and their environment with a rosy glow."

Alloy and Abramson noted that depressive realism, as this phenomenon came to be called—and, by the way, this work has never been refuted; cognitive theory, as we will see in later chapters, chugs along as if it never happened—raises a "crucial question": Does "depression itself [lead] people to be realistic, or [are] realistic people more vulnerable to depression than other people?" They did not mention that Freud had already posed this question when he wondered why we have to get sick in order to have access to the truth. But then again, Freud wasn't bringing up the question in order to answer it. He was making a point: that an excess of truth is bound to make a person suffer. Just ask Job.

It would be tempting to see Freud's increasing pessimism about the prospect for escaping ambivalence—or, to put it more directly, to achieve happiness—as his response to the excess of truth imposed by World War I, in whose shadow he wrote "Mourning and Melancholia." That cataclysm, as he put it in "Transience," an essay published in 1916, "brought our instincts to the surface, unleashed within us the evil spirits that we thought had been tamed by centuries of education." It was in the aftermath of the war that Freud developed the idea of the tripartite self, an ego stretched between id and superego, never quite up to the task of mediating between these protean forces. Eventually he would liken the ego to a garrison occupying a rebellious city, one whose walls would sooner or later be breached by the rest of the unruly psyche.

But historian Eli Zaretsky reminds us that the war's depredations showed up for Freud and other clinicians in a very specific way: the veterans on both sides of the conflict returned home plagued by nightmares and agitation and depression, by what the DSM now calls post-traumatic stress disorder, but which Freud knew as shell shock. Shell shock, Zaretsky argues, forced Freud to reconsider the significance of trauma in mental suffering. Although he once had seen external events—specifically, childhood sexual abuse—as the culprit in the hysterias he was treating, he had come to think of the memories of abuse as fantasies spun out by the psyche as it manufactured a Manichaean reality in an attempt to come to terms with its own divided nature. But the whole world had witnessed the horror that gave rise to shell shock; there was no use denying that the trauma was real.

Still, it took an active mind to turn shells into shell shock—which it did, according to Freud, by what he came to call the "repetition compulsion." The veterans' psyches forced them to repeat their experience in flashbacks and dreams and in the unending anxiety—so much like their lives in the trenches—that plagued them. Freud had no question that this was an attempt to master the experience in fantasy if not reality, but he also saw something darker at work. It wasn't only wars and sexual abuse and other overwhelming experiences that breached the garrison; it was a desire, built into our animal nature, to return to our inorganic origins, to obliterate life—or, as Freud named it, Eros—before it obliterates us. "What lives, wants to die again," wrote Freud's earliest biographer by way of explaining this. "Originating in dust, it wants to be dust again." A veteran relives the trenches for the same reason that another man recreates the trench warfare of his family—not only to revisit and "work out" that formative trauma, not only to have the opportunity to play out the drama with the odds evened up a bit, but also out of an inborn and perverse attraction to horror itself: the *Todestrieb* ("death instinct"), Freud called it.

Even as he formed these dour theories, however, Freud retained some optimism that the war, and the truths about human nature that it seemed to reveal, would be the occasion for mourning more than for melancholia:

> Once mourning is overcome it will be apparent that the high esteem in which we hold our cultural goods has not suffered from our experience of their fragility. We will once again build up everything that the war has destroyed, perhaps on firmer foundations and more lastingly than before.

By 1930, however, even this pale optimism—which biographer Peter Gay describes as "far more a matter of duty than conviction"—had nearly disappeared. "I can no longer get along without the assumption of this [death] drive," he wrote to a fellow analyst. And he was deeply worried about its implications for a people in possession of even the prenuclear version of weapons of mass destruction:

> The fateful question for the human race seems to be whether, and to what extent, the development of its civilization will manage to overcome the disturbance of communal life caused by the human drive for aggression and self-destruction . . . Human beings have made such strides in controlling the forces of nature that, with the help of these forces, they

will have no difficulty in exterminating one another, down to the last man . . . And now it is to be expected that the other of the two "heavenly powers," immortal Eros, will try to assert himself in the struggle with his equally immortal adversary.

These were the final sentences of *Civilization and Its Discontents* when Freud sent it to the printer on October 29, 1929, the day that the New York Stock Exchange collapsed. The following year nearly one hundred Nazis were elected to the Reichstag, and in 1931 Freud added a new last line, one that made it clear he no longer considered a victory for Eros a safe bet: "But who can foresee the outcome?"

As much as the book (whose German title, *Das Unbehagen in der Kultur* Freud once proposed translating as *Unhappiness in Culture*) keeps returning to the large scale, it also considers the fate of individuals, and here Freud is, as Peter Gay puts it, "pitiless" about our prospects. "The life imposed on us is too hard for us to bear," Freud wrote. "It brings too much pain, too many disappointments, too many insoluble problems." The worst by far of those insoluble problems is what Freud calls the "sense of guilt," which he says differs from guilt itself in that we, rather than some external authority, inflict it upon ourselves. Developing a conscience, we control our instincts and use the channeled energy to erect a civilization; keeping ourselves in line, we can dispense with the dictators in favor of modern nation states and all the liberties they bring. But the price of freedom is high: "the threat of external unhappiness . . . has been exchanged for an enduring inner unhappiness." Modern civilization requires a mutilation of the self, less barbaric than circumcision or ritual scarification, but still leaving a permanent mark.

Your mum and dad, whatever trauma they inflict upon you, are only determining the particulars of how the conflict between your boundless instinct and the demands of civilization is going to make you suffer. The garrison of the ego is under constant assault from within and without; how and where the shells penetrate its walls is a matter of your personal history, but in the war between Eros and Thanatos you are mere collateral damage.

Our existential condition, then, is one in which both Job and his comforters are correct. The game is rigged against us, as Job complained, but our misery is a function of how we have played the game, as Eliphaz insists. Which means that a very peculiar form of comfort is available: psychoanalysis, which is not a technique for correcting those negative thoughts and achieving happiness, but merely the crucible, some would say the ordeal, from which the patient emerges with the strength and courage to face the truth about himself, to resist what he must resist and enact what he can enact, to live without the illusion that he is not guilty of what he accuses himself of and yet to stop accusing himself. You don't have to surrender to the whirlwind. If you face it, which is to say if you are an assiduous patient, you can trade in your neurotic misery for normal human unhappiness and carry on with a better story, one in which self-knowledge is your only consolation.

You have to marvel at the fact that this unrelentingly bleak philosophy, and its expensive, inefficient, and not terribly warm and fuzzy therapy, ever caught on in the United States. The full story of how this happened is too long and complex for my purposes, but Eli Zaretsky has done a very good job of telling it in *Secrets of the Soul*, and Joel Kovel has given an alternate account in *The Age of Desire*. What you have to know is that some things did change— notably Freud's conviction that the ego would never do better than to hunker down behind some imaginary green line while civilization and instinct relentlessly lobbed their mortars. Within a decade of his death in 1939, Freud's followers, including his daughter Anna, having left a broken Germany for England and the United States, saw to this, decreeing that the ego was at least potentially a commander-in-chief capable of leading the psyche beyond discontent. By the 1950s, psychoanalysis was dominated by the ego psychologists, whose job was no longer to acquaint people with the source of their discontents, to clue them in to the endless conflict that raged in their psyches, and to console them as they faced their inescapable suffering, but to help them manage their impulses and adjust successfully to civilization. And by the 1950s, psychoanalysis had become what Kovel sneeringly called a "psychology for winners," which of course made it a much easier sell in the land of opportunity.

Even before this transformation, however, psychoanalysis had made deep inroads into American popular culture. Zaretsky is among the historians who argue that Freud came along at just the right time: in the first half of the twentieth century, when "personal life"—the interior world that had once been the province of philosophers and priests—became a sphere in which everyone had a stake. "Freudianism helped construct a new object—personal experience," Zaretsky says. "It gave [people] a new sense, according to which individuality was rooted in one's unconscious, one's desire, and, above all, one's childhood." As John Calvin had done for an earlier generation, Freud provided the template for the new story that Americans were telling themselves about themselves, their mental hardships, and the means for overcoming them.

Those hardships were substantial, the discontents widespread. Leaving America after an eight-month lecture tour in 1927, Sandor Ferenczi, a Hungarian analyst and close colleague of Freud, told a *New York Times* reporter that "life in America is so strenuous that the people are naturally driven into neurotic conditions." Criminality and insanity and "incorrigible children" demanded our attention, Ferenczi added, but "another issue . . . is the psychological readjustment that thousands upon thousands need in their relation to family, profession and society in general." Which meant that the whole world could be put on the couch.

But that didn't mean that the whole world was insane—or, for that matter, that it needed doctoring at all. Ferenczi may have sounded like George Beard in flagging the dislocations of modernity as a widespread pathogen, but the Freudian diagnosis was at once more universal and less medical than Beard's—not neurasthenia but neurosis, not the rest cure but restless exploration of the incurable human condition. This difference helps to explain something remarkable about *Civilization and Its Discontents:* that in this sometimes bitter, often mournful, and always melancholy lament about who we are, what kind of world we have created, and where we are headed, Freud never once mentions melancholia.

My doctors at Mass General would no doubt have interpreted the absence of melancholia from Freud's book as yet more historical evidence that depression is not unhappiness itself but an illness that has unhappiness among its symptoms, and that Freud himself was trying to make that distinction. They would claim that now that they have returned to a Kraepelinian diagnostic scheme, they are able to sort out the diseased from the merely discontented. Far from indicating a flaw in that logic, or at least that their diagnostic net was cast too wide, my presence in their study meant that their science worked. It told them something about me that I didn't know: that if I thought I was only suffering from *das Unbehagen,* that was my depression talking.

But Freud would have had a different explanation. *Civilization and Its Discontents* may not have been an account of a medical condition—his own or anyone else's. ("I believe that I have not given expression to any of my constitutional temperament or acquired dispositions," he wrote, perhaps forgetting how this denial would sound to a Freudian.) But that didn't mean it wasn't scientific. The book's gloomy conclusions were the only ending that the evidence allowed: "My pessimism appears to me as a result," he wrote, "the optimism of my adversaries as a presupposition." Science, after all, sometimes tells us what we would prefer not to hear.

But what kind of science was psychoanalysis? Ferenczi's interview with the *Times* ended with a coded message that answered this question. All that psychological readjustment, he said, "opens a tremendous field for the analytically trained social worker." The general reader might have heard the good news that help was on the way for the discontent he evidently was bound to suffer; young people looking for career opportunities may have heard a different encouraging message. But to doctors, especially those with a professional interest in psychoanalysis, Ferenczi's words had another meaning: social workers, not doctors, were going to reap this bounty. Medical science was one thing, psychoanalytic science another.

Ferenczi was in fact firing a salvo in a battle that had already been raging in his profession for a few years and that had just recently surfaced in a *Times* article that had run just a couple of weeks before the interview. An American doctor by the name of Newton Murphy, the story went, had gone to Vienna for analysis with Freud. The master was too busy to see him and referred him to a student analyst, Theodor Reik. After several weeks of treatment, Murphy, according to the *Times,* "declared that his health was worse rather than better." He complained to Freud but evidently received no satisfaction, because he then approached the Austrian authorities, claiming that because Reik was not a physician, he was guilty of quackery.

The trial was attended by the elite of medicine and psychoanalysis in Vienna, among them Julius Wagner-Jauregg, a Nobel Prize–winning psychiatrist. Testifying on behalf of thirty-one of his colleagues, Wagner-Jauregg warned that psychoanalysis was "dangerous when practiced by a man not educated in medical science." Freud countered that medical science was not only irrelevant to his treatment, it might actually get in the way. "A medical man cannot practice psychoanalysis because he always has medicine in his mind," he told the court. The judge decided that Freud knew what he was talking about when it came to psychoanalysis and dismissed the case against Reik.

But a court in Vienna couldn't stop what had already happened in America. In 1926, the New York Psychoanalytic Society declared that only physicians could practice analysis. Freud's response came in *The Question of Lay Analysis,* in which he spelled out why doctors were ill-suited to psychoanalysis: their education was exactly the wrong preparation for the job. "It burdens [a doctor] with too much . . . of which he can never make use, and there is a danger of its diverting his interest and his whole mode of thought from the understanding of psychical phenomena." Doctors are subject to the "temptation to flirt with endocrinology and the autonomic nervous system," as if psychic suffering was just another illness whose cause and cure were organic.

Some mental suffering may indeed be organic in origin. A few cases of melancholia, Freud wrote in "Mourning and Melancholia," "suggest somatic rather than psychogenetic diseases," but his own interest lay in the "cases whose psychogenetic nature was beyond a doubt." These maladies—the neuroses—were the proper object of his therapy, and because the mind was shaped by history and culture, the education of analysts must "include elements from the mental sciences, from psychology, the history of civilization and sociology, as well as from anatomy, biology, and the study of evolution." There was no time to teach medical students these subjects in addition to all they had to learn about medicine. But without this breadth of knowledge, doctors would make poor analysts. Perhaps even more important to Freud—whose "self knowledge," he wrote, "tells me that I have never been a doctor in the proper sense"—their ignorance would lead them to turn psychoanalysis into a "specialized branch of medicine, like radiology."

That was the last thing Freud wanted. "As long as I live," he declared, "I shall balk at having psychoanalysis swallowed by medicine." But even Sigmund Freud could not control the fate of psychoanalysis. The New York Psychoanalytic Society continued the policy that Freud, his already rabid anti-Americanism inflamed by his quarrel with American doctors, called "an attempt at repression." Mental suffering may have been democratized, but it was still an illness, its understanding and treatment still firmly in the hands of the medical elite. The social workers would have to find some other way to save the world.

Staking the territory of *das Unbehagen* for medicine, the rene-gade analysts opened the way for the depression doctors eventu-ally to corner the vast unhappiness market—a debt of which my doctors at Mass General were most likely unaware. But the New York psychoanalysts left their future colleagues with a problem that would only deepen as more and more of the mysteries of the human organism fell to the microscope and the scalpel: the diseases the psy-chiatrists were claiming as their own were problems of the mind, their origins in culture and history, their treatment in the refash-ioning of biography. But the authority behind the doctors' claim derived from treating diseases that were biochemical in origin for people who, as Freud grumbled, "expect nervous disorders . . . to be removed." It was only a matter of time before the obvious con-tradiction between form and content became an embarrassment.

The solution to this problem is obvious in retrospect: to swallow psychoanalysis and all the psychotherapies it spawned, to turn them into a specialized branch of medicine, the depression doctors had to turn away from biography and back to biology. They eventually had to declare that the mind does not exist except as a property of the brain. Which meant that doctors could have it both ways: domin-ion over our discontents *and* a claim to scientific knowledge about them. Only then would *das Unbehagen* be folded into major depres-sive disorder, the disease at which my doctors, funded by the federal government and employed by the most prestigious university in the country, could aim their magic bullets.

CHAPTER 7

THE SHOCK DOCTORS

The most fun part of my clinical trial came on my fifth visit. By then I'd been dutifully taking my pills—five glistening amber gelcaps a day—for six weeks. I'd been asked the same questions, filled out the same forms, gotten my parking ticket stamped by the same receptionist four times. I knew the combination to the lock on the men's room door by heart.

I'd also been told that I was improving. And maybe I was feeling a little better, a circumstance that I would normally have attributed to some minor successes or to the relative ease of life in the summer or to the random nature of emotional life or to increasing maturity and wisdom or to the fact that I was finally getting my book off the ground or indeed to nearly anything other than a daily dose of three grams of omega-3 fatty acids.

Or maybe I wouldn't have ventured any explanation at all. After a while, you just start to think that depression and its remission just can't be explained, not fully anyway. You look at your immediate circumstances and see if there is one you can change, some trouble to manage or irritant to eliminate, some loss to mourn. You take the steps, make the change, spill the tears or voice the rage, and if that doesn't make you feel better, if you still wake up nauseated and afraid and spend your day that way, you contemplate other measures, therapy maybe or some distraction or maybe even psychiatric drugs, but all the time you are doing this you are also just waiting for it to pass like bad weather. Maybe you regret that you are built this way, the same way that you regret that your musical talent is limited, that you are losing your hair, that you drive away some people and attract others, and maybe you stand in alternating awe and resentment at just how narrowly the margins are drawn around what you can change, and you take all this as a reason to develop your humility before the indignities really catch up with you. But mostly you can't really know what made life turn ugly any more than Job could.

Because you can't live an experimental life and a control life at the same time. You just take your best guess at what causes what and try to live accordingly. And by the time you're fifty, you like to think that the few things of which you are certain—beyond, of course, the increasing impossibility of being certain about any-thing—are also true, that you haven't just snared yourself in some unjustified faith, some ideology, held against science and, increas-ingly, common sense, that posits that consciousness has to be more than the sum of its parts, that history is important, that self-examination is, if not a cure, then surely more than a mere consola-tion. So when the doctors start not only to tell you that they know what is wrong with you better than you do, but also to show you the proof that you are actually getting better in exactly the way they predicted, when they add up their numbers and the survey says you have improved, when their certainty about where your depression resides and what ought to be done about it has the ring of scien-tific truth, you really have to wonder about the conclusions you've arrived at. Maybe you have to face the possibility that you are like Schopenhauer, in William James's version, barking at the moon. Maybe you have to choose between being right about the ways of the world and being happier.

But I couldn't decide. I kept taking my pills, but I never totally got on board with the doctors, no matter what the numbers said. This was partly my native orneriness, a pigheaded clinging to my worldview over theirs. But it seemed to me there was a big problem even within their world. It was those tests, the ones they were using to measure my depression and to tell me that I was improving.

In addition to the HAM-D and a questionnaire filled out by the doctor about constipation and fevers and other possible side effects, I was completing a battery of forms on every visit—the Q-LES-Q, which rated my life enjoyment and satisfaction, the Quick Inventory of Depressive Symptomatology (QIDS), which asked me to circle a number on an item like this one:

View of Myself

0. I see myself as equally worthwhile and deserving as other people
1. I am more self-blaming than usual
2. I largely believe that I cause problems for others
3. I think almost constantly about major and minor defects in myself

And the Ryff Well-Being Scale, which measured my emotional state by asking me to rate on a scale of one to six how much I agreed with statements like "For me, life has been a continuous process of learning, changing, and growth" or "My daily activities often seem trivial and unimportant."

What was bothering me about the tests wasn't only that they seemed inane and puny compared to what they were trying to measure. It was also their logic—or their lack of it. It's the burden the depression doctors took on when they revived Kraepelin: you have to assume that the patient is depressed in order for his feelings to be considered symptoms, but the symptoms are the only evidence of the depression. Wondering if "life is empty" or "if it's worth living," may be, as the QIDS insists it is, a thought of suicide or death—but only if you're depressed. Otherwise, it's just a common, if disturbing, thought. To logicians, this is known as assuming your conclusion as your premise, or begging the question.

The depression doctors know about this problem. Even the best doctors are skeptical of the ability of these tests to parse inner life. On an early visit, I complained to Papakostas about having to choose from one of four options, or worse, a yes or no, to describe what I thought were complex, sometimes even incomprehensible experiences. "I'm sorry to seem dense about this," I said, "but it's just not how I usually think about things."

Papakostas was reassuring. "You know, this question condenses a lot of areas of life into just a number. It doesn't work well," he said. "Some questions we just don't like."

Since condensing life into a number seemed to be more or less exactly what we were supposed to be doing here, and since the results were the basis of my diagnosis and the claim that the drugs were treating it (not to mention of the whole antidepressant industry) this seemed like a startling admission—sort of like a priest telling me from his side of the confessional that he's not so crazy about this venial sins business. And later, when we got to the question about my naps (I had snoozed four times for thirty minutes or more that particular week) and Papakostas said, "See, some of the questions are really nice in terms of being objective," it seemed like the right time to speak up, to remind him that when doctors and drug companies tell people that drugs cure the disease of depression, they don't add, "But by the way, the tests that allow us to say so are really bullshit."

But I didn't protest. Quite the opposite, I sympathized. "I suppose it would be easier if there were biochemical markers," I said. "Otherwise, you're just stuck with language."

And even when Papakostas said, "Hey, we're psychiatrists. Language is good," as if this entire enterprise weren't an attempt to avoid the uncertainties of language, I still didn't speak up.

Maybe it was the abrupt change from Papakostas to Dording, or just the fact that I didn't like her so well, but by the time of my fifth visit, I was over my attack of Stockholm syndrome and ready to stop giving the depression doctors a free ride.

My chance came when Dording, administering the HAM-D, asked (as Papakostas had already asked four times), "In the past two weeks, have you been feeling excessively self-critical?" There's no doubt that I am a very self-critical person. If there's a problem somewhere in my vicinity, if someone I care about is unhappy, I assume that it's at least partly my fault. I don't particularly enjoy this about myself. But is it excessive? Or is it what makes me caring, responsible for myself, a conscientious citizen, an effective therapist, a decent writer? And to what or whom am I supposed to compare my self-criticism to determine its excessiveness? To another depressed patient? To the way I wish I were or think I ought to be? So I asked.

"If there's a comparator implied, it's always to when you're not depressed," she answered crisply, as if no one had ever asked such a silly question, as if it was as plain as the nose on my depressed face. She seemed so sure of herself that I began to wonder if her answer really was as circular as it sounded, if it meant more than saying self-criticism is a problem when it's a problem and not when it's not, and if it wasn't a call, if ever there was one, for some self-criticism on the psychiatrists' part. It seemed like a denial of the basic assumption of this whole clinical trial—that they were the experts about my mental health, that depression isn't something I'm equipped to detect in myself, because if I was, I'd have been in the other study, the one for the minor depression I thought I had in the first place. I began to wonder if this was really the old Kraepelinian problem or if all this wondering and my resulting inability to blurt out a yes or a no was just another example of my excessive self-criticism.

But I was staying on my side that day, on the side of language and meaning. So I asked her if she really thought self-criticism is pathological.

"Pathological?" she asked, as if she'd never heard the word. "I don't know if I'd call it pathological."

"Symptomatic, then," I offered.

"Well, it's certainly not optimal."

"Optimal," I said, deploying the therapist's repeat-and-pause tactic, hoping she would tell me exactly how much self-criticism is optimal and how she knew.

"Certainly not optimal." She did her own pause.

"But being self-critical is something that helps people achieve, isn't it?"

"Sometimes yes, sometimes no. I don't think being excessively self-critical is ever a great thing. No." She started turning pages again, trying to resume the interview.

But I didn't want to let it drop. I went back to the question I should have asked Papakostas a long time before. The numbers aside, I wanted to know, just between us pros, did I really seem depressed to her? *Majorly* depressed? I couldn't quite get myself to ask it this way, so instead I asked her what she thought the difference was between minor depression and dysthymia, a DSM-IV mood disorder that, at least until minor depression makes it into the diagnostic big leagues, comes closest to capturing my melancholy.

"You're getting into close quarters here," she said.

I think she really meant to say that I was getting into fine diagnostic distinctions here. In another world, one in which psychiatrists actually liked language, we might have explored this unintended revelation of discomfort at my intrusion into her professional space. In this world, however, there was no room for discussing such slips. But that doesn't mean she didn't make one, and as she explained that "dysthymia is more low-level chronic; minor depression may or may not be long term, but it's typically less criteria than major depression," and then closed her notebook to walk me out, I was feeling vindicated.

And, of course, guilty.

I'm not sure what it says about me that my little quarrel with Dording was fun. But I'm not the first person to make this kind of mischief, to enter the belly of the beast and give it a little heartburn.

I've already told you about David Rosenhan and his seven friends who, three decades before my Mass General caper, infiltrated mental hospitals across the country. Their biggest mischief, of course, was placing their write-up in *Science,* but there were other little pleasures along the way, like catching the attendants rousting the patients in the morning by screaming, "Come on, you motherfuckers, out of bed!" or keeping track of the time doctors actually spent with patients and determining that it amounted to an average of 6.8 minutes per day. My personal favorite moment in "On Being Sane in Insane Places" comes when "a nurse unbuttoned her uniform to adjust her brassiere in the presence of an entire ward of viewing men. One did not have the sense that she was being seductive. Rather, she didn't notice us." That must have been fun to watch while jotting in a notebook what the unsuspecting nurse thought was just another manifestation of your insanity.

But Rosenhan wasn't the first guy to pull this kind of stunt either, and his results and mine put together can't hold a candle to those of the man who was. To be fair to us, even if we'd had the moxie, neither Rosenhan nor I could possibly have done what Joseph Wortis did in 1934. We were born too late to show up at Berggasse 19 in Vienna, lie down on the most famous couch in the world, and prank Sigmund Freud.

That's not how Wortis, who was born in Brooklyn in 1906, described what he did. His account starts with the suicide of a wealthy Harvard art historian, Kingsley Porter, who threw himself off a cliff in Ireland in 1933 when his lover, Alan Campbell, rejected him. Porter was married and, Wortis recalled, "the bereaved widow went to Havelock Ellis, who was a friend of Kingsley Porter, saying she wanted to use her wealth to do something for the cause of homosexuality." Ellis, a British psychologist, was famous for his matter-of-fact research on human sexuality, which included not only the deviance studied by Richard von Krafft-Ebing in his 1886 *Psychopathia Sexualis* or the polymorphous perversity Freud was so interested in, but also your normal day-to-day man/woman sex.

Perhaps for his candor, unusual in the late Victorian era, or simply because his work was exciting to read, Ellis was, according to Wortis, the "literary and scientific hero of my college days." Hero and acolyte met in 1927, and the good feeling was evidently mutual, for when Mrs. Porter approached Ellis six years later, Wortis's name came to mind. Ellis relayed his interest to Adolf Meyer, by then at Johns Hopkins, who asked around at Bellevue Hospital, where Wortis was beginning his psychiatric training. The Bellevue staff told Meyer that Wortis was "very unusually talented," and the next thing Wortis knew, two of the most prominent medical men in the world were throwing Mrs. Porter's money at him.

There was only one problem. "I had no wish . . . to become a sexologist," Wortis said. "I also had some doubts and misgivings about a project that might be intended to involve special pleading on behalf of homosexuals." But he didn't let that dissuade him. "I would be glad to accept a fellowship of the sort described," he told Ellis, "if it allowed me to pursue my general psychiatric training, with a view to later turning my interest to special studies in the field of sex." (Wortis never did turn his interests in Mrs. Porter's desired direction—at least in part because after seven years he concluded that he couldn't agree with "the views of the widow . . . who thought her husband was born this way, couldn't help it, that his rights needed to be defended, and that science should come to his defense.")

If you had an active mind, a command of German, a love of Europe in general and Vienna in particular (Wortis got his M.D. at the University of Vienna), the sponsorship of the world's leading psychologists, and access to a rich widow's money, you'd probably at least consider doing exactly what Wortis did next. He took the money and ran, using it to fund a little research project of his own. "Though I am myself skeptical of the dogmas and claims of the psychoanalysts," he wrote to Meyer, "don't you think it would be worthwhile to learn something of the subject at first hand?" Meyer and Ellis were skeptical about Freud, so they did think this was a good idea and pledged sixteen hundred of Mrs. Porter's dollars to Wortis's "training." Wortis approached Freud in September 1934 and after two meetings Freud told him that his bankroll would pay for four months in analysis, which they could begin presently.

Freud knew that Wortis was associated with Meyer and Ellis and that neither of those men held psychoanalysis in high regard. He must have suspected that Wortis shared their views, but the game was on. "He [Freud] would have thrown me out because he got impatient with me, but he didn't want to acknowledge his failure," Wortis told an interviewer sixty years later. "I came under the grand auspices of Havelock Ellis and Adolf Meyer . . . So he had to put up with me." What Freud didn't know was that after every session, Wortis was going to a nearby café and writing down as close to a verbatim transcript as he could and sharing the highlights (and eventually the transcripts) with Ellis and Meyer. They wrote back to congratulate him and egg him on. (In this, Wortis falls on the transparency spectrum somewhere between Rosenhan and me: Rosenhan deceived the hospitals outright; I told my doctors that I was recording our interviews and writing about my experiences and that I had published critical articles on the subject of antidepressants and clinical trials.) In the 1994 interview, Wortis, who first published his account of his analysis in 1954, had to admit that he had had some fun on the couch. "I was taunting Freud," he said.

Freud probably was aware that he was being taunted. But there was something else that he couldn't know—and that Wortis couldn't know either. Freud, by now a cranky old man, suffering from mouth cancer, struggling to talk with his prosthetic jaw, bitter and scared about the rise of the Nazis, was intemperate ("It is true you have no palpable symptoms, but you have no right to be too proud of your health") and even downright mean ("You know shit about psychoanalysis"). He was sometimes pathetic ("He seemed to be a bit hard of hearing," Wortis wrote, "but did not admit it. On the contrary he continually criticized me for not talking clearly and loudly enough"), occasionally pithy ("Dreaming is nothing but the continuation of waking thought." "No man could tell the truth about himself"), often doctrinaire ("You have not yet completed the transition from the pleasure principle to the reality principle"), and always engaged with Meyer and Ellis at least as much as with Wortis ("A person who professes to believe in common sense psychology [i.e., Meyer] and who thinks psychoanalysis is 'far-fetched' can certainly have no understanding of it, for it is commonsense which produces all the ills we have to cure." "I feel sure . . . that Ellis must have some sexual abnormality, else he would never have devoted himself to the field of sex research").* All of this was predictable, if entertaining. The surprise, however, came when, in the midst of his analysis, Wortis made a visit to another doctor in Vienna, an encounter that would change not only his professional life, but the course of psychiatry, and especially the treatment and understanding of depression, in the United States.

Ten weeks into his analysis, in the middle of December 1934, Wortis told Freud about a demonstration he'd attended over the previous weekend. In front of a group of doctors and students, a psychiatrist named Manfred Sakel had injected insulin into a schizophrenic, brought him to the point of death, and then revived him with glucose. After this ordeal, the patient seemed transformed—quiet, oriented, and calm.

Sakel, whose real name was Menachem Sokol, said he was a descendant of Moses Maimonides. He also claimed that he had tested his treatment on animals before administering it to people. The *New York Times* was satisfied enough that the first claim was true to repeat it in Sakel's obituary when he died in 1957. No one, not even Sakel himself, was ever able to substantiate the second. Nor could he say exactly why he thought to use hyperinsulinization

---

* In this at least Freud proved to be right. Ellis was married to an openly gay woman, was impotent until he was sixty, and was turned on by the sight of urinating women.

in the first place. Sometimes he spoke of vague theories—something about toxins, the digestive tract, the "vagotropic nervous system," and the "restoration of balance." He didn't mention that psychiatrists had already discovered that when they gave insulin to asylum patients who were refusing to eat, it not only boosted their appetite but also improved their psychological state. But then again, a would-be maverick genius generally can't afford to acknowledge his predecessors.

He can, however, flaunt his departures from orthodoxy and turn his ignorance about why his method works into a virtue. He wasn't even looking for an answer, he said, but rather had "deliberately abandoned the normal scientific procedure which first seeks to establish the cause of a disease and then formulates a treatment accordingly." If he had plodded along scientifically, worrying about comas and convulsions and the causes of disease, he would never have had his "accidents" and seen the "dramatic psychological changes" that occurred when patients slipped from mere hypoglycemia into comas and lived to tell about it. (In this respect, Sakel was a harbinger: twenty-five years later the accidental improvement of psychological states led to the antidepressant revolution—and to the theories cobbled together to explain those accidents.)

The idea that a doctor could practically kill someone and thus make him better—a common feature of ancient medicine, with its bloodletting and mercury treatments—was not unheard of in modern medicine. In fact, Julius Wagner-Jauregg, the psychiatrist who testified against Theodor Reik in the Vienna malpractice trial, won his Nobel Prize for infecting neurosyphilitics with malaria—an idea he hit upon when he noticed that high fever often relieved their symptoms. (Actually, he tried tuberculin first, but tuberculosis, while perhaps preferable to general paresis, is still a pretty devastating disease.)

Pioneer or not, Sakel's colleagues hailed him for having "courageously persisted in his experiments," but the accolades should probably have gone to his patients for enduring such a gruesome procedure. Once you get a big dose of insulin—assuming you aren't a diabetic or didn't just down a thirty-two-ounce Slurpee—it takes about forty-five minutes before the symptoms of hypoglycemia come on. Then you start to become disoriented, your speech slows down until all you can do is murmur or groan, and you may well start to hallucinate. If your doctor, wanting to show an assembled audience just how his treatment works, pricks your arm with a pin, you will wipe the spot over and over, and if he claps his hands next to your ears, you will jump even if you are nearly unconscious—which you probably are. As you slide into a stupor, you might have a convulsion or two, and sooner or later, your whole body will be racked by spasms that will leave your feet extended and your toes curled, your arms outstretched and your fists clenched. You will sweat like crazy, drool from your mouth, and drip mucus from your nose in strings. Your face will get pale, your heart will slow down, your breathing will become irregular and labored and maybe even stop, and your eyes will stop responding to light, their pupils fixed. "Beyond this point," as a how-to manual for psychiatrists put it, "the changes are likely to be irreversible," which is doctor talk for "you will die."

But if you are schizophrenic, and if you survive the four or five hours that it takes to induce the coma, and the hour or so for which it lasts (unless your doctor has decided you are a hopeless case, in which case he might extend the coma to twenty or thirty hours), and then you are brought back to life courtesy of some sugar water, chances are good you're going to feel a lot better. "One is frequently surprised by the patient's changed attitude immediately on awakening," the manual says. "He asks for help, he is friendly, accessible, interested in his comfort and in the little things of daily life, especially food . . . The schizophrenic patient becomes *gemütlich*."

Unfortunately, this state lasts for only thirty minutes or so. But if you get treated six times a week for two or three months, you will very likely be less disturbed and more tractable, less prone to hallucination and delusion, better able to get along with people, able to give and receive affection—in general, an easier patient to manage.

You may be really lucky and have so complete a remission that you can leave the asylum altogether, although you will not necessarily be like this:

> One of our patients, paranoid for 17 years, first improved in every way, but after planning to leave the hospital he declared that he was not able to stand health, that his sickness was a protection from something much more dangerous.

But especially if you were unfortunate enough to be schizophrenic in 1934, you, and anyone around you, would have been amazed. All the advances in medicine of the past fifty years had brought virtually no improvement to schizophrenics—or to manic-depressives. These two major forms of insanity—the diagnoses that no one wanted to hear and that no doctor wanted to deliver—still amounted to a living-death sentence.

This was the news that Wortis was announcing to Freud: that the age of therapeutic nihilism was over, that mental illness could be the subject of proper doctoring, the kind that involved doing things to people's bodies, and not just the mind-cure malarkey Freud was offering.

You could call the era that insulin coma therapy kicked off—which continues today—the age of therapeutic exuberance. But the doctors of the 1930s didn't bother naming it. They were too busy finding new therapies.

You can't blame them for this. To turn psychosis into *gemütlichkeit* was no small accomplishment, even if you didn't really understand how it had happened. One doctor—the Hungarian psychiatrist Ladislas von Meduna— even had an actual theory for his own contribution to this new zeitgeist. The theory would eventually turn out to be wrong, but at least for a while it gave Meduna the ability to claim that his method was "the result of research based upon a previously developed working hypothesis," while Sakel's was "developed in a purely empirical manner." Sakel, in other words, was still practicing the old way, while Meduna, at least according to Meduna, was leading psychiatry into the modern era.

Meduna's hypothesis was based on an old observation: that epilepsy and schizophrenia rarely occurred together. Some doctors had tried to use this apparent antagonism to their advantage. For instance, Paracelsus, the sixteenth-century Swiss physician, used camphor to bring on fits in psychotic patients in hopes of curing them. But camphor injections caused excruciating pain and sometimes even life-threatening infections, which made the procedure even more forbidding. Although one doctor went so far as to transfuse schizophrenic patients with the blood of epileptics in 1932, most psychiatrists by then had given up on inducing seizures to cure psychoses.

Meduna thought he knew the reason for the antagonism between seizures and psychosis. As a psychiatrist in a small hospital in Budapest, he had occasion to autopsy the brains of many people with mental disorders, including epilepsy. He noticed that the epileptics' brains contained an excess of the glial cells that surround and nourish the brain's neurons like worker bees around a queen, while the schizophrenics' brains had a dearth of them. He also found epidemiological studies that not only confirmed the ancient observation but also showed that epileptics who became schizophrenic often stopped having seizures, and that in at least two cases, the reverse was true: schizophrenia remitted when epilepsy developed. Meanwhile, he talked a coworker into secretly biopsying the brains of living epileptics and schizophrenics (of course, it wasn't a secret to the patients, only to other doctors, who might have objected to the procedure) and found the same results. Armed with what he thought was biological as well as empirical justification, Meduna began to search for a chemical to induce seizures in animals. He tried strychnine, caffeine, and absinthe, among other drugs, but finally decided that the old standby camphor was the most effective and least dangerous of them all.

In January 1934, at just about the time that Joseph Wortis was hatching his plan to drop in on Freud in Vienna, Meduna was in Budapest giving camphor to a man who had been in a catatonic stupor, unmoving and tube fed, for four years. Forty-five minutes later, the patient seized. After the seizure was over, Meduna later wrote, the patient continued to lie in bed "like a wooden statue, oblivious to his surroundings." But the doctor persisted, and eighteen days later, two days after the fifth injection,

> the patient got out of his bed, began to talk, requested breakfast, dressed himself without any help, was interested in everything around him, and asked how long he had been in the hospital. When we told him, he did not believe it.

Meduna treated five more patients with similar results, and by the end of 1934 had found a better drug than camphor: Metrazol, a newly marketed cardiac drug that at high doses caused seizures. The Budapest medical establishment called Meduna "a swindler, a humbug, a cheat," he wrote later. "How dare I claim that I cured schizophrenia, an endogenous hereditary disease!" But in the larger world, Metrazol therapy joined the insulin treatment to offer hope where there had previously been none.

Of course, as with insulin treatment, it came at a price. Metrazol didn't work by nearly killing people as insulin did, but it didn't exactly make for a pleasant afternoon. A drug-induced grand mal seizure is bad enough—limbs extending and contracting violently two to four times per second for up to one minute, the thrashing intense enough to break bones and dislocate joints, breathing so violent and irregular that it might lead to aspiration and pneumonia before it stops altogether for a minute or so at the end of the seizure, not to mention the indignity of vomiting, urinating, and sometimes even ejaculating on the table—but at least the patient didn't remember most of it. (And as doctors later found out, many of these problems could be avoided with a judicious shot of a paralytic drug like curare, famously used as the poison on the end of a dart.) What patients could not forget, however, was the awful dread they felt as the drug came on and the amnesia, the individual memories that disappeared and sometimes never came back.

Meduna and Sakel soon came to loathe each other. The Hungarian missed no opportunity to tweak the Austrian for his empirical approach and for missing the fact (in his view) that the convulsions his patients had on their way to unconsciousness, and not the comas themselves, were curative. Sakel, apparently unaware that Meduna was a fellow Jew, grew convinced that he was part of a vast anti-Semitic conspiracy to give all glory to the racially pure.

But they did agree on one thing: each called his therapy "shock treatment." The name referred to its putative mechanism, a powerful shock to the system, but it took on a new resonance when Ugo Cerletti, an Italian doctor, decided to replace the drugs with real shocks, the kind you get if you plug yourself into a wall socket.

Cerletti was using electricity to induce seizures in dogs in his Genoa lab. He wasn't trying to cure anything, just to understand the neurology of epilepsy. His technique was simple, if crude: he put an electrode in a dog's mouth, another one in its rectum, and turned on the juice. If the dog wasn't killed outright—the current passed through its heart, resulting in a mortality rate of 50 percent—it would generally have a seizure. Later, when the dog was sacrificed, Cerletti could compare its brain to one taken from a dog that hadn't been shocked, and he began to amass a body of data about the effect of convulsions of the mammalian brain.

Cerletti had heard about Meduna's work in 1936 and wondered why the Hungarian was "not using the much simpler method of inducing fits with electricity." The next year, at a psychiatric summit in Switzerland, he floated the idea of trying this approach on patients (he was by then running a psychiatric hospital in Rome), and no one raised an objection. Of course, they may not have been so accepting had they known about that 50 percent mortality rate, so he went to work on figuring out a method of delivering the shock that was not so dangerous.

Cerletti had heard reports that butchers were using electricity to kill pigs in Rome's slaughterhouses. He sent his assistant, Lucino Bini, on a fact-finding mission. It turned out that the butchers weren't electrocuting the pigs, just stunning them into unconsciousness before slicing their throats—exactly what Cerletti wanted to do to people (except for the throat-slitting part). Even better, they had figured out how to bypass the whole rectum-heart-mouth nexus by using a pair of electrified forceps to deliver the shock through the pigs' temples. The abattoirs soon became an impromptu lab, where Bini determined, among other things, that the margin between the seizure-inducing dose and time (120 volts for about a tenth of a second) and the lethal dose (400 volts for a minute) was so great that it was safe to start using electricity to do shock treatment.

In April 1938, a patient wandered into the Clinic for Nervous and Mental Diseases at the University of Rome, incoherent and babbling, and was quickly diagnosed with catatonic schizophrenia. He couldn't tell the doctors his name or anything else about himself. He was, in other words, a perfect subject for an experiment.

Over the next ten days, Cerletti tried to put his new patient—whom he named Enrico X—into seizures. He tweaked the dose and duration of the shock until finally, at 92 volts and a half second, Enrico seized for more than a minute. He ejaculated, stopped breathing for 105 seconds, went pale, and lapsed into unconsciousness for about five minutes. After eleven of these treatments over the next three weeks—many of them witnessed by colleagues summoned from all over the hospital by trumpet blasts—and another month of hospitalization, Enrico was released "calm, well oriented," with "thought and memory unimpaired." The treatment was announced to the public with the usual miracle-cure fanfare. A new word entered the Italian language—*zapare*, which means exactly what you think it does—and a new idea began to take hold in the public consciousness: that if you let psychiatrists take extreme measures, they can actually cure insanity.

*   *   *

These miracle cures all turned out to be too good to be true. Insu-lin coma therapy proved too dangerous; every treatment was an experiment, as no patient's reaction to insulin could quite be pre-dicted, and when the experiment failed, the doctor had a dead patient on his hands. With the advent of Cerletti's device (which even Meduna hailed as an improvement), Metrazol fell out of favor; it eventually lost its FDA approval. Doctors are still known to *zapare* their patients—it's known today as electroconvulsive therapy, or ECT—but reports of nasty side-effects like permanent amnesia and depictions such as Ken Kesey's in *One Flew Over the Cuckoo's Nest,* have made it not only infamous but subject to tight legal control in many places, including California. Doctors have for the most part moved on to quieter methods.

But even in their heyday in the 1930s and 1940s, shock thera-pies raised eyebrows. One scandalized British doctor said to his col-leagues, "Our patients seem to be in danger of having a very thin time. First we Cardiazolize them [Cardiazol was the British version of Metrazol], then we insulinate them, and now we are proceeding to electrocute them." But you didn't have to be a critic to be shocked by the therapies. One of Meduna's colleagues noted that their treat-ment amounted to "driv[ing] the Devil out of our patients with Beelzebub." The authors of a manual on shock therapies referred to them as attempts to "bedevil the psychotic into a state of nor-malcy," and the doctor who introduced their book started his essay with this ringing endorsement: "Shock therapy has thrust its none-too-pretty form into the field of psychiatry. Whatever the method of producing 'shock,' the process itself is distasteful." For their part, patients, at least some of them, agreed:

> One . . . patient, when coming out of coma, usually asked
> how many times she would have to die, and added that she
> was happy to be alive again . . . Another schizophrenic asked

us, "Why do you kill me every day?" One of our paranoid female patients compared the physician forcing her into coma with someone pulling the wings off a fly.

It's hard to know whether the doctors were bragging or complaining about having all this power—and whatever their enthusiasm it didn't compare with that of Walter Freeman, a doctor who had perhaps the worst case of therapeutic exuberance. He figured out how to perform a frontal lobotomy in his office with a tool modeled on an ice pick, and then packed his gear into a suitcase to take psychosurgery into the American hinterlands. And then there were the doctors who used drugs to keep patients asleep for a week at a time or had them breathe nitrogen until they turned blue or refrigerated them until their body temperature dropped below 85 degrees. The really scary part is that none of the shock doctors had any idea, at least any scientific idea, of why their treatments worked.

Cerletti didn't even try to explain it. He was content to describe himself as no more than an inventor improving on Meduna's methods. But the theoretical coattails he was riding—Meduna's vaunted theory—never proved out: there are indeed people with both epilepsy and schizophrenia, and the neurological findings about glial cells remained (and still remain) inconclusive. Maybe, as Sakel thought, insulin coma provoked some kind of biological regression:

> The various reflexes disappear during hypoglycemia in the order of their evolutionary development and they reappear in reverse order . . . In mental processes too those components of mind which happen to be most dominant and active are most quickly and effectively eliminated . . . In cases which progress favorably, repeated and correctly managed hypoglycemia states finally serve to produce a permanent dominance of those psychic components which have hitherto been repressed.

Or "the therapeutic effect . . . may be due to the destruction of great numbers of nerve cells in the cerebral cortex"—new brain damage undoing the old. Or maybe the treatment simply focused the mind wonderfully, the improvement "due to the patient's experience of the treatment as a threat to his existence, or as punishment, or as death and rebirth."

But, as Ladislas von Meduna might have said, you can't argue with results—even if they aren't exactly what you expected. That's the beauty of not having a theory about something like shock therapy: even when you're wrong you can be right. Or, as Manfred Sakel—who first discovered the benefits of insulin when a heroin addict he was treating with lower doses of insulin slipped into a coma and, upon being revived with a shot of glucose, emerged gemütlich—once wrote, "the mistakes in theory should not be counted against the treatment itself, which seems to be accomplishing more than the theory behind it."

So let's say the treatment you theorized was a cure for schizophrenia turns out to cure something else. Let's say you aimed at an elephant and brought down a rhino. That doesn't mean that your bullets are no good—a principle that remains in effect, happily for men of a certain age whose Viagra began life as a not-so-good heart disease drug with a very interesting side effect.

That's why when Sakel noticed (or says he noticed; he was known for revising his autobiography to suit his needs) that depressions seemed to lift in patients who had convulsions while being insulinized, or when Cerletti concluded that he was getting better results with depressed patients than with schizophrenics, or when an American doctor wrote that he was using Metrazol to cure depressions, or when Philadelphia psychiatrists reported that 70 to 85 percent of their depressed patients were recovering (and none of their schizophrenics) after electroshock therapy, or when a controlled study in 1945 found that 80 percent of the ECT-treated depressives improved and their average length of hospitalization was cut from twenty-one months to five months, or when suicide rates among the depressed who received ECT decreased dramatically, and all the while shock treatment's effect on schizophrenia, the disease it was theoretically supposed to cure, proved more and more disappointing—when all this happened, psychiatrists were happy to skip the theorizing and get on with the treating. Not of schizophrenia, of course, but of depression.

Those 80 percent improvement rates, by the way, are way better than anything that any antidepressant, no matter how cooked the books, has delivered, and they have been replicated often. But before you wonder why ECT is not the treatment of choice, you have to remember one thing: these depressives were very sick. They had *affective psychoses,* which meant that they were immobilized, delusional, nonfunctional—much as you would want people to be before you start shocking them into convulsions.

It's not that doctors didn't try to use their methods on the walking wounded. Unhappy people can be every bit as desperate as disabled people. But the shock doctors discovered that, as Lothar Kalinowsky, one of ECT's major proponents and the man who did the most to spread it in the United States, put it, "the results [with neurotics] are as a whole disappointing"—adding that especially if the patients were anxious as well as depressed, ECT was not indicated.

These results would not necessarily have stopped the shock doctors from continuing to make their miracle-cure claims. That's the other great advantage of having no explanation for why and how your therapy is effective. Because your doctor knows how antibiotics work, he knows which one to prescribe for your particular infection. But if there was no theory, and more to the point, if your doctor was a psychiatrist who knew only that a treatment was getting results with patients who somehow remind him of you even though they have a different diagnosis from yours, then why not try it on you anyway? Or, to put it another way, why not shoot first and ask questions later?

That's how Kalinowsky explained it to the *New York Times* in

1949. "As treating physicians," he said, "we cannot wait for satisfactory theories . . . We psychiatrists, like other physicians, will learn to select the right therapeutic techniques for our patients." Kalinowsky quickly reassured readers that even if doctors didn't know exactly what they were doing, they would be careful, that shock therapy would be "applied with discrimination." Talking to his colleagues, on the other hand, he emphasized the importance of discrimination, given the unprettiness of their techniques. After all, something even more important than patient well-being was at stake. "Indiscriminate use in neurotics is particularly likely to discredit the method," said Kalinowsky. You wouldn't want to kill the goose that laid the golden egg.

Shortly after Joseph Wortis observed Sakel at work in Vienna, and six weeks from when his therapy was scheduled to end, he got a letter from Havelock Ellis:

> I am pleased to hear the Freud analysis has been going well, even though you will be glad to reach the end of it. Not surprising that it has yielded no new revelation of yourself, and you can hardly have expected that it would. But it must certainly yield a revelation of Freud and his technique, and that is what you want.

Wortis was not the only eyewitness to report that Freud was querulous, combative, imposing, and dogmatic. Nonetheless, he did provide one small but important revelation about him. When it came to the insulin cure, and the prospect of a biological psychiatry, Freud, the self-assured pessimist, was surprisingly sanguine and more than a little modest.

Wortis described the insulin therapy to Freud "with great enthusiasm"—and a little bit of taunting:

I said incidentally that it was now theoretically possible to produce a paranoia in the course of a morning with insulin and stop it in a few minutes with sugar, which seemed to disprove the psychoanalytic explanation of its etiology.

Freud didn't accept this diagnosis.

Psychoanalysis [Freud said] never claimed there were no organic factors in paranoia, it simply indicated the psychic mechanisms behind it. A mere organic explanation would explain nothing, any more than you could explain why one drunk became manic and another remained quiet.

The fact that a doctor could induce and curtail a psychological state with a biological intervention was not proof that the doctor had discovered the cause of the psychological state. All he could say with certainty was that he had found the way—or perhaps one of the ways—that the body (and presumably the brain) provided the experience. It was still possible—probable, in Freud's view—that the psyche needed what the brain was doling out, that the symptom had some meaning.

Freud didn't dispute the fact that insulin made some people better, but he argued that this didn't rule out analysis as a cure. "'Analysis never claimed a prerogative over organic forms of treatment, if such a treatment is more successful,'" he told Wortis. And besides, he reminded his young patient, he had always granted that there may be illnesses of the body that manifested themselves as problems of the mind, and "analysis never undertook to cure [them]."

Wortis persisted. "I said that in New York one often saw purely organic cases that had been treated in vain for a long time by psychoanalysts, at great expense to their patients." Here he finally got a rise out of Freud, who was perhaps still smarting from losing the fight over lay analysis. "'What your American *crooks*'—Freud used the English word—'do is certainly not representative or typical of the science of psychoanalysis.'" But his bile soon subsided.

> Analysis is not everything. There are other factors . . . what we call libido, which is the drive behind every neurosis; psychoanalysis cannot influence that because it has an organic background. You very properly say that it is the biochemists' task to find out what this is, and we can expect that the organic part will be uncovered in the future.

Freud did get in his digs—for instance, when Wortis told him that he'd dreamed that the Sakel method was a failure and Freud responded that "what [Wortis] really wished was that Freud would fail in his method." But he remained firmly, if blandly, ecumenical even as his patient brought the subject up for the third time in three weeks, insisting that there was no reason that psychoanalysis and biological psychiatry couldn't fashion a peaceable therapeutic kingdom. For that matter, Freud went on, even these two approaches didn't exhaust the possibilities. "As Charcot [Freud's early mentor] always used to say, 'We cannot compete with Lourdes'; and many cases were actually sent there."

The analyst never lay down with the shock doctor (or the priest)—although, as Edward Shorter and David Healy point out, throughout the 1940s and early 1950s, ECT was the "secret love" of many analysts, who would quietly send their patients for treatments even while denouncing it in public. Freud may simply have been angling to keep a place for analysis in the temple of a biologized psychiatry. He may have understood immediately what might happen now that doctors had found a reliable biological route to relieving suffering. Perhaps that's why his first response to the news was to remind Wortis—already the representative of Freud's antagonists, now going proxy for new challengers—that organic explanations explain nothing, because he knew that the shock doctors were about to claim not only that they could make you better, but that they had explained what was wrong with you to begin with: that something had gone wrong in your brain, that when it comes to psychological suffering, the psyche is only another side effect.

Joseph Wortis didn't bring any of this caution back with him when he returned to America from Vienna in 1935. Instead, he brought insulin therapy to Bellevue Hospital, from which it spread to the rest of the country. He made his own contribution to the method, perfecting a technique for keeping people in comas for up to twenty-four hours—the program was terminated in 1942 after a patient died—and eventually became the editor of the journal *Biological Psychiatry*. Manfred Sakel arrived in the United States shortly after Wortis's return. He administered insulin treatments to his wealthy clientele in hospitals, hotel rooms, and private homes and acquired a reputation, at least with the *New York Times,* as the "Pasteur of psychiatry." In 1937, a pair of Upper West Side doctors started treating depressed New Yorkers with Metrazol, and in 1940, just as Lothar Kalinowsky set sail for America, an epidemic of ECT spread to asylums in New York, Chicago, Philadelphia, and even Cincinnati. Headlines like "Insanity Treated by Electric Shock" began to appear in the *Times,* and by 1950 the age of biological psychiatry had begun in earnest.

If anyone was worried about the irrationality of all this therapeutic exuberance—other than the analysts whose livings it threatened—they weren't saying. But then again, the guinea pigs in this experiment were terribly sick, which made it easy to justify desperate measures taken on their behalf. Had the shock doctors' methods been less extreme and unpretty, had they been, say, gaily colored pills with friendlier names than *electroshock therapy,* remedies that just tweaked consciousness a little bit, that could be taken in the privacy of one's own home, that had only a few side effects, and that were held out to cure a disease afflicting 20 percent of the population, there might have been a little more worry. In this sense, the depression doctors are in infinite historical debt to the shock doctors. They softened up the market for them, getting people used to the idea that doctors could mess with their heads even if they didn't know exactly what they were doing.

They also got people ready for the idea that Freud warned Wortis about and that my clinical trial made a fetish of: that our discontents and their cure are in our brains. The shock doctors, starting with the observation that biological interventions relieved psychic suffering, began to build the case that mental illness is fundamentally a biological problem. They took Kraepelin's brilliant, build-it-and-they-will-come strategy—claim scientific authority by speaking in scientific language—one giant step farther.

Kraepelin had in effect issued a promissory note: eventually, he promised, an explanation would emerge that would validate his taxonomy; on that assurance, the taxonomy, which *sounded* scientific, should be accepted now. The shock doctors realized that so long as they did something dramatic to a patient's body, so long as what they did was plausibly biological, and so long as they got results, they could further claim that they had proved what they were still only assuming. They could have the capital without even making the promise. They also identified the market: not schizophrenia, which often remained unaffected by their treatments, and which rendered its victims nearly inhuman, but depression.

CHAPTER 8

THE ACID AND
THE ECSTASY

M y first depression lasted for a couple of years. When drugs made it go away, it was an accident. Not the taking drugs part—that I meant to do—but the depression-lifting, anxiety-erasing, total-revolution-in-my-head part. I'm not sure what I was expecting when I took Ecstasy for the first time—although the name should have been a clue—but what actually happened is the last thing I would have predicted.

When it comes to taking drugs like Ecstasy—whose official name is MDMA, or methylenedioxymethamphetamine—I am an educated and careful consumer. Before my first LSD trip, when I was eighteen, I prepared myself like a Boy Scout. I read about other people's experiences, learned about what made them good trips or bad trips, sat with friends through theirs. On the appointed day, I made sure I trusted the people I was with, that we were in a safe place, that our nutrition and hydration needs could be met. I did one last walkaround of my own psychic state (as much as one can when one is so young) before I popped that little piece of paper in my mouth. No one told me to do all this; it just seemed prudent, the same way that it is prudent to learn how to drive a car before you get on the expressway. I guess that when it comes to taking dangerous drugs, I have a self-preservation instinct.

And MDMA can be a dangerous drug. At high doses—not too high, maybe four or five times the normal human dose—it's neurotoxic.* Of course, so is Prozac, or for that matter Pepto-Bismol (bismuth causes brain damage), although in both cases you need more like a hundred times the normal dose to cause problems. But that's not why I hadn't taken it before 1990. My reluctance wasn't about the drug itself, but about me. I was concerned that my emotional state was too fragile to risk any disruption. I had, in fact, declared a moratorium on psychedelic drugs when, not long after I first got depressed, I took a dose of psilocybin mushrooms and spent a few hours resisting the urge to dash my brains out on a rock.

Like I said, educated and careful.

On the other hand, my circumstances seemed in some ways perfectly suited to MDMA. Its reputation among psychotherapists, at least of a certain stripe, was stellar. They gave it rave reviews, so to speak, for its ability to foster open, fearless communication—a mainline from the heart to the mouth and the ears, they said. It could, they went on, accomplish in an afternoon what years of therapy could not and was especially effective for couples with relationship issues.

And boy did we have some of those.

Susan had moved across the country to be with me. She'd left behind a marriage to a man whom she had been with for fourteen years, and all the life that went with that—house, friends, money—to move to my little New England village, whose inhabitants' Yankee reserve registers with anyone from outside as disdain. She arrived to find me in the midst of my first depression, so morose

---

* It's hard to say exactly what the toxic dose is because virtually all MDMA research is designed to prove how dangerous the drug is, rather than to establish whether there is a safe dose.

and sour that I couldn't even give her a kiss with her Valentine's Day gift.

We'd agreed not to live together, but that didn't stop me from feeling responsible for her, a feeling I responded to by keeping her at arm's length, by being cold, by letting her know in every way that I wasn't going to take care of her, that I'd had enough of that in my first marriage and in my life and it wasn't going to happen again—speeches that, had I been able to think clearly about it, I would have realized were unnecessary because that was the last thing in the world she wanted.

Susan was, and is, too humane and proud to pressure me (and maybe too clever; she wasn't about to fulfill my prophecy), but after a year of my ambivalence, I knew that she was getting impatient. Which I helpfully responded to by being even more standoffish. And more depressed.

My therapist at the time asked, "But in your core, in your heart of hearts, do you love her?" That's one of the all-time stupidest questions a therapist has ever asked, but I couldn't blame her. She was tired of my dithering too.

"Love is beside the point," I told her. "I'm too old for that." I was thirty-two, but I felt ancient. "I'm just not sure I want to marry her."

"She's wants you to marry her?"

"Well, she hasn't said that. But I'm pretty sure."

"I wouldn't be so sure. If someone treated me the way you've been treating her, I'm not sure I'd want to marry you."

Which was one of the all-time smartest things a therapist has ever said. It didn't do anything for my depression, at least not directly, but I did resolve to figure this out—if for no other reason than to spare further misery to someone whom, in my core, in my heart of hearts, I did love. I gave myself a deadline—six more months and then I'd have to call it quits.

Around that time, a friend of mine sent me some MDMA in the mail. (I should acknowledge that various felonies were committed here, and unjust as this is, it's not something I am recommending that you try at home.) Susan and I decided to take it with us when we traveled to a grimy Rust Belt city to see the Grateful Dead. We took the capsules—125 milligrams each of pure, lab-certified MDMA—in our hotel the morning of the show.

We lay down on our bed and waited. Even the best psychedelic drug experiences are an ordeal, especially at the beginning. So we're lying on this bed in the Rust City Holiday Inn, looking up at the ceiling, close but not quite touching, talking about nothing in particular, certainly not about the fact that at this moment I'm remembering my last psilocybin trip and wondering why in the world I've done this to myself. I'm just about to resign myself to four or five hours of avoiding the rocks, and I turn to say something to Susan, maybe even something honest about how scared I am, and she just happens to turn toward me at the same time. Our faces are about a foot apart at the nose, and I look into her eyes, which are a perfect shade of violet, soft—*limpid* is the word that comes to mind, a word that has rarely, if ever, crossed my mind before—and inviting, so inviting, like a calm blue sea. I dive in. I am looking back and seeing me as she does—literally, viscerally, Vulcan-mind-meldedly. My God, she loves me. I can't say that I understand this, but whatever: I am forty fathoms deep in love.

Which is cool enough, but nowhere near so cool as what comes next. In one moment, a nanosecond maybe, the dread, the self-loathing, the sadness and despair and nausea, all the dark, twisted thoughts that have black-dogged me for two solid years, keeping me up nights, ruining days, driving away opportunity by the carload and hurting the people I care the most for, including especially the one right here next to me, the one who has so graciously lent me her eyes—it all drains away, every last polluted ounce. I feel it leave me, like dirt washed off in a shower. I don't know where it has gone, and I don't care. Because for the first time in two years, and maybe in my life, the world feels like a hospitable place, and more than hospitable: welcome, friendly, full to the brim with love. Whatever poison has been running through me, I believe I have found the antidote. Not the drug, of course, but this person next to me, this steadfast, patient, kind woman who, beyond all reason, loves me. Who is the channel through which all this ecstasy flows.

I'll spare you the rest. Well, no I won't. I have to tell you about Grace and Angel.

Susan and I are lying on the bed inside this numinous bubble, it's been an hour or so and we must be positively glowing as we stare into each other's eyes. By now, I'm on top of her, but our clothes are on—as we will find out later (and I really will spare you this), the kind of lovemaking that involves genitals is a dicey proposition on this drug, probably because it's totally redundant—and we're talking gently to each other about this newfound world, when there is a knock at the door, followed quickly (or maybe not so quickly, our sense of time not being all that it could be) by the turning of the key and the entry of the maid. Whose hotel is full of Deadheads, so this can't possibly be the weirdest thing she's walked in on today—not to mention she's a chambermaid, a professional walker-in-upon— but she is nonplussed anyway. She backs out quickly, starts to close the door. We stop her, tell her to come on in.

It turns out that the Ecstasy bubble expands, and Angel—I'm not making these names up—soon calls her fellow housekeeper Grace into the room. We love her too. We love the janitor they fetch, whose name is probably Gabriel, but I'm not asking. We love the Irish whisky that we toast them with—and so does Gabriel—and the salsa music they find on the radio, and the beige uniforms they wear, and especially the rolling carts they've left by the door. They are brimming with fresh white linens, tiny bottles of shampoo and lotions, clean cups and bars of soap and rolls and rolls of toilet paper, and all this *stuff* designed to cater to our needs seems like an embodiment of all the inexhaustible supply of love in the world. And as we dance—Gabriel all sinewy and loose (and I don't think this is the first snort he's had today), the women bemused and happy, the joy jumping out of our skins—I can't imagine what in the world that depression could have been about. All I know is that it's like someone has finally shut off a jackhammer outside my window, one that has been banging for so long that I've almost forgotten it is there, let alone what the world sounds like without it. And I resolve never to feel that way again. Because it seems so totally unnecessary.

On the other hand, I also find myself saying—because the man on the radio mentioned something about the local team—that it might be fun to go to a hockey game someday, that hockey must be a wonderful sport, and even in my disbelief-suspended state, I realize that this might be a bit much, that whatever valve has closed, and whatever humor it has shut off, the one that tempers love with judgment, that I probably can't get along without it forever. The world is too dark and mean. But for the moment, I've stood on the other side and looked back. I've lived in the warm embrace. I've seen worry and despair for the folly that they are. I am healed.

Susan and I were married within eighteen months. We're still married, and I am sure we always will be. Even more germane, it was years—years!—before I was ever again depressed.

Now, that's my idea of an antidepressant.

It turns out that I'm not the only one with that idea. Or at least something like it.

MDMA has a very powerful effect on the brain. It rouses the nerve cells in a part of the brainstem called the dentate gyrus into a frenzy. Those cells produce electrical signals that travel through axons—fibers that project from the cell body into the farther reaches of the brain and end at the axon terminal, in which serotonin is stored. When a signal reaches the end of the line, the axon releases its serotonin into a synapse, a gap between nerve cells. On the other side of that gap are receptors attached to fibers (dendrites) that lead to other cell bodies. The MDMA-enhanced signaling causes massive amounts of serotonin to be launched into the synapse, where it stimulates the hell out of those dendritic receptors. Meanwhile the drug also suppresses the enzymes that would normally whisk the serotonin out of the synapse, break it down into its constituent parts, and recycle it for further use, which means that the increased amount of serotonin stays in the synapse longer than it usually would.

To a literary guy like me, the fact that all the time you are bathed in the MDMA lovelight, your synapses are being bathed in serotonin is irresistible. You can't help but want to cite it as evidence that the universe is a benevolent place, or at least that there are pockets of benevolence in an otherwise merciless world. You want to write a panegyric to your neurotransmitters, send them a thank-you note for making your experience possible, gush about this unmistakable and beautiful isomorphism, use this brilliant conjunction of the inner and outer worlds for all its metaphorical worth. And you'd consider taking the drug again, maybe see if it can do for your writing or your musicianship or your tennis game what it has done for your mood and your love life.

On the other hand, if you're a scientist, let's say a scientist interested in curing mental illnesses, this conjunction is irresistible for other reasons. The fact that changing serotonin metabolism has such profound and positive results on experience and behavior makes you wonder what would happen if you figured out other ways to accomplish that goal—maybe ways that wouldn't involve an overnight at the Holiday Inn or the risk of brain damage or imprisonment or a sudden wish to go to a hockey game. Even harder to resist would be the conclusion that the alteration of serotonin activity is what causes the alteration in experience, that the ecstasy is somehow in the serotonin. Basking in the glow of these results, you might well forget the basic rule of science: correlation is not causation. To discover that two things happen together is not necessarily to discover the cause of a phenomenon.

That's what was wrong with Hippocratic medicine in the first place: if the fact that a syphilitic gets better shortly after taking mercury leads you to the conclusion that mercury cured him, you might unleash a few centuries of using a bad poison as a medicine. Maybe giving the patient mercury was what we call a *necessary* cause of his cure—perhaps for no other reason than the placebo effect created when a doctor offers a remedy to a sick and desperate person— but until you find the mechanism of the response, or at the very least you replicate that response many times under controlled conditions, you can't conclude that mercury is also the *sufficient* cause of the remission or that syphilis was a mercury insufficiency in the first place. The "cure" could just be a coincidence.

This is exactly what has happened with neurochemistry, especially the neurochemistry of serotonin and depression. Scientists have made too much of coincidence in fashioning depression as a brain disease to be treated with antidepressants. It's an understandable mistake. The series of coincidences about serotonin with which they were confronted starting in the late 1940s were as compelling as, well, a drug trip.

That's how the serotonin story begins—with a bicycle ride taken in 1943 by a Swiss scientist who was high to the gills on a then-unknown drug.

Albert Hofmann was a thirty-seven-year-old medicinal chemist working for Sandoz, a pharmaceutical firm. For seven years, he had been trying to figure out what was going on, pharmacologically speaking, with ergot, a purple fungus that grows on rye. Ergot was infamous as the cause of St. Anthony's fire, an affliction that causes a burning pain in the extremities, followed by gangrene and, eventually, after the patient's toes and fingers start to drop off, a very painful death. (Ergot also induces hallucinations, and some historians think that witchcraft epidemics began with bread made with contaminated flour.) Ergot also had a history of medical use—as a labor-inducing drug that, according to one nineteenth-century physician, "expedites lingering parturition and saves to the accoucheur a considerable portion of time." Sandoz was hoping that Hofmann could find out why it hastened labor and use this knowledge to patent a derivative of ergot for medical purposes. What better way to sell a medicine than to promise a doctor that his schedule won't be held hostage to a woman's "lingering parturition"?

Hofmann proceeded the old-fashioned way used by Paul Ehrlich and William Perkin before him—tweaking the compound, mixing together this and that to make new molecules and then seeing if they did something promising in the test tube or to an animal. He eventually invented a way to assay ergot's presence in animal tissue, discovered its active ingredient, and synthesized it and a series of variations, but no clinical applications emerged. In 1938, he thought he'd found a drug, *Lysergsaüre-diäthylamid* (lysergic acid diethylamide), or LSD, that would be a good stimulant, but the guys in pharmacology concluded that it had no clinical promise. Five years later, Hofmann wanted to take another crack at convincing them. He whipped up a batch and suddenly found himself feeling unsteady and weak. He thought he was coming down with a cold, so he went home and got into bed, whereupon, as he put it in his report to his boss, he experienced "an uninterrupted stream of fantastic images of extraordinary plasticity and vividness accompanied by an intense kaleidoscopic play of colors."

That was on Friday. By Monday, Hofmann had determined a course of action. "I decided to conduct some experiments to find out what was the reason for that extraordinary condition I had experienced," Hofmann told an interviewer fifty years later. Perhaps because it had been a "not-unpleasant" experience (certainly more pleasant than John Hunter's self-inoculation with syphilis), he decided to run the experiment on himself.

At first, he suspected that he had inhaled the solvent he'd used at the end of the synthesis. But when he tried directly exposing himself to it, nothing happened. Then he turned to the LSD. "I was open to the fact that, maybe, some trace of the substance had in some way passed into my body. That, maybe, a drop of the solution had come onto my fingertips and, when I rubbed my eyes, it got into the conjunctival sacs. But, if this compound was the reason for this strange experience I had, then it had to be very, very active." He decided to proceed with "extreme caution" by taking what he thought was only a tiny dose—a quarter of a milligram.

Hofmann quickly discovered that he had underestimated LSD. Within forty minutes of swallowing it, he could no longer write in his notebook. ("Beginning dizziness, feeling of anxiety, visual distortions, symptoms of paralysis, desire to laugh" was the last entry he managed to make.) That's when he decided to go home—by bicycle, because the war was restricting the use of cars. It was a terrifying trip—"Everything in my field of vision wavered and was distorted as if seen in a curved mirror. I also had the sensation of being unable to move from the spot. Nevertheless, my assistant later told me that we had traveled very rapidly"—but that was nothing compared to what happened when he arrived home.

> Every exertion of my will, every attempt to put an end to the disintegration of the outer world and the dissolution of my ego, seemed to be wasted effort. A demon had invaded me, had taken possession of my body, mind, and soul. I jumped up and screamed, trying to free myself from him, but then sank down again and lay helpless on the sofa. The substance, with which I had wanted to experiment, had vanquished me. It was the demon that scornfully triumphed over my will. I was seized by the dreadful fear of going insane. I was taken to another world, another place, another time. My body seemed to be without sensation, lifeless, strange. Was I dying? Was this the transition?

Hofmann didn't die. Nor did he go insane, at least not permanently. He did call a doctor, but by the time he arrived, "the horror softened and gave way to a feeling of good fortune and gratitude," and Hofmann began "to enjoy the unprecedented colors and plays of shapes that persisted behind my closed eyes." After a few hours, he fell asleep, and the next day awoke

> refreshed, with a clear head, though still somewhat tired physically. A sensation of well-being and renewed life flowed through me. Breakfast tasted delicious and gave me extraordinary pleasure. When I later walked out into the garden, in which the sun shone now after a spring rain, everything glistened and sparkled in a fresh light. The world was as if newly created.

It didn't take long for Hofmann's story, which he wrote up in a report to his boss, to filter back to his colleagues. They were incredulous at Hofmann's tale, particularly about the possibility that 250 millionths of a gram of anything could wreak such havoc, but then they helped themselves to some LSD. "The effects were still extremely impressive, and quite fantastic," Hofmann wrote. "All doubts about the statements in my report were eliminated."

Sandoz had no idea how LSD worked and no immediately obvious commercial application for it. But by 1947, perhaps because the son of Hofmann's boss, a psychiatrist, had started experimenting with the drug on himself and his patients, the company saw enough potential to give it a name—Delysid. Psychiatrists soon began receiving samples from Sandoz. Enclosed was a note suggesting that they use it as a psychotomimetic—a drug that would induce "model psychoses" and thus allow firsthand study of the "pathogenesis of mental illness."

The recipients wasted no time in investigating LSD. They had all sorts of ideas. They gave it to therapy patients in lower and higher doses and found that a single dose could bring about "spectacular, and almost unbelievable results," according to one enthusiastic review, "in which an individual comes to experience himself in a totally new way." They gave it to psychotic people, who didn't like it much. They gave it to normal people to see if they really became psychotic. Which they did, in a manner of speaking, but not in a way that was terribly helpful to investigators like Max Rinkel, who complained that "subjects appeared more interested in their own feelings and inner experiences than in interacting with the examiner."

Rinkel got some of his research money from the CIA, which, upon hearing of LSD had tried to buy up Sandoz's entire supply—twenty pounds by the agency's estimate. By then, the suggested dose was around one hundred micrograms, which meant that the CIA thought Sandoz had enough LSD to turn on the entire population of the United States at least once. That wasn't what the CIA had in mind, of course. They were more interested in driving the entire Soviet Union crazy, or maybe in loosening up captured spies, but in any event, the agents whom they dispatched to Sandoz headquarters in Basel with $240,000 in cash quickly discovered that intelligence estimates about this weapon of mass destruction had been faulty. Sandoz only had 40 grams on hand—a mere 400,000 doses. Not to be deterred, the CIA prevailed upon Indianapolis-based Eli Lilly to provide a homegrown version of LSD, which they provided to experimenters like Rinkel. The Company also funded the dosing of unsuspecting soldiers (one of whom flung himself out a window when he thought he'd gone permanently nuts) and even paid prostitutes to give it to their johns so that agents could observe the results.

LSD was a failure as a research drug, largely because its effects were so dependent on circumstances. It turned out that patients tripping in hospital beds, surrounded by men in white coats asking them questions and taking notes didn't have a very good time of it. They failed to achieve the "happy and dreamy feeling of ecstasy" or "seeing something of extraordinary beauty, as it was stated in early reports." Even worse, "subjects became hostile when treated in a cold, investigative, unsupportive or hostile manner"—in other words, when they were treated as subjects.

But in the meantime, many doctors were taking Sandoz up on another suggestion the company had sent with the free samples: "By taking Delysid himself, the psychiatrist is able to gain an insight into the world of ideas and sensations of mental patients." The result wasn't exactly what Sandoz had in mind. Doctors glimpsed the inner world all right—not their patients', but their own. And not only doctors. By 1955, LSD had found its way to literary figures like Aldous Huxley, who was so impressed by LSD (and by the mescaline he'd recently taken) that he proposed that it belonged to a new class of drugs. He even suggested a name for it, based on the Greek for "making the soul visible" and announced it in a couplet sent to Humphrey Osmond, a British psychiatrist who was among the first to use LSD therapeutically:

> *To make the trivial world sublime*
> *Take half a gramme of phanerothyme*

Osmond responded with his own coinage. "To fathom hell or soar angelic," he wrote back, "take a pinch of psychedelic"—a word that drew on a more familiar Greek word for soul, one that, thanks to psychoanalysis, was much more doctor friendly.

LSD turned out to be a little too doctor friendly, at least when it came to doctors like Timothy Leary and Richard Alpert, who, starting in the early 1960s, scandalized Harvard University by turning its psychology department into a psychedelic Animal House, terrified parents by urging their kids to "turn on, tune in, and drop out," and eventually provoked the U.S. government into making LSD illegal, thus bringing scientific research into the drug to a near halt. But there was another, much less famous, doctor, a British pharmacologist named John Gaddum, who took the Delysid challenge. And while Gaddum's acid trips don't seem to have changed his life much, let alone the course of American popular culture, his effect on us was in a way more profound than Leary's.

If Humphrey Osmond thought that LSD could illuminate the soul, Gaddum had a different idea—that it could illuminate the flesh. Specifically, he thought it could shed light on what serotonin did to blood vessels and how. He wanted to use the drugs in the old-fashioned way: not as a therapy but as a means of investigating how biology works by treating tissue with a substance and watching what happened.

Like Betty Twarog, Gaddum had gotten samples of serotonin to play with in the early 1950s. He knew about its effects on circulation, and to find out more about them he was introducing other chemicals to animal tissues in tandem with serotonin and comparing the result to what would have happened with serotonin alone. From this he could deduce the chemical properties of serotonin and get a clue about what was going on with blood pressure. Among the drugs he tried was LSD (Gaddum knew that ergot derivatives dampened vasoconstriction), and when he dosed rat uteri with both LSD and serotonin, he found that LSD almost completely blocked the effects of serotonin. It was, in other words, a serotonin antagonist.

Gaddum wasn't entirely uninterested in the effect of serotonin on the mind—or at least on the brain. Indeed, he found serotonin in the mammalian brain in May 1952, just a month after (and independently of) Twarog. And in 1953, he began giving LSD to cats and making behavioral observations. "The cats became for a time unreasonable and intolerant," he wrote, but that wasn't really the cats' fault—their eight hundred microgram dose was, on a body weight basis, forty times even Hofmann's initial huge dose. It was also more than three hundred times the dose Gaddum gave himself in the midst of the cat experiments, on Good Friday 1953. Not much happened, and when he reported on LSD and serotonin to Britain's Physiological Society a week later, he didn't mention his own experience. But he did point to a very intriguing convergence of data:

> Lysergic acid diethylamide has a powerful action of the brain . . . There is evidence that HT [Gaddum's shorthand for serotonin's chemical name, 5-hydroxytryptamine] is present in some parts of the brain. The molecular structures

of these two substances are similar; lysergic acid contains tryptamine as part of its molecule. This last fact suggested that these two powerful drugs might interact.

Gaddum wasn't ready to connect the dots until, in the months after the meeting, he took more LSD, ramping up the dose, and adding a small dose of amphetamine, until he finally "experienced some of the known effects of this drug, such as a feeling of irresponsibility and euphoria." Soon, he was ready to make quite a bit more of that convergence. If LSD changes the way the brain works and also blocks the action of serotonin, and if serotonin, which is chemically related to LSD, is present in the brain, then it is possible, Gaddum wrote in 1954, "that the mental effects of lysergic acid diethylamide are due to interference with the normal action of this HT."

Gaddum remained the careful scientist, letting hang in the air the obvious implication—that serotonin plays an important role in our mental lives. But a couple of his American colleagues, scientists at the Rockefeller Institute in New York whose LSD status is unknown, were less cautious. Gaddum's work, they wrote in a note at the back of *Science* in 1954, suggested

> that the mental changes caused by the drugs are the result of a serotonin deficiency which they induce in the brain. If this be true, then the naturally occurring mental disorders— for example, schizophrenia—which are mimicked by these drugs, may be pictured as being the result of a cerebral serotonin deficiency arising from a metabolic failure rather than from drug action.

It's hard to overstate just how far out on a limb these scientists were climbing. They really had no idea what serotonin was doing in the brain; some researchers thought it was a waste product. Much more important, the idea that chemicals played a role in the central nervous system was complete heresy. It's tempting to think that it was LSD itself that led to these bold insights. Gaddum is mostly mum on the subject, his notes about his trip mostly limited to accounts of what happened when he tried to dictate a passage from a book or touch his finger to his nose. (Although he does write—with impressively legible penmanship, given the circumstances—about the way that his hand looked "queer like a monstrous picture of a hand—that writhes about until I fix it with a look.") But according to his daughter, his experiences were powerful enough to make him think that serotonin was the key to normal mental functioning. This world-in-a-grain-of-sand, everything-fits-together revelation is the signature of the LSD experience, and one researcher who worked with the drug——France's Jean Thuillier, who administered the drug to patients and took it himself—left no doubt about what he thought inspired this theory:

> We have all had wonderful dreams. At one stroke we thought we knew everything . . . We had mental illness in microcrystals, delirium in homeopathic suspension. An active, calculated dose of a few hundred million LSD molecules, thrown at our fourteen billion nerve cells,* was the detonator . . . Surely, the mechanism would be found and dismantled. One day the action of LSD on the neuron was discovered, the next day the action on the barrier between the brain and the meninges, and the following week on nerve transmission . . . It was thought that everything had been demonstrated.

On the other hand, all of my nerve cells once sang out in unison to me that I would like hockey. You have to be careful with drug-inspired revelation.

---

* This turns out to be a vast underestimate. Scientists now think we have 100 billion neurons in our brains.

CHAPTER 9

GETTING HIGH AND MAKING MONEY

This chapter is going to be full of words like *catecholamine* and *iminodibenzyl* and *phenothiazine*. I'm not exactly apologizing for this, but I am warning you: this might be a good time to take a stretch and get some coffee or whatever drug you use to stay alert. It will be worth it, because if you get through this stuff, you'll be prepared when your doctor starts to tell you about your chemical imbalances. You'll understand what he or she is talking about, where that idea came from, and what its strengths and weaknesses are. And you might even be able to respond to your doctor better than I did when George Papakostas and I talked about these things.

I performed especially poorly when I told Papakostas about my adventures with mind-altering drugs. The topic came up three times. The second time was when he asked me if I'd had any prolonged remissions, periods when I just didn't feel depressed. I recounted my MDMA experience and its aftermath. I left out the part about Angel and Grace. You have to know your audience, and I already knew that my Harvard doctor was no Timothy Leary.

This had become clear in our first conversation on the subject, which took place in the first half hour of our first day together. Your drug-taking habits are an important part of a clinical trial intake interview. Partly that's because of the possibility of interactions between the study drug and whatever drugs you might be taking on your own initiative. Serotonergic drugs like MDMA and Prozac, for instance, can on rare occasions add up to too much of a good thing, specifically to serotonin syndrome, a condition in which your brain loses control of vital functions like body temperature. You can die that way, and researchers don't want your blood on their hands.

But even more important to the clinicians taking your drug inventory is the question of just exactly how sick you are—and with what. The DSM offers them seventeen diagnostic categories to choose from, including *substance-related disorders*. That section, at 106 pages, is the longest in the 700-page book. That's not because the doctors who wrote the DSM have something against drug use. Indeed, with substance use disorders as with all diagnoses, they have taken great care to maintain neutrality. For reasons that will become clear in chapter 12, they want to make sure that no one can say that they are pathologizing mere deviance, making value judgments about people's behavior rather than identifying actual illness.

In general, this means defining *mental disorder* as a condition that causes you clinically significant impairment or distress. For the most part, it's up to you to decide if you are impaired or distressed; a mental illness is an illness only if it is a problem *for you*. It's a strange way to think about disease. If a routine chest X-ray inadvertently turns up the fact that I have cancer, I'm still sick whether or not the condition was actually causing me problems before I heard about it; my distress may well *begin* with the diagnosis. On the other hand, ensuring that the disorder is a problem for you imposes an important safeguard: psychiatrists can't become gulag commissars by making pathological what they (or their bosses) merely find distasteful or dangerous.

So, theoretically anyway, one man can use MDMA to save his relationship and fix his depression and not be guilty of substance abuse, while another can take it so much that he ends up with a diagnosis. In real life, of course, things aren't quite so simple. Sometimes doctors have to convince patients that the reason they keep getting fired or divorced or going broke is because they are drinking too much, and that their failure to comprehend this is denial. I myself have done this, and I think I've been correct in most cases. But under this logic, it remains easy to diagnose away disagreement, to call behavior sick when it merely fails to conform to standards that have little or nothing to do with what we consider health.

For instance, among the clinical impairments listed in the DSM are *legal issues,* which means that after 1986, when it was made illegal, MDMA use could earn a diagnosis in a way that it previously could not. The DSM, committed to neutrality, can't comment on the political or social dimensions of this symptom. Instead, it can only refer to a patient's run-in with the law as a health problem—as if the only reason to break the drug laws is that you are mentally ill. Similarly, if you get arrested for drunk driving, the DSM is going to diagnose your difficulty as substance abuse rather than the misfortune of living in a country where mass transit barely exists and where the focus on individual responsibility is so great that lawmakers don't even bother trying to require cars to be impossible to start if a driver is intoxicated.

Driving, in fact, is what led Papakostas to tell me that I was a substance abuser. He was following the script of the Structured Clinical Interview for the DSM-IV (SCID), asking me about my history of substance use. As I listed the drugs I had taken, he blinked rapidly, maybe even blanched a little. But he refrained from comment until I answered his question about whether or not I'd had impairment or distress because of my drug use.

"Not really," I said. "Oh, I suppose there have been times when I'm high that I wish I wasn't, but I'm pretty careful about the circumstances. I only went to school stoned once. Seemed to me a waste of my time, a stupid place to be high. Same with work. I've never been arrested or fired or any of that. I've never operated heavy machinery under the influence."

Papakostas was silent.

"Except once," I added. "The first time I took LSD. I was eighteen. I tried to drive the half mile home from my friend's house. I got about three blocks. The road was converging ahead of me, like railroad tracks do when you look into the distance, only it was happening just a few feet in front of me. I was sure I was going to get squeezed into a single atom if I kept going, and that's when I realized that driving was not such a good idea. So I parked and walked the rest of the way and fetched the car in the morning."

"Well, you know, that's drug abuse," he said.

"It is?"

"Yes. You used in a way that was dangerous."

It was, in fact, a textbook case. Substance abuse, the DSM says, includes "recurrent substance use in situations in which it is physically hazardous (e.g., driving an automobile . . . when impaired by substance use)." Well, nearly a textbook case. It did only happen once.

But I didn't remember the "recurrent" part at the time. I was too hot with shame to think clearly. Looking back at myself through Papakostas's eyes I did not like what I saw, and I wanted to set the record straight. I wanted to tell him that I thought I had done a hell of a job being a responsible citizen, especially for an eighteen-year-old kid, that given the fact that I'd just spent six hours on the beach watching the gulls and waves and boats consort under a pulsating winter sky, that I had seen myself—indeed I could right now, more than thirty years later, see exactly what I looked like, what my friends and I were wearing; I could smell the day like a madeleine—dancing with them toward the threshold of my adult life, that for maybe the first time it had felt not like the edge of a cliff but a portal to a better world, that maybe, just maybe, I was going to get up and fly away, that given all this, I thought leaving the car showed good judgment, was a sign of incipient mental health even.

"But it was more than thirty years ago," I squeaked out. "And it was only *three blocks*," and when I heard myself pleading my case, my shame only deepened. I'd never been called a substance abuser before, and certainly not by a Harvard doctor. It was one thing to get pigeonholed the way I'd been prepared for—as a depressive—and quite another to be told that something I considered to be an important and mostly positive part of my life, something that I thought I handled pretty well, was a problem. And not just a problem but a mental disorder.

But even if I'd had the presence of mind to cite the DSM, or, for that matter, to remind Papakostas that one of the greatest neuroscientists of the twentieth century had taken LSD, and chased it with amphetamine to boot, there was no explaining my way out of it. Partly that was because Papakostas wasn't really trying to cast judgment on my drug habits. He was just trying to make sure that I was diagnostically pure, that I didn't have a disease that would add an unwelcome variable to his statistics, and the way to prove that surely wasn't to convince him that I was a *responsible* substance abuser.

At the time, it seemed as if we were standing on opposite sides of an ontological divide that would swallow up any attempt at conversation. On the other hand, maybe our worldviews weren't all that different. Freud wrote in *Civilization and Its Discontents*, "It is precisely those communities that occupy contiguous territories and are otherwise closely related to each other that indulge in feuding and mutual mockery. I [call] this phenomenon the 'narcissism of minor differences.'" Freud had in mind the traditional warring cousins: the Spaniards and the Portuguese, the Scots and the Brits. But he could as easily have been writing about the Prozackers and the Potsmokers or whatever you want to call the communities to which Papakostas and I belong, which occupy territories that are not only contiguous but often overlapping: the domain of consciousness-altering drugs. Antidepressants (which, interestingly, are not listed as possible drugs of abuse in the DSM, despite the fact that they cause both withdrawal syndromes and dependence) are not only, chemically speaking, the spawn of LSD, one of the most notorious recreational drugs ever to come down the pike. They also, as you'll see shortly, owe their entire existence to the fact that people taking drugs for conditions other than depression—tuberculosis, allergies, schizophrenia—suddenly and unexpectedly felt a whole lot better. Or, as we drug abusers say, they got high. Maybe not as high as John Gaddum or Albert Hofmann got, but high enough.

If you're a psychiatrist or a drug company, this uncomfortable closeness places a great premium on dividing up the territory, on separating your chemicals from theirs, on making sure that yours are medicine and theirs are drugs, that you are treating illnesses while they are abusing substances. You have to disown your embarrassing cousins, even if it means buying an escutcheon from a different family and claiming a scientific pedigree that you don't really deserve.

This brings me to the third conversation that I had with Papakostas about psychedelic drugs. A few weeks after my clinical trial got underway, Carlos Zarate, a National Institutes of Health scientist, released the results of his own trial. Zarate had given one intravenous dose of the anesthetic drug ketamine to seventeen depressed people who had not responded to antidepressants. Within a couple of hours of the injection, most of the patients felt much better; a day later twelve of them were significantly improved, and five of those met the criteria for remission. Half of the responders continued to feel better after a week.

These numbers were remarkable for two reasons: first, that the response was so strong—71 percent, a number that SSRIs can only dream of—and second that it was so rapid. The ten-day to three-week lag between starting to take SSRIs and when they kick in is not only inconvenient, it's also the period during which patients seem to be at most risk of becoming suicidal. Not surprisingly, Zarate's study was hailed as a breakthrough in the national media.

In his report, published in *Archives of General Psychiatry,* Zarate took credit not only for solving the biggest problems of the SSRIs— their weak performance in clinical trials and the lag period before they take effect—but for running a study that, unlike most antidepressant research, was based on a secure theoretical foundation. Like Ladislas von Meduna, who argued that the theoretical nature of Metrazol therapy made it superior to Sakel's empirical insulin therapy, Zarate informed his readers that all our current antidepressants owed their existence to "serendipitous discovery," while his treatment was anything but accidental.

Zarate built his case on two lines of evidence. First, he cited the psychological research. Normally, subjecting animals to inescapable stressors leads to behavioral despair. But, Zarate wrote, "a single dose [of ketamine] interferes with the induction of behavioral despair for up to 10 days"—which means, presumably, that the animals won't go on strike no matter how bad you make their working conditions. Employers of the world, take notice!

And then there is the data from the biochemists. Zarate recounted the evidence that the glutamates, a group of neurotransmitters affected by ketamine, "play an important role in the pathophysiology of depression." The glutamate theory is a state-of-the-art account that states that glutamates, and especially N-methyl-D-aspartate (NMDA), are the ultimate beneficiaries of antidepressants, which means that changing serotonin metabolism is only the indirect means to changing NMDA metabolism. Zarate didn't consider what this means for all those patients who thought that the reason they had to take antidepressants for the rest of their lives is that they have a serotonin imbalance, but then again, among neuroscientists the serotonin-imbalance theory has long been a thing of the past. Besides, Zarate's point was more parochial: ketamine is precision manufactured, a bullet designed for a specific target rather than a shot in the neurochemical dark. Unlike what has come before, his was a truly scientific approach to depression.

But speaking of the narcissism of minor differences, the distinction between ketamine and its accidentally discovered cousins is not all that Zarate cracks it up to be. Indeed, there is a "serendipitous discovery" in Zarate's closet: anesthesiologists and pain doctors have long noted that depressed people given the drug for surgery or pain management—its usual uses—just happen to get less depressed. All that stuff about NMDA and pathophysiology may well be true, but the reason doctors like Zarate were looking there in the first place and spinning these elaborate stories is exactly the reason that, as we will see in a moment, doctors were looking at serotonin: that a drug that unexpectedly made people feel better was found to have an effect on certain neurochemicals.

That's not the only embarrassment. Zarate also fails to mention that people who come out from under ketamine anesthesia some- times have an emergence reaction. They experience themselves as awake and aware and yet disembodied, a "consciousness without an I," as one person told me, pure energy traveling freely through the cosmos. This may be an unwanted side effect to doctors, but to psychedelic warriors it's a glimpse of the universe in all its glory and indifference—which is why they've been using the drug therapeu- tically, mostly underground, for forty years. (Naturally, ketamine has also found its way into the recreational drug scene, something else Zarate left out.) Zarate did mention that his patients suffered certain adverse events, including "perceptual disturbances, con- fusion . . . increased libido . . . euphoria [and] derealization." But just as he erased this part of ketamine's history, he never, not once, mentioned what the experience was actually like for his patients. Perhaps he figured that if he didn't ask them, he wouldn't have to tell the world whether or not they got high.

Or maybe Zarate just thinks that inner life is entirely a side effect. That was what I really wanted to talk to Papakostas about when I brought up Zarate's research—my increasing sense that I was only the middleman, the guy he had to go through to get to the truth of the matter, which lay not in my thoughts and feelings but in my biochemistry, not in my spirit but in my meat. It was one thing to ask questions about my subjectivity that seemed designed to erase it. It was quite another to study the effect of a psychedelic drug on consciousness without ever talking about consciousness itself. I wanted to know, do depression doctors really think that subjectivity, especially the experience of transformation, is irrelevant? And if so, then aren't they making an enormous and controversial claim about the mind: that it is merely the biggest side effect of all?

Our conversation was short. Papakostas knew about Zarate's research, he told me. But he didn't seem to know that ketamine was a consciousness-altering drug. I explained what I knew to him. "Sort of like ECT," he said. "The way it's supposed to reset your neurotransmitters."

"But isn't there a difference between ECT and ketamine?"

"Well, of course ketamine works mostly on glutamate pathways . . ."

"No. I mean that you're conscious when you take ketamine and unconscious when you get ECT."

The distinction seemed lost on Papakostas. He looked at his watch, said we were out of time. For a second I thought I caught a whiff of regret, as if maybe he wanted to talk about the nature of consciousness and its relationship to what we were doing. But I might have been making that up. Mostly what I felt was an immense frustration, a version of the same frustration I felt when he told me I was a substance abuser: that when it came to *Homo sapiens,* we simply weren't talking about the same animal.

But then again, I was the depressed person and he was the doctor, and maybe he knew something I didn't. Maybe the insuperable indeterminacy of consciousness is really not a supreme mystery that will always beg for and elude solution in nearly equal measure, but rather the sign that the question isn't worth asking, that if you want to feel better the last thing in the world you should do is inquire into your inner life. Unless by that you mean the amino-acid-rich neurochemical soup that roils in dumb silence inside your head.

That idea—that Chemicals 'R' Us—is the conceptual backbone of the depression industry. Before you can accept that your mood is merely the symptom of a brain disease, you have to at least implicitly believe that consciousness is nothing more than the steam that rises off the soup. This notion has been around for more than a century—it was 1874 when Thomas Huxley, Aldous's grandfather, first compared consciousness to steam (in his case the steam that pours out of a whistle)—but it began to take its current shape as an article of scientific faith in the 1950s, with a series of accidental discoveries about brain chemistry, many of which, as it happens, came about as the result of antidepressant research.

Many of those accidents involved people, like Albert Hofmann, who unexpectedly experienced altered states of consciousness. To a scientist interested in the workings of the mind, these experiences naturally beg for explanation. But if the scientist in question is working for a drug company, more is at stake than mere curiosity. Drugs like LSD and MDMA, with all their psychic fireworks, don't fit very well with the pharmaceutical paradigm, in which a drug with predictable effects is used—preferably on a frequent basis—to treat a specific illness. To understand the mechanism of these drugs is to harness their power, to domesticate them—which, in the course of the 1950s and 1960s, industry scientists were able to do. In this respect psychopharmacology in general, and antidepressants in particular, are the bastard offspring of mind-altering drugs and the pharmaceutical industry, or, to put it another way, of getting high and making money.

If that sounds like the ravings of a substance abuser, then consider the case of Madame X, a woman who went into a Paris hospital in 1948 for a nose job. For the obvious reason, she couldn't take anesthetic through a mask, nor was she eager to stay awake for the procedure, so she was given a cocktail containing a newly discovered drug. She remained conscious as the surgeon worked, but she remained perfectly calm. "I felt the hammer striking and the scissor cuts, but as if it was happening to someone else's nose: to me it was indifferent."

Madame X's strange state of consciousness was not entirely accidental. Her doctors got the idea for the anesthetic potion from none other than Paul Ehrlich. In his search for a cure for malaria, Ehrlich had tried a dye that lit up that parasite beautifully on slides: methylene blue. It didn't cure the disease, but in the course of trying, he found something curious. If he injected the dye into frogs, it stained their nerve cells much more vividly than other types of tissue. Methylene blue and nerves, in other words, had a strong affinity for each other; their receptors—structures that Ehrlich had hypothesized to exist but had never seen—were a good match.

Ehrlich was impressed enough to try the dye as an analgesic. It didn't work, but in 1899, a Genoan doctor, Pietro Bodoni, started to give methylene blue to psychotic people. The dye calmed them down, and Bodoni's colleagues in Italy began to use it in their asylums. Thirty-five years later, an American doctor used methylene blue successfully with schizophrenics, and as late as the 1970s doctors were giving it to manic-depressives. So why haven't you ever heard of this promising drug? Well, according to psychiatrist/historian David Healy, it's partly a case of bad luck. Every time doctors got interested in methylene blue, it was eclipsed by more dramatic therapies: barbiturates, which came into vogue just as Bodoni was doing his experiments, the shock treatments of the 1930s, and the more powerful antipsychotic drugs of the last half of the twentieth century. But Healy thinks methylene blue is a victim of a business decision: "Patents had been obtained on newer agents and no drug company would market an old drug even if it worked."

But what about a new drug based on methylene blue? Now, *that* you could patent. And in the late 1940s, a French drug company, Rhône-Poulenc, was very interested in expanding its market share of antihistamines, drugs like Benadryl, which had been profitable since 1933. The company had figured out how to extract the nucleus of methylene blue—which it called phenothiazine—in order to make new drugs. None was more effective at treating malaria than methylene blue had been, but they had an overall effect that was interesting. If you gave them to rats that had been trained to climb ropes either to get food or to avoid a shock, they stopped climbing; even stranger, they didn't seem bothered by the shock. Upon further study, the Rhône-Poulenc scientists concluded that the rats were neither sedated nor impaired; they just didn't give a rat's ass.

The drug that helped Madame X keep her head while doctors chiseled away at her nose was a phenothiazine. The psychiatric implications of such a drug were obvious, and soon French doctors were using a particularly strong phenothiazine—RP4560, or chlorpromazine—on their psychotic (mostly schizophrenic) patients. At first, chlorpromazine was administered as an adjunct to a shock treatment popular in France—packing psychotic patients in ice, on the theory that the drug enhanced the hibernation effect, which allegedly cured patients by putting them in a somnolent state. One day, so the story goes, there was no ice on the ward, but the nurses administered the drug anyway. The patients improved, and soon nurses were relieved of the burden, and patients of the discomfort, of the ice treatment.

The drugs became known as tranquilizers, and their effects were immediate and profound—on both the patients and their keepers. One asylum doctor reported:

> In the corridors . . . one no longer passed patients in shirts walking with their straitjackets open with straps undone on the way to the toilets, but patients dressed in heavy blue cloth, strolling decently and quietly . . . The most evident sign of this extraordinary therapeutic result could be appreciated even from outside the building—there was silence.

Like Rhône-Poulenc's rats and Madame X, the patients were calm but not sedated. Indeed, previously unresponsive people seemed to wake up, and nearly everyone who took the drug became more oriented and able to engage with others. It was as if chlorpromazine had tamed their inner demons sufficiently to allow them to reenter the world. Or, as another doctor put it, the drug acted like a "pharmacological lobotomy." He meant that in a good way.

Chlorpromazine was slow to catch on in the United States, where 80 percent of psychiatrists were private practitioners providing therapy to neurotic patients rather than asylum doctors presiding over the psychotic. Smith Kline and French did license the drug in 1953, but only because it happened to reduce nausea and SK&F had just introduced a heart drug that made people vomit; the idea of selling one drug to counteract the side effects of another was irresistible. The company, which named the new drug Thorazine, did manage to prevail upon some leading psychiatrists to try it; they were unimpressed with the results and worried by strange side effects. Finally, however, SK&F got the drug into the hands of the head of New York's state hospital system, and in 1955, the company sold $75 million worth of Thorazine, most of it for use in asylums across the country. Having found the right target for the phenothiazine effects—whatever they were; scientists couldn't explain Thorazine any better than shock therapies—psychiatry had its first magic bullet.*

That $75 million figure did not escape the notice of the rest of the pharmaceutical industry. Even before Thorazine hit the market, Geigy Pharmaceuticals, one of the drug companies that started its life as a dye manufacturer in the mid-nineteenth century, had been looking into antihistamines, and SK&F's sudden success only ramped up the effort. Geigy wasn't interested in a me-too drug, a version of the compound modified just enough to steer clear of patent infringement. Instead, they looked at another dye that they had on their shelves—summer blue, from which they derived iminodibenzyl, a molecule that, like phenothiazine, showed an affinity for nerve tissue and could be easily modified.

---

* Unless you count penicillin, which, in the late 1940s, Pfizer figured out how to mass-produce. Doctors gave it to neurosyphilitics. The results were dramatic and certainly buoyed hopes that mental illnesses could be cured with a drug. But Thorazine, unlike penicillin, was a psychiatric drug from the beginning.

Geigy's team soon synthesized what they thought was a promising compound, G22355. The youngest member of the team, British pharmacologist Alan Broadhurst, was the first test subject. This was in the days when drug research was almost entirely unregulated. That's why a company could bring a drug to market just a few months after its first use in humans. It's also why the trial didn't come to a sudden halt when Broadhurst fainted as soon as he received the injection.

Once they got the dosage problem straightened out, the Geigy scientists sought out patients on whom to test their drug. Fortunately for them, Roland Kuhn, who ran an asylum in Switzerland, was just then getting tired of paying top dollar for chlorpromazine. He had worked for Geigy before, and when he heard about G22355, he offered his patients as subjects in return for a free supply of the drug. It quickly became apparent that G22355 wasn't going to compete with chlorpromazine, no matter how cheap it was. If anything, it made schizophrenics more delusional. Broadhurst remembers one case in which a patient, in an unwitting reprise of Albert Hofmann's bicycle trip, "rode, in his nightshirt, to a nearby village, singing lustily, much to the alarm of the local inhabitants. This was not really a very good PR exercise for the hospital." Broadhurst added, "And I can't say it endeared the hospital to Geigy either."

The scientists, according to Broadhurst, "stumbled around considering a variety of unlikely hypotheses and mechanisms" to explain these untoward results until finally they began to realize—with what Broadhurst now calls the "most naïve scientific reasoning . . . something which I now look back on with a kind of embarrassment"—that if they switched their focus, they just might have a valuable drug on their hands. "If the flat mood of schizophrenia could be lifted to hypomania by the drug," they reasoned, "then could not in a similar fashion a depressed mood be elevated also?" If G22355 got people high, in other words, then why not use it with people who were low?

In 1956, Geigy decided to answer that question. The company

asked Roland Kuhn to give G22355 to his depressed patients, and within weeks, he reported back that depressed patients taking the drug felt much better. In 1957, he sang the drug's praises to the readers of a Swiss medical journal and to the International Congress of Psychiatry at their meeting in Zurich. A year later, he repeated the performance at Galesburg State Hospital in Illinois, a talk published by the *American Journal of Psychiatry* at year's end. His data was overwhelming, his enthusiasm even stronger, but G22355—by then known as imipramine—was hardly taking the world by storm, and Kuhn was roundly ignored. Indeed, according to one doctor active in the late 1950s, psychiatrists compared him to Ichabod Crane, an outsider with an inflated sense of himself and his ideas.

Reading Kuhn's paper now, it's easy to see why his colleagues isolated him. At a time when psychoanalysis was the mainstay of psychiatry, he was arguing that imipramine could "bring a complete change in the situation within a few days, which could not be achieved by intensive prolonged psychotherapy." In the immediate wake of Thorazine's success, he had advanced the claim that a closely related drug would have the opposite effect—stimulating rather than tranquilizing patients. These notions were heresy enough, but even more jarring was Kuhn's idea that imipramine cured an illness that few doctors had even heard of: endogenous depression, in which unhappiness seems without external cause and thus can be presumed to be biological in origin. This "vital disturbance," Kuhn claimed, is what imipramine was uniquely suited to treat.

Endogenous depression was easy to miss in part because its primary symptoms—"a general retardation in thinking and action, associated with fatigue, heaviness, feeling of oppression, and a melancholic or even despairing mood"—were less florid than the delusions and mania of the affective psychoses. But the disease was also insidious. It hid in the plain light of more obvious problems—anxiety, phobias, hysteria, insomnia, impotence, homosexuality; indeed, according to Kuhn, "Almost any neurotic symptom can be caused by a depressive state or be maintained because of the simultaneous occurrence of a depression." It could even be the cause of reactive or, as the Freudians were calling it, neurotic depression—the reason that those external factors kindled a melancholic response to a loss in the first place. Worst of all, endogenous depression came on slowly and imperceptibly, manifesting in "mild disturbances, which hardly appeared pathological at the time." You could be endogenously depressed—indeed, if you were suffering from a "neurotic symptom," you very likely were—and not even know it.

Until you took the drug, that is, and realized, as many of Kuhn's patients told him, that you "had not been so well for a long time." This was no mere euphoria, Kuhn assured his skeptical colleagues, and imipramine was no simple amphetamine, which doctors had been using for thirty years to lift people's moods; people didn't develop tolerance and craving for it. Imipramine "completely restore[d] what the illness . . . impaired—the power to experience." Imipramine didn't simply improve your mood; it changed something underlying it, the endogenous, and presumably biochemical, problem that gave rise to your mood—in short, your disease. Amphetamines only made you feel *better*, but imipramine made you feel *well*—which meant that you must have been sick all along. The drug had revealed the contours of a disease.

The suggestion that doctors had misunderstood the nature of depression, and the controversy it inspired, were not new. But by 1958, the arguments among Kraepelin and his colleagues, and between them and the rest of the psychiatric profession, had been forgotten. Freud's nosology and his contention that psychological suffering arose out of personal history had sent Kraepelin into eclipse—a "world-wide ignorance," that Kuhn bemoaned.

Kuhn's affinity for Kraepelin is obvious. His theory was Kraepelinian logic on drugs. Endogenous depression was exactly the disease that imipramine cured, and the proof that you had been sick was that imipramine cured you. Imipramine wasn't only a cure; it was the chemical that assayed for the presence of the disease. Kuhn was pushing psychiatry closer to fulfilling Kraepelin's promise that mental illnesses would be found in nature. After all, if a chemical could identify and fix the problem, then doesn't it stand to reason that it was chemistry that was broken in the first place?

You would think that, especially in a permissive regulatory climate, a drug company would sit up and take notice when a doctor informs it that its new drug is a magic bullet for a widespread, if hitherto under-recognized, disease, puts that assertion on what appears to be a firm scientific footing, and then adds that "therapy must be maintained as long as the illness lasts"—which, since the problem was "endogenous," may well be a lifetime. But Geigy was mostly unimpressed with Kuhn's reports; the company's marketing experts said that depression was too small a market to bother with. It would take thirty years before the ideas really caught fire that a disease could show up in nearly any malady and be a problem before the patient even knew it, that not feeling well enough was an illness that could be remedied by a prescription.

By then, of course, marketing experts had gotten more savvy and drug companies more determined. They wouldn't, for instance, have let the fact that an antidepressant drug was derived from a substance called summer blue pass them by, or be dissuaded by ethical concerns about convincing people that they are sick. But that's not to say that the first generation of antidepressants were without their hucksters. It's just that they weren't to be found in the Old World, but in the United States.

In 1952, a story appeared on the front page of the *New York Times* reporting on a drug for tuberculosis that was being tested on patients at Sea View Hospital on Staten Island. The drug—iproniazid, or as Hoffman–La Roche called it, Marsilid—was having some remarkable effects. Not only was it healing people's lungs, but according to doctors, it was also relieving their pain enough for them to give up narcotics and, at least sometimes, instilling in them a "euphoria" that after three or four weeks leveled off into a "normally optimistic instead of a depressed attitude." In fact, as doctors soon discovered, the drug was capable of making patients feel better than their X-rays said they were. "This suggested," wrote the *Times* reporter, "the use of the chemical in conditions other than tuberculosis."

Iproniazid was invented when the stocks of hydrazine, Germany's primary fuel for their V-1 and V-2 rockets, went to Roche at the end of World War II. There was some evidence that hydrazine could be modified into an antihistamine that might be particularly effective against tuberculosis, but once the psychological side effects made themselves known, it didn't take a rocket scientist to figure out what to do next. It also didn't take a drug company executive— Roche was as lukewarm as Geigy about the commercial prospects of an antidepressant. What it took was Nathan Kline, an American psychiatrist known as "an international wheeler-dealer, a flamboyant, buccaneering fellow [who] would try anything on his patients." And that was according to one of his friends.

Kline was honest about his own ambitions and the love of power from which they came. "Research scientists are wide-eyed manipulators," he once wrote.

> When an observant brat discovers for the first time that he can push buttons, turn faucets, open doors, dial phone numbers and exploit his parents, he is astonished and delighted to uncover and control the physical and social environment. Some of us never recover.

Iproniazid wasn't the first button that Kline had put his finger on. In 1954, he'd gotten hold of reserpine, a drug that Geigy had recently synthesized. Reserpine was the active ingredient of *Rauwolfia serpentina,* or snakeroot, an herb known for centuries to India's ayurvedic physicians, who called it *sarpagandha,* as a cure for many complaints, and to holy men as *pagal-ki-dawa,* a remedy for insanity. Kline gave it to more than seven hundred psychiatric inpatients and found that it calmed down most of them—enough, he said, that the glaziers at his hospital reported that they had fewer shattered windows to replace.

Given what John Gaddum and other researchers were saying about LSD and serotonin, Kline and other scientists speculated that reserpine was increasing serotonin levels in the brain. Experiments carried out at the National Institutes of Health, however, showed that reserpine *lowered* levels of serotonin in rabbits' brains even as it made them sleepy and immobile. But the idea that a simple brain glitch was the cause of mental illness was too pleasing to be discarded because of one experiment, and soon scientists were talking about serotonin *imbalances* rather than *deficiencies;* insanity could now be a problem of excess *or* deficient serotonin.

As you'll see in a moment, this new theory was even more wrong than the one it replaced, but before anyone knew that, Kline had already taken it a step further, using the same "naïve" reasoning that embarrassed Alan Broadhurst. If a drug could depress mental functioning—a good thing in the case of people who were agitated and hallucinating—what about a drug that did the opposite? A "psychic energizer," as Kline named it,

> would relieve simple depression and . . . the sadness and inertia of melancholia . . . reduce the sleep requirement and delay the onset of fatigue . . . increase appetite and sexual desire . . . Motor and intellectual activity would be speeded up. It would heighten responsiveness to stimuli, both pleasant and noxious.

In 1956, Kline heard that scientists had given iproniazid along with reserpine to lab animals, which, far from being sedated, had become hyperactive. The possibility that the drug could reverse the reserpine effect, Kline said, "immediately led me to speculate whether this was the psychic energizer for which we had all been looking."

That's how Kline told the story in 1970 anyway. Like Zarate thirty-five years later, he mostly left out the part about how the drug was already known to make people high in favor of a story about scientists reaching conclusions through sober reasoning. It's not clear which version Kline told to the research director at Roche, but by then the company had become disenchanted with iproniazid as a tuberculosis treatment—those darned side effects!—and they remained uninterested.

"Here indeed was a fairly unique situation!" Kline wrote years later. "A group of clinical investigators were trying to convince a pharmaceutical house that they had a valuable product rather than the other way around." Kline was in a good position to mount his campaign for iproniazid. The drug was already on the market, and he was the director of research at New York's Rockland State Hospital. In 1957, he put together a team that quickly launched a study of iproniazid treatment for seventeen hospitalized schizophrenics and nine of Kline's office patients who were depressed. When the depressed patients seemed to be getting better, Kline arranged to testify to Congress on the subject of iproniazid and to have his work published in the *Congressional Record*—a move that at once avoided peer review and guaranteed publicity. He also used his position as the head of the American Psychiatric Association's committee on research to secure a spot at a regional conference in Syracuse, New York, to present his findings. He parlayed this relatively obscure gig into a preconference interview with a *New York Times* correspondent, who dutifully reported that "a side effect of an anti-tuberculosis drug may have led the way to chemical therapy for the unreachable severely depressed mental patient."

Kline's showmanship paid off, at least temporarily. In the next year, Roche, which had all but abandoned iproniazid as an antitubercular, sold the drug to four hundred thousand presumably happy, or at least less unhappy, customers.

*   *   *

It is impossible to know how many of these patients would have impressed Kuhn as endogenously depressed, or, for that matter, how many American doctors rendered an official diagnosis of depression before reaching for the prescription pad. In part, that's because diagnostic specificity and the record keeping it allows were still a thing of the future. The DSM in use in 1958 was a mere 132 pages and counted among its 125 diagnoses inadequate personality and imbecility, diseases whose eye-of-the-beholder vagueness could only support the prevailing view that, as one doctor put it in 1960, "a disease is what the medical profession recognizes as such." It's enough to make you grateful for the faux objectivity of the more recent DSMs.

But there's another reason to wonder about what was wrong with those four hundred thousand iproniazid patients in the first place. Like imipramine, iproniazid seemed to change something underlying mood; as an anonymous "New York industrialist" told the *New York Times*, "On the antidepressant drugs, something has happened. I just feel entirely different." The drug, according to the *Times*, worked by bringing about a "state of well-being and happiness"—*"eudaemonia,"* the reporter called it, explaining that this was Greek for "Aristotle's conception of a life of activity in accordance with reason as constituting human felicity."

Where Roland Kuhn had been careful to tie imipramine to a disease—albeit a disease that seemed tailor-made to the drug—Kline showed no such restraint. Neither was he interested in turning over the "moral and social implications" of a drug that altered people so deeply to philosophers and clergy, as Kuhn had suggested. Instead, by the end of 1957 Kline was urging doctors to move ahead with a project to determine whether drugs like iproniazid "could improve ordinary performance."

You could scour the medical journals from 1957 and 1958 (or you could let me do it for you) and you would find virtually no indication that the new antidepressant drugs were being used for any-thing other than the "severely depressed mental patient." But then again, you could also take a quick look at the *New York Times* from April 1958, just a year after Kline first reported his findings about iproniazid, and conclude that Kline's experiment was under way, courtesy of those four hundred thousand patients—and with some untoward results.

The problems began in April 1958, when Frances Simpson, a fifty-five-year-old San Franciscan, met an untimely demise, which the coroner blamed on liver failure due to iproniazid use. Her doctor claimed that Mrs. Simpson had a "persistent moderate depression," but the *Times* had a different diagnosis, rendered as a headline: "DEATH OF WOMAN LAID TO 'PEP PILLS.'" By the next day, Roche's public relations department had kicked into gear, telling the paper that Mrs. Simpson was taking the 150-milligram dose intended for "severe depression," that in this role it "had repeatedly proved to be a life-saving drug," that there was no such thing as a "completely safe drug," and, perhaps most important, that while iproniazid may be an energizing drug, the "very opposite of a tranquillizer, it is not . . . a so-called 'pep pill.' Such pills give a quick lift. Iproniazid's action is slow and cumulative."

The good news for the company was that these denials, unlike the original headline, were printed on the paper's front page. The bad news, however, was that the new article—"CITY RESTRICTS SALE OF ENERGIZING DRUGS"—was mostly about the forty health-department agents who were at that very moment fanning out across the city to impound supplies of iproniazid. And three days later, after the inspectors had seized 2,671 bottles containing nearly 324,000 tablets, the city medical examiner, under the headline "DRUG INVESTIGATED IN 2 DEATHS IN CITY" disclosed that two recent deaths had a "suggestive relationship" to the drug.

Within a few years, reports of jaundice and other liver-related ailments in iproniazid users began to pour in. In 1960, a doctor—and not Nate Kline this time—testified to Congress ("PHYSICIAN

WARNS ON WONDER DRUGS") about the dangers of the drugs. Despite a spirited defense by Senator Everett Dirksen, who accused his colleague Estes Kefauver of "headline hunting" when he convened hearings on these questions, Roche voluntarily withdrew iproniazid from the market in 1961, the pharmaceutical house by then no doubt ruing the day that it had listened to the likes of Nate Kline.

Iproniazid succumbed to liver problems, but plenty of drugs before and since have been a lot worse for their patients and still stayed alive in the marketplace. One clue about what turned a little jaundice into a terminal condition for Roche can be found in those headlines, in their pep-pill innuendo, their implication that doctors and patients and pharmaceutical firms were up to something not quite savory with these mind-altering drugs. Another can be found in the roster of "wonder drugs" hauled before Kefauver's committee in 1960: iproniazid, chlorpromazine (Thorazine), norethandrolone (Nilevar, an early anabolic steroid), and Diabenes, an anti-diabetes compound. Can it be a coincidence that three of these drugs are for conditions that, especially in 1960, were not considered strictly medical, and that two of them were mind-altering substances?

Kline's proposal to use drugs merely to "improve ordinary performance" may simply have come too early, before Americans were willing to wring their hands about this idea, rather than simply rejecting it outright. But the blitheness with which he tossed it off indicates an ignorance—perhaps willful—of a deep confusion in the American character on the subject of using drugs simply to make life easier.

It's not that Americans don't like to cheat nature—consider, for example, the Interstate Highway System, which renders mountains nearly irrelevant. But when it comes to changing our inner landscape, our destiny isn't quite so manifest.

On the one hand, Americans have always enjoyed a good buzz.

Even the Puritans, the same people who once outlawed the celebration of Christmas on the grounds that it was sacrilegious, kept their larders stocked with rum and ale. Indeed, while John Winthrop was giving his shipboard sermons about a life of hard work consecrated to God, barrels of booze were rolling around in the hold and one of his shipmates was no doubt figuring out where to put the pubs in the City upon a Hill.

On the other hand, Americans have also always been suspicious of getting high. They once amended the Constitution to outlaw drinking and currently spend something like $14 billion a year on a "war" to keep the country drug free and to round up those who would cheat in the pursuit of happiness.

But there is a third hand, which becomes obvious when you realize that $14 billion is only a little more than the national expenditure on antidepressants, and if you throw in tranquilizers like Valium and the uncountable volume of opioid analgesics like Vicodin that are used long after the pain from surgery wears off, you've dwarfed the war-on-some-drugs budget by an order of magnitude. Apparently, some ways of getting high are acceptable after all.

Confused? Well, Gerald Klerman, a Harvard psychiatrist who was an early antidepressant researcher, has a concept for you: pharmacological Calvinism.* According to this doctrine, Klerman wrote in 1972 (at just about the time that it seemed like the whole coun-

---

* I'm not sure why Klerman called this Calvinism. I think he was trying to get at the Protestantism embedded in the antipathy to unearned pleasure, which probably goes back at least to Luther, and maybe ultimately to St. Paul. Calvin, I suppose, is the post-Reformation figure with the sternest reputation, and the godfather of the Jonathan Edwards–led first Great Awakening, which spread the suspicion of earthly pleasure across the New World, but if it were up to me, I'd have called it pharmacological Methodism. Because according to Calvin, you are predestined to salvation; nothing you do can stop you from going to Heaven or the other place. But Klerman makes it clear that drug users can be saved from perdition—not by faith alone, but by the works of the doctors and scientists who turn feeling better into feeling well.

try was teetering on the brink of turning on, tuning in, and dropping out), "If a drug makes you feel good, it must be morally bad." "Psychotropic hedonism," as he called this something-for-nothing degeneracy, is bad because you haven't *earned* your happiness.

But there is an exception to the rule. "The dominant American value system," Klerman continued, "condones and sanctions drug use only for therapeutic purposes and then only under professional supervision by physicians and pharmacists." Make the drug a medicine, get it from a pharmacy on the advice of a doctor, turn feeling good into feeling well (and feeling not-so-good into a disease), and you've found the way to steer your drug-laden camel through the eye of a just-say-no needle. So long as you're sick, you're not cheating if you take drugs. That's what all those pep-pill poppers, and their doctors and druggists, were guilty of—seeking the "quick pill," as Everett Dirksen put it, when they weren't really sick in the first place.

This was Nate Kline's big mistake: failing to reckon with pharmacological Calvinism, he forgot to make "ordinary performance" into a disease.

Iproniazid's demise was not by any stretch the end of Nate Kline's career. He went on to win the Lasker Award in 1964 for his role in developing antidepressants, to grace the cover of *Fortune,* and eventually to become THE ULTIMATE SPECIALIST ON DEPRESSION—or so the cover of his best-selling *From Sad to Glad: Kline on Depression* shouted. Even as iproniazid was crashing and burning, imipramine and a host of related compounds were on a slow and steady ascent. And as Kline and other doctors began to advance the claim that, as *From Sad to Glad* promised, "You can conquer depression without analysis"—i.e., with drugs—they were also creating the means of redemption, the scientific works that, by tying the drugs to a biochemical disease, would save those drug users from the sulfurous pit of psychotropic hedonism.

I've already mentioned Kline's first foray into explaining what was going on with these drugs—his experiments with reserpine, which supported the hypothesis that serotonin imbalances caused mental illness. According to David Healy, the 1955 *Science* article that publicized this result was the "inaugural article" of the antidepressant era. "A bridge had been built from neurochemistry to behavior, and it remained for other researchers to cross this bridge and establish a beachhead on the other side."

And pour over they did. Leading the charge were scientists wielding their knowledge that iproniazid blocked the action of monoamine oxidase (MAO), an enzyme that broke down a group of brain chemicals that included serotonin, known collectively as monoamines. If iproniazid was a monoamine oxidase inhibitor (MAOI), and if at least one monoamine—serotonin—had been implicated in mental illness, then it stood to reason that iproniazid worked because it stopped the destruction of serotonin, thus increasing its presence in the brain. The reserpine experiments, the ones that showed less serotonin activity in the brains of animals that had suffered a reserpine-induced "depression," gave credence to this emerging hypothesis about how iproniazid worked, and, in the bargain, to the serotonin theory of mental illness.

War soon broke out among the troops, however. A young Swedish researcher, Arvid Carlsson, soon proved that Kline, and Carlsson's own bosses at NIH, were wrong about reserpine. He put serotonin back in the brains of the reserpine-treated rabbits, but they remained lethargic—the opposite of what should have happened if reserpine's depressant effects were due to serotonin depletion. Only when he treated the rabbits with L-dopa—a chemical that in the body became dopamine, another newly discovered monoamine—did the rabbits become frisky again. This led Carlsson to theorize that dopamine, not serotonin, was the culprit in the rabbits' torpor. He wasn't interested so much in the psychiatry of this finding as in its implications for Parkinson's disease—the rabbits were suffering from motor problems, he thought. Soon he had demonstrated that the reserpine effect was indeed caused by a depletion not of serotonin but of dopamine—work that led to a treatment for Parkinson's disease and a Nobel Prize for Carlsson.

By the late 1950s, the biological psychiatrists had decamped for a new neurochemical territory: not just dopamine but its cousins in a subgroup of monoamines, the catecholamines—which did not include serotonin. In particular, interest focused on norepinephrine, a chemical related to adrenaline, whose role in the fight-or-flight response had been known since the early twentieth century. Norepinephrine had recently been discovered in the brain. It was known to be depleted by reserpine. MAOI drugs like iproniazid, on the other hand, countered the reserpine effect and raised levels of norepinephrine. These observations led scientists to conclude that antidepressants worked by boosting levels of norepinephrine, providing some kind of stimulation that worked against depression.

There were at least two problems with the *catecholamine hypothesis,* as this theory came to be known. First of all, in 1955, two British scientists did what no one else had done. They ran a controlled clinical trial of a prospective antidepressant, one in which all variables were eliminated except the treatment. The trial was a success, but the drug was reserpine, which, according to the theory should have been anything but an antidepressant. Second, imipramine does not block the action of monoamine oxidase, and yet it still has an antidepressant effect.

Scientists never even tried to explain the first problem. The reserpine study was simply ignored, probably because it ran counter to the increasingly prevailing wisdom. The second problem, on the other hand, yielded an answer much more amenable to the emerging story about how antidepressants worked. There is an excellent account of how the imipramine mystery was unraveled between 1959 and 1963 in David Healy's book *The Antidepressant Era.* I'd love to tell that story here because it involves some of my favorite things: dyes (both William Perkin's old friend aniline and newfangled radioactive reagents), many dangerous drugs and a host of dead lab animals (no coincidence there), basic discoveries about such things as how the liver works, serendipitous drug company bonanzas (the research led to the invention of Tylenol), and scientists fighting over a Nobel Prize. But it would require a digression too long and rambling even for me. Suffice to say that the result was the uncovering of a second way that the brain conserves the chemicals secreted by nerve endings. In addition to being broken down by enzymes like MAO, those chemicals are also reabsorbed whole by the same nerve endings that released them. Julius Axelrod, the doctor who eventually won the fight for the Nobel, called this a *reuptake mechanism*, and soon neuroscientists had two ways to make a brain chemical more available: stop its destruction by inhibiting an enzyme or prevent its reabsorption. And when Axelrod showed that imipramine worked the latter way, blocking the reuptake of norepinephrine, it was easy to believe that the catecholamine hypothesis was correct.

Or so thought Joseph Schildkraut, an NIH scientist. In 1965, he reviewed all the evidence—the reserpine research, the clinical trials with drugs like imipramine and iproniazid, the studies showing evidence of increased catecholamines in the blood and urine of people taking antidepressants, the discoveries of enzyme and reuptake inhibition—and concluded, "There is good evidence to support the thesis that the antidepressant effects of both the monoamine oxidase inhibitors and the imipramine-like drugs are mediated through the catecholamines." So much for serotonin, at least until it returned with a vengeance twenty years later.

But Schildkraut wasn't only interested in settling some parochial argument about which brain chemical was being influenced by which pharmaceutical chemical. If he had been, he probably would have called his paper, which appeared in the *American Journal of Psychiatry*, something like "The Catecholamine Hypothesis of Antidepressant Action." Instead, however, he called it "The Catecholamine

Hypothesis of *Affective Disorders*" (emphasis mine), as if all this scientific work was somehow about the cause of depression rather than the effects of drugs—and as if figuring out how antidepressants worked was the same thing as figuring out how depression worked.

But the paper never addressed the obvious scientific and logical problem: just because you know that a drug makes you feel better, and that feeling better is correlated with a change in brain chemistry, you can't claim that the problem was a lack of those brain chemicals in the first place. All you've explained is how the drug affects the chemicals you're looking at. For all you know, by tweaking the metabolism of one chemical, you've changed something else, something you haven't even examined.

But even if your explanation turns out to be correct, scientifically speaking, you still haven't proved that you've repaired an underlying problem, let alone explained how that alleged chemical deficiency creates the experience we call depression. You may have explained antidepression, but not depression itself.

Schildkraut acknowledged as much. The catecholamine hypothesis, he wrote, was "at best, a reductionistic oversimplification of a very complex biological state." But that didn't stop him from making the claim anyway. He even had a justification for his overreaching. The catecholamine hypothesis, simplistic and speculative as it was, still had "considerable heuristic value, providing the investigator and the clinician with a frame of reference integrating much of our experience with those pharmacological agents which produce alterations in human affective states." So long as scientists framed their investigations of drugs as investigations of the biochemistry of our affective states, they couldn't help but generate knowledge about their causes. To put it another way, although probably not the way Schildkraut would have, if you know what you're looking for, you're bound to find it.

In this case, what scientists were looking for was that single target, the one that they reckoned their bullets must be hitting. Their conviction that such a target must exist got an enormous boost from another scientific development of the early 1960s. The debates about catecholamines and other monoamines were skirmishes in a larger scientific war—over the hypothesis, first advanced by Betty Twarog, that the brain, like the peripheral nervous system, worked by means of neurochemical transmission. The leading neuroscientists of the day refused to accept the concept, preferring to cling to the old idea that sparks flew in the brain, perhaps because it preserved some semblance of the ethereal, of the spirit, amid the gathering knowledge of just how carnal humans are.

Caught in the crossfire between the "soups" and the "sparks," as the factions came to be known, most of the depression doctors hedged their bets: those monoamines might be neurotransmitters, but it was also possible that they were toxins, or antitoxins, or hormones, or fluids that, like Hippocrates' humors, somehow maintained the proper ambience in the brain. But then scientists started to find ways to catch the monoamines at work within the brain. In 1958, a British team inserted recording electrodes directly into a small group of neurons, introduced various alleged neurotransmitters, and listened in as the firing rates of the nerve cells increased in the presence of the chemicals. Three years later, John Gaddum figured out how to inject a saline solution into a cluster of neurons and then withdraw it after stimulation so that the resulting changes in the cells' chemistry could be assayed. And then in 1962 Arvid Carlsson and a colleague used fluorescent dyes that reacted differently with different monoamines to show that dopamine, norepinephrine, and serotonin each had their own pathways in the brain's tangle of cells and fibers. The brilliant, beautiful photos (along with images, captured a few years later, of chemical activity at the synapse) did not prove the principle of chemical neurotransmission, and a few dead-enders held out until the end of the 1960s. But once scientists could hear those recordings and hold those pictures in their hands, once they had visual and auditory evidence that what goes on in the brain is no different from what happens elsewhere in the nervous system—that the inner life is the life of carbon, oxygen, nitrogen, and hydrogen mixing and colliding and recombining—the war was all but over.

So when Schildkraut suggested that even if the catecholamine hypothesis was simplistic, it was still a worthy frame of reference, it didn't occur to his colleagues to complain that he was assuming his conclusions and making vast claims not only about depression but about humanity in the bargain, or that they should do anything at all but rush headlong to the next beachhead. Indeed, "The Catechol-amine Hypothesis" quickly became one of the most-cited papers in the medical literature. And even as the catecholamine hypothesis, which had replaced the serotonin hypothesis, gave way in the 1980s to a new serotonin hypothesis, which itself was replaced a couple of decades later by other hypotheses about other neurotransmitters, even as drug company scientists have packed their tents and rushed to successive new fronts, even as the depression doctors have confidently told their patients about the molecule, whatever it is, that is the source of their woes, the idea itself—that depression is caused by chemical imbalances—has only gathered strength.

Because that was the point all along: to find something inside your head on which doctors could work their magic *and* relieve you of your unhappiness. By the time Schildkraut wrote his paper, that wish had been building for nearly a century, since Paul Ehrlich first set medicine on the course of manufacturing bullets. Kraepelin's notion that mental illnesses would be found in nature, Adolf Meyer's idea that we all could have them, the shock therapists' discovery that they could relieve mental illnesses by doing something to the body, Gaddum's speculation that the problem was in the chemicals of the brain, Kuhn's assertion that imipramine had shown those chemicals at work in one of those diseases—all of these ideas came together in Schildkraut's neat and irresistible package. The fact that even he knew that it wasn't exactly true—and that scientists would soon prove it false—couldn't stop his idea from taking hold. It was a juggernaut, the culmination of decades of yearning for a way to set people free from their psychic afflictions, a way to comfort the

Jobs of the world without accusing them of sin or forcing them to reckon with a whirlwind.

Of course the depression doctors were exuberant. Anyone would be. But science and enthusiasm are an uneasy pair. No scientist labors at his bench without the wish that his work will pay off, and no good scientist allows his excitement to run roughshod over the facts. That's what the scientific method is for—to rain on the researcher's parade—and that's why we believe its results: because they are the facts pure and simple, not subject to the distortions of credulity or ambition.

Or so we hope. But even as the drug makers were gearing up to treat depression, the method for determining whether drugs worked was still under construction. And the solution that the Food and Drug Administration finally settled on could not have been better for the industry. Soon enough, the United States government had not only given its imprimatur to the drugs, it had also created a huge incentive to invent a disease for them to treat—a disease that would also receive the government's stamp of approval.

CHAPTER 10

DOUBLE BLIND

E very two weeks, the research assistant at Mass General, a tall young woman fresh out of college, handed me a new supply of pills. They came in a brown paper bag, the kind you carry your lunch in, its top neatly folded over. Inside were two plastic bottles stuffed to the brim with fat golden capsules. The bottles had childproof caps and two labels: the standard prescription label telling me to take four from one bottle and two from the other every day and another, smaller one which read DRUG LIMITED BY FEDERAL LAW TO INVESTIGATIONAL USE.

That wasn't exactly true. In fact, I could have bought my "drug" at the Whole Foods Market right next to the offices of the Depression and Clinical Research Group, and I could have taken it, as many people do for many reasons, as a supplement. (For that matter, so long as I didn't mind a big dose of mercury with my omega-3s, I could have just eaten a lot of salmon.) Taking fish oil without all the medical folderol would not have felt anywhere near so momentous, and it certainly couldn't have contributed to science, but going the commercial route would have had at least one major advantage: I would have known what I was getting.

Specifically, I would have known whether my pills actually contained omega-3s. They might have been placebos—in this case, soybean oil. No one involved in the experiment knew which was the case, except the pharmacist who filled the bottles according to the condition I'd been assigned by a random number generator. This ignorance was intentional and crucial to the business at hand. Indeed, the double-blind, placebo-controlled, randomized clinical trial (RCT), the kind of experiment I participated in, is known in the industry as the gold standard for investigating the efficacy of drugs. RCTs are the tool that the FDA uses to determine the validity of most drug treatments, including antidepressants.

The RCT is a simple method, or so it seems at first. My first day of testing established the baseline of my depression—eighteen points on the HAM-D, which is considered the threshold for major depression. On subsequent visits, I took the same tests. The score from my last day was then compared to the baseline. (Intermediate scores help to verify that the final score is not an outlier, a score whose divergence from the others makes it suspect.) After two hundred people are recruited and run through this procedure—which investigators reckon will take two to three years—the study will be unblinded, the subjects sorted into a treatment group and a placebo (or control) group, and the aggregate differences between their baseline and endpoint scores will be computed. If the treatment group shows a greater decrease in symptom severity than the placebo group, then that will indicate that omega-3s are effective antidepressants. The mutual ignorance of researcher and subject about who is in which group will increase confidence that the difference between groups reflects not persuasion or gullibility or hope or some other artifact of human credulity but something *real:* the drug and not just the pill.

Prior to the advent of scientific medicine in the mid-nineteenth century, credulity was the active ingredient in most cures. Aside from aspirin, quinine, morphine, digitalis, and a handful of other drugs, most of what doctors handed out was hokum, except for the stuff like mercury and calomel that was downright dangerous. This was the state of affairs that led Oliver Wendell Holmes to comment on how bad it would be for the fish if the nineteenth-century pharmacopoeia were thrown into the sea. According to Lewis Thomas, the great twentieth-century physician/essayist, Holmes wasn't alone in this opinion. Doctors turned to science, Thomas wrote, after "it was discovered, sometime in the 1830s, that the greater part of medicine was nonsense."

As nonsense goes, however, placebo effects are pretty impressive. Patients taking those ancient remedies—poisonous and inert alike—routinely got better. In part, that was because so many illnesses remit on their own, and the potion's reputation was only coincidence trumped up by post hoc reasoning—superstition, in short. But after years of giving placebos in virtually every clinical trial, it is now a matter of scientific fact that there's more to these cures than nature running its course. People given a pill, any pill, will do better than those for whom nothing is done. Researchers have figured out how to allow for this in their calculations: a drug's effect is the treatment group's response minus the placebo group's. But despite the fact that placebos are without a doubt the most widely studied medical treatment in human history, and the hidden subject of every placebo-controlled trial, scientists haven't figured out why they work. In part, that's because science in general has a hard time grappling with irrationality, with cases that blur the bright line between sense and nonsense. But science, at least the variety of science bought and paid for by corporations like drug companies, also has a hard time getting interested in sugar pills— which, after all, can't be patented.

So the placebo effect has been relegated to the role of stalking horse for drugs that can turn a profit. Sometimes it is entirely outmatched. Give a person with bacterial pneumonia a placebo and he is nowhere near as likely to survive as the person given streptomycin. But sometimes the dummy pill gives the drug a run for its money. Sometimes it even wins. In fact, more often than not, that's just what happens in trials for antidepressants. A total of seventy-four trials have been submitted to the Food and Drug Administration for the twelve leading antidepressants. Of those trials, only thirty-eight showed an advantage of drug over placebo. That advantage, when it is there at all, is small: another analysis of clinical trials showed that drugs improved HAM-D scores by an average of ten points, placebos by an average of eight, which means that 80 percent of the effect of the antidepressants is due to placebo effects. The drugs that are in the pills, in other words, don't work very often or very well.

That's not to say that this two-point difference is negligible. It doesn't reach the threshold recommended by some governmental bodies—notably Great Britain's National Institutes for Health and Clinical Excellence—and the FDA's own director of clinical research in the psychiatric drug department once questioned whether antidepressants should continue to be approved based on such weak numbers (only to conclude that it would be unfair to change direction at this stage of the game). But the FDA data do show that the higher a subject's initial HAM-D score, the greater the difference between drug and placebo response. And while some researchers think this only means that severe depressions blunt the placebo effect, rather than that drugs are more effective with the severely depressed, still it is clear that those two points may well have saved lives or at least made people feel better enough to get out of bed and go to work and rejoin their families. This is not a trivial effect.

It is, however, much less than what the drug companies claim. You wouldn't know from a Prozac ad that the drugs have failed almost half of their tests, or that even their successes are well short of miraculous.* But then again when the FDA says a drug is scientifically proven to treat a disease, its manufacturer is well within its rights to take that ball and run with it; that is what the United States government has issued it a license to do. Especially if the company's

---

* You wouldn't know this from reading the scientific literature, either. Of those thirty-eight trials considered successful by the FDA, thirty-six were published in professional journals. Only fourteen of the unsuccessful trials saw print, however. And, according to a team of reviewers, the papers reporting eleven of those studies were written in such a way as to convey a "positive outcome," despite what the FDA said. A doctor reading every paper published would therefore be correct to conclude that 94 percent of antidepressant trials were successful.

best marketing strategy is to sell not only the drug but the disease that it treats, and if its best proof for the existence of the disease is the effect of the drug, then getting this approval is an enormous boon.

That's why the manufacturers of depression, as if they hadn't had enough good luck already, must have pinched themselves to make sure they weren't dreaming when, in 1962, at just about the time that the catecholamine hypothesis was catching on, the FDA suddenly got into a business that it had never been in before: assuring the American public that drugs did what their makers said they did. Government-sanctioned science: as marketing tools go, when it comes to convincing the citizenry that depression is as common as feeling not quite well, that you have the fever and they have the cure—well, it just doesn't get much better than that.

Depression doctors didn't invent the strategy of using science and government to sell their wares. Their professional ancestors got the ball rolling for them in 1847, when the American Medical Association, in its first code of ethics, forbade its member physicians to advertise. The ban was part of doctors' attempt to distance themselves from apothecaries, barbers, medicine-show men, and all the motley assortment of healers to whom people traditionally turned with their suffering. The embargo covered the drugs sold by doctors, which came increasingly to be the drugs produced by the budding pharmaceutical industry and it helped both parties drape themselves in the mantle of modernity. Scientists, the AMA was telling the public, didn't have to advertise; the results of scientific inquiry spoke for themselves, and it was up to real doctors to inform you of those results. That meant that when it came to drugs, there were two kinds in the world: patent medicines like Cuforhedake Brane-Fude (30 percent alcohol) whose ads you could read in any newspaper, and ethical drugs like heroin (Bayer's cure for headache, which was mostly acetylated morphine), which were the only kind good doctors recommended.

As that last example might indicate, the drug industry's claim to possess the facts about disease and cure was really no more justified than that of Emil Kraepelin, who at exactly this time was using the language of science to create the impression that he had discovered mental illnesses in nature. The United States Pharmacopeia, the official compendium of ethical drugs, consisted largely of the same drugs that Holmes hesitated to feed to the fishes. But the branding was brilliant—after all, if your drugs are ethical, then it's pretty clear what the other guy's are—and the U.S. government, just then on the cusp of its Progressive Era, ratified this distinction as it turned its attention to the wide-open medical marketplace.

Congress's attempt to impose some order on that frontier began in the 1890s as part of an overall effort to ensure the safety of America's supply of food and drugs. Congress was reluctant at first. Despite ongoing suspicions about the drug industry, despite impassioned oratory, despite even a highly publicized stunt—Harvey Wiley, a Department of Agriculture chemist, organized a Poison Squad, whose volunteer members dined for six months on nothing but foods containing suspect additives and got really sick—the bills died in committee. Until the winter of 1905–6, that is, when Samuel Hopkins Adams, in a series of *Collier's* articles called "The Great American Fraud," blew the whistle on the patent drug companies, revealing, for instance, that Radam's Microbe Killer was 99 percent water with a little bit of sulfuric and hydrochloric acids thrown in, and that infant cough syrup was likely to contain cocaine or opium, or maybe even both. Also that winter, Upton Sinclair's *The Jungle* grossed out Americans with its barely fictionalized account of the meatpacking industry. By June, the Pure Food and Drug Act had been passed and signed into law.

The ethical drug manufacturers were ecstatic. The new law, said Mahlon Kline, vice-president of Smith, Kline and French, would be "the greatest instrument for the moral uplifting of the average businessman that had ever been bestowed upon the American people." The fact that the Bureau of Chemistry, the Agriculture Department agency charged with enforcing the act, chose his company's chief chemist as its new head probably figured into Kline's enthusiasm. But even without this favor, the regulations were helpful to his industry. They would give it something that the makers of Brane-Fude and Microbe Killer couldn't have: the legitimacy conferred by the United States government.

The bureau obliged the industry by imposing standards for drug preparations with which only a large company possessing the resources to control its supply and manufacturing process precisely could comply. Even better, while it forbade manufacturers to place on the label "any statement . . . which shall be false or misleading in any particular," the bureau didn't establish the method by which those statements could be verified. It was left to the drug makers to engage in a war of rhetoric in which the products backed by the most convincing experts prevailed. A company like Smith, Kline, which could muster an army of respected, science-talking doctors to attest to the truth of their labels, had a huge advantage over patent drug makers and their hucksters. Especially after the Harrison Anti-Narcotics Law had limited the amount of opiates allowed in medicines and the Internal Revenue Service began to impose taxes based on alcohol content, the makers of Muco-Solvent, Humbug Oil, and Swamp Root were on the run.

The situation only got better for the ethical drug industry in 1911, when the bureau tried to shut down the company that made Dr. Johnson's Mild Combination Treatment for Cancer, on the grounds that it didn't really, as the label claimed, cure cancer at home. Johnson fought the law all the way to the Supreme Court, and the law lost. Oliver Wendell Holmes—the jurist son, not the physician father—delivered the bad news to the bureau. He wrote in his majority opinion that Congress had only meant to regulate drugs "with reference to plain matter of fact, so that food and drugs should be what they professed to be." That was easy enough to establish, by comparing what the label said was in the bottle with what was actually there. But drug efficacy was a different story.

Congress, Holmes said, surely did not mean "to distort the uses of its constitutional power to establishing criteria in regions where opinions are far apart." And whether or not a drug actually did what the label said it did, whether or not Dr. Johnson's Mild Combination Treatment actually cured people's cancer—this, Justice Holmes said, was a matter of opinion.

The United States Congress was in a reforming mood in those days, and lawmakers were not content to leave public health to the invisible hand. They tried to give the Bureau of Chemistry the power to require that drugs be effective without running afoul of the Supreme Court, but after a few bruising floor battles, the best they could do was to outlaw "any statement . . . regarding the curative or therapeutic effect . . . which is false and fraudulent." This law actually discouraged research by making ignorance a defense; falsehood is not fraud unless a company has taken the time to figure out the truth. The "greatest instrument" was only getting better for Mahlon Kline and his friends.

That may seem startlingly obtuse, but it makes sense when you remember that with a few exceptions—like Paul Ehrlich's Salvarsan—there was no way to say with certainty why a sick person who took a particular drug got better; there was still precious little understanding of the biochemistry of disease or cure. In the absence of that knowledge, Holmes thought, a drug's ability to cure illness was a matter for the marketplace to decide. You bought the potion, decided whether you liked it, and told your friends (and your doctor) what you thought—the same as you did with any other consumer product. The government's role was simply to make sure that the public got what it paid for and didn't get harmed in the process.

That turned out to be harder to accomplish than it sounds. In 1931, the Food and Drug Administration was spun off from the Bureau of Chemistry. It didn't take long for the new agency to discover just how limited its powers were. In 1937, S. E. Massengill Company introduced Elixir Sulfanilamide. Sulfa drugs had burst onto the medical scene in 1936, when Franklin Roosevelt, the president's son, had been cured of a streptococcus infection (in those days, strep infections were potentially fatal) by a timely injection of sulfanilamide. The patent for the drug—Prontosil, derived from a dye that Paul Ehrlich had fooled around with and which was used to redden leather—had long ago expired, so drug companies sought market share by inventing new preparations. Massengill hit upon the idea of putting the antibiotic into a raspberry-flavored syrup. The problem was that the drug refused to dissolve in the syrup—until Massengill's chemists added diethylene glycol. The company's chemists may have been the only chemists on earth who didn't know that this compound, used as an industrial antifreeze, was a fatal poison.

After the bodies were counted (105), after the remaining 234 gallons of Elixir Sulfanilamide had been rounded up, and even after a government official determined that chemists at Massengill "just throw drugs together and if they don't explode they are placed on sale," the FDA found that its only recourse was to fine the company $26,100 for mislabeling—*elixir,* the agency said, was a term reserved for alcohol-based preparations. Inadequate as this punishment was to the scope of the tragedy, it could have been even worse. Had Massengill named its drug something else (Sulfa-Freeze, perhaps), the company might have been entirely beyond the reach of the law.

The Elixir Sulfanilamide debacle provoked Congress to action. In the Food, Drug and Cosmetic Act of 1938, it banned the interstate shipment of harmful substances, which in turn gave the FDA the authority to require drug companies to prove that their new products were safe before they could be brought to market. The label remained the focus of the law. It now had to disclose not only the contents of the bottle but accurate information about proper dosage and potential dangers. The law exempted some drugs from this requirement—those whose effective use and safety hinged on conditions too complex for a layman to assess. For these drugs, no label was necessary because people weren't going to be buying them on their own say-so, but rather only on a doctor's orders. Along with the prescription, doctors would dispense the information necessary for safe use—information that they got from the drug companies, and mostly from the army of salesmen, or detailers, that the industry now began to deploy.

Like earlier pure drug laws, the 1938 version was a hit with doctors and drug companies. It gave more power to physicians and it left it up to the industry, whose in-house research staffs were still tiny, to recruit the doctors who would submit the safety information to the FDA, a cozy relationship that couldn't hurt either party. Even more important, the law continued to steer a wide berth around the question of efficacy. That was still, as it had been in 1911, a matter of opinion, not something on which the FDA was going to weigh in.

But that didn't mean that efficacy was off the table completely. Drug safety is generally not a straightforward question; cases of outright poisoning like Elixir Sulfanilamide are rare. More common is the problem that arose when effective drugs caused unforeseen problems as a function of, or in addition to, their therapeutic action—in other words, side effects. Sulfa drugs, for instance, could deplete white blood cells, but that risk, especially if managed by a skilled doctor, was clearly outweighed by the infection-killing benefits of the drug. Safety could only be assessed as part of a cost-benefit analysis in which efficacy had to play a part. A few articles in a journal, testifying to the effectiveness of a drug and written by doctors whose credentials as researchers weren't necessarily sterling and whose methods were haphazard, would suffice.

The FDA did make at least one attempt to face the question squarely—with drugs that seemed to have no therapeutic effect whatsoever. These drugs were by definition unsafe because there was no benefit against which to weigh the cost. The agency focused on "glandular substances"—hormone-like remedies intended for women of a certain age that had no discernible pharmacological action—and proposed a fix: a label that acknowledged "there is no scientific evidence that such products . . . possess any therapeutic activity."

The industry howled in protest. The FDA, a trade journal complained, was telling drug makers that they "must undertake to educate physicians"—a function they were perfectly happy to fulfill when the news was good. And when the agency declared in 1948 that it couldn't certify the safety of such a preparation, industry hauled out the heavy guns. The FDA, it said, was interfering with the sacred doctor-patient relationship. One of Congress's own doctors brought the warnings close to home. This edict, he told lawmakers, would make it impossible for him to offer glandular remedy to "the wives of my Congressional group."

That worked. By 1950, the FDA, reminding Congress that the Supreme Court had long ago tied their hands, was officially out of the efficacy business, but it remained in the drug-certifying business. And once again, regulation uplifted the businessmen. The prescription drug industry, which already claimed science for its side, and whose products were officially only understandable by experts, could now obtain the government's imprimatur for its products without ever proving that they worked.

This was the regulatory environment in which the antidepressant discoveries I described in the last chapter took place. The FDA could only comment on safety, it had to take doctors' word about efficacy, and the agency had only a short time to respond to a new drug application (sixty days, after which the drug, in the absence of a response, was automatically approved) and a small staff (1,065 in 1956, the height of the industry's postwar boom, and only 117 more than it had ten years earlier). The FDA's $6 million budget was dwarfed by the $140 million that the pharmaceutical companies were spending annually on research and development. A drug could make it from bench to market in just a few months.

In some respects, this laxity didn't seem to matter. The 1950s were a time of true wonder drugs—not only the antibiotics that followed on the sulfa drugs and penicillin, but also chlorpromazine, corticosteroids and other hormone treatments, and diuretics for high blood pressure. The results, in lives saved or transformed, seemed to speak for themselves. Armed with its government-sanctioned success, the drug industry had gained the confidence of the public as a reliable supplier of magic bullets.

The industry wasted no time in exploiting its achievements. Sometimes this was too much of a good thing, as Johns Hopkins doctor Louis Lasagna complained in 1954:

> The doctor of today is under constant bombardment with claims as to the efficacy of drugs, old and new. It is difficult, if not impossible to read a journal, attend a medical meeting, or open the morning mail without encountering a new report on the success or failure of some medication.

As if to prove Justice Holmes correct, medical journals were chock full of opinion. Doctors, it seemed, were to be guided in their prescription choices by whoever among their colleagues sounded most trustworthy to them. Or, for that matter, by the advertising in their journals: the number of pages of *JAMA* devoted to drug company ads doubled in volume in the 1950s, even as the AMA stopped requiring advertisers to earn its Seal of Acceptance before they could hawk their wares. And if doctors were too busy to read the journals or look at the ads, there were always detail men to regale them with the latest lab results over a round of golf.

The resulting therapeutic chaos alarmed Estes Kefauver and his Anti-Trust and Monopoly Subcommittee—the same body that listened to testimony about iproniazid and other "quick pills." A Tennessee Democrat who was Adlai E. Stevenson's running mate in 1956, Kefauver was a liberal populist, an early champion of racial equality, a fierce opponent of Joseph McCarthy. He fought the pharmaceutical industry to the death—his own, in 1963, the fifth year of his hearings. By that time he had infuriated doctors, the drug industry, and lawmakers like Everett Dirksen with his repeated attacks on the way that the prescription drug scheme had tilted the market toward the drug companies and turned doctors into their shills. "He who orders does not buy," he said, "and he who buys does not order." Consumers were at the mercy of drug companies; they couldn't even evaluate advertising claims on their own because the targets of the ads, and the only people who saw them, were their doctors.

At the very least, Kefauver argued, drug companies should have to prove the merits of their drugs to the government's satisfaction before they began their advertising onslaughts. He proposed a law giving the FDA the power to require drugs to be proven "safe and efficacious in use"—a question that he thought science had finally made into a plain matter of fact. Although a few renegades, like Lasagna, supported Kefauver, the AMA strongly opposed his proposed law warning that it was a step toward socialized medicine. The Judiciary Committee, to which Kefauver's subcommittee belonged, took out the efficacy provision; the bill had been so defanged by the time it reached the Senate floor that Kefauver refused to manage it. It was headed for a quiet death in mid-1962 when lawmakers suddenly became aware of a magic bullet that had turned lethal, and that the citizens of the United States had only barely dodged.

The drug was thalidomide, and it had been invented in the late 1950s by Chemie Grünenthal, a German company that was hoping to get into the psychiatric drug business with a new tranquilizer. The company hawked the drug, known as Contergan and available over the counter, not only as a tranquilizer, but as a sleep aid, a flu remedy, and, at least according to a doctor on retainer to Grünenthal, a suppressor of young men's desire to masturbate. Grünenthal also claimed, based on animal studies, that the drug was completely nontoxic, which meant that it was safe to give to pregnant women.

Frances Kelsey, an FDA physician-bureaucrat, was not so sure. In 1960, when Richardson-Merrell applied for a license to sell thalidomide in the United States, she asked the company to supply more information. She was concerned about reports of peripheral neuritis, irreversible nerve damage in patients taking thalidomide, and pointed out that there were contradictions in the safety data that might bear on this effect. She also noted that the company had not provided information on the drug's effect on the developing fetus—not even on whether or not it crossed the placental barrier—a crucial absence given the fact that the company was pushing the drug as a remedy for women made jittery by their newly discovered pregnancies. Merrell was much slower to respond to Kelsey than they were to distribute 2.5 million doses of the drug to more than 1,200 doctors for them to use on a trial basis. Twenty thousand American patients received the drug.

Working for Merrell, as for most pharmaceutical companies at the time, was pure gravy. The drugs were free, the doctors didn't have to report results if they didn't want to, and even if they did, they wouldn't have to go to all the bother of gathering data or writing up the results. Merrell's medical director, like medical directors at most drug firms, was glad to provide them with completed manuscripts attesting to the drug's effectiveness and ready for their signature and to send them on to the medical journals, where they would become part of the record establishing the value of the drug.

Even as Merrell was ramping up its marketing efforts, however, trouble was brewing. Doctors in Australia, England, and Germany were seeing not only peripheral neuritis, but something much more disturbing in the offspring of their thalidomide patients: a sudden increase in cases of phocomelia, a birth defect in which limbs fail to develop, and which leaves infants with hands and feet growing directly from their shoulders and hips. By the time epidemiological and animal studies, conducted over Grünenthal's objections, had confirmed the link between thalidomide and phocomelia, thousands of European children had been born with massive deformities. In March 1962, Merrell, after two years of heated argument with Kelsey, finally withdrew its application for thalidomide.

The European tragedy might have passed unnoticed in the United States, where fewer than twenty thalidomide babies were born. But Kefauver's staff recognized the opportunity in the debacle and, three days after his proposal hit the Senate floor in July 1962, they informed a *Washington Post* reporter about what had happened overseas. The story was reported on the front page. In short order legislators were falling over one another to do something about the drug industry, and there just happened to be a bill ready for their approval. The Kefauver-Harris Drug Amendments to the Pure Food and Drug Act, their efficacy clause intact, passed in 1962.

The new law had absolutely nothing to do with thalidomide. Even Roman Hruska—the Nebraska senator who had once defended a Richard Nixon Supreme Court nominee who had been called mediocre by insisting that there was a place for mediocrity in public life—could see that "thalidomide was already barred and the public was protected under the 1938 act." But no matter. Kefauver had gotten his way. For the first time pharmaceutical companies were required to prove to the FDA that their drugs worked in order to get a license to sell them.

Turning this requirement to corporate advantage was easier than you might think, thanks in part to Justice Holmes. The new law had to address his original worry about congressional reaching into the realm of opinion. This meant that it wasn't enough for Congress to say that science had made it possible to sort out fact from opinion; it had to specify how those facts would be established. The answer was that "substantial evidence . . . consisting of adequate and well-controlled investigations . . . by experts qualified by scientific training and experience" would establish the efficacy of a drug.

That seemingly innocuous phrase—"substantial evidence"—contained a huge break for drug companies. Lawmakers had considered a different standard—the *preponderance of evidence*. The difference, as one senator put it, was that to require only substantial proof meant that a drug could be deemed effective "even though there may be preponderant evidence to the contrary based upon equally reliable studies." Especially after the FDA determined that two independent trials with statistically significant results in favor of the drug constituted substantial evidence, this meant that a drug up for approval could have as many do-overs as a drug company wanted to pay for. So long as the research eventually yielded evidence of efficacy, the failures would remain off the books. This is why antidepressants have been approved even though so many studies have shown them to be ineffective.

That wasn't the only way that Kefauver-Harris turned into a sweet deal for the drug companies. They also had in mind a way to address the requirement for adequate and well-controlled investigations: the randomized clinical trial, the method used by my doctors at Mass General. This approach, as the industry soon figured out, could easily be made to say more than it really said and do something quite different from what it was intended to do. Both the RCT and the statistics used to assay its outcome are much better at telling scientists when a treatment *doesn't* work than when it does, to *disprove* rather than to prove drug efficacy.

The eagerness among drug doctors to get more out of the RCT than it is equipped to provide features in the earliest attempts to sell it as the method for verifying drug efficacy. In explaining why he thought regulators should adopt the RCT, Louis Lasagna cited a momentous event in medical history. In 1747, Lasagna recalled, the ship's doctor on the HMS *Salisbury*, James Lind, decided to check out an old and unproven theory that acids would cure scurvy. Since the *Salisbury* was returning to England after a long time at sea, it had no shortage of subjects. Lind divided a dozen scurvy sailors into six pairs. "Their cases were as similar as I could have them," he later wrote. "They lay together in one place and had one diet common to all." Lind randomly assigned each pair to one of six treatments, which included a dose of vinegar and a garlic concoction—and, fatefully, oranges and lemons. Most of the sailors stayed ill, but within a week, one of the citrus-eaters was so well that he "was appointed nurse to the rest of the sick," and by June 16, when the *Salisbury* pulled into Plymouth, the other was fully recovered.

That was a clever experiment. But even if he had randomly chosen the sailors who received the fruit and tried to keep their other
conditions equal, Lind's trial was not really controlled. He did not
account for a crucial variable—the possibility that the placebo effect
had cured the sailors. He knew who was getting which treatment,
and he had a stake in the outcome. Even if his reports were honest,
his belief might have been contagious, his enthusiasm the cause of
the cure's success. He couldn't say with certainty that something in
the fruit had cured the sailors because he did not control for credulity—his or his patients'.

That might seem like an unfair criticism—after all, doctors of
the time didn't know that most of their medicines were placebos—
but the confounding power of the placebo effect was understood by
at least one eighteenth-century scientist. In 1784, Benjamin Franklin was living in Paris when Louis XVI tapped him to head a scientific commission investigating a claim that had all of Europe in
a stir. Franz Anton Mesmer was telling people that he had discovered a force in the universe as real and important as gravity. He
called it "animal magnetism," and in parlors across the continent he
was demonstrating how a physician could harness it in the service
of healing. Patients swore that their rheumatism, skin ailments,
asthma, and nervousness had been cured by Mesmer. The mesmerism craze alarmed the king, and he charged Franklin with the task
of determining whether or not animal magnetism really existed.

Franklin and Mesmer had different ideas about how to answer
this question. Mesmer suggested an experiment much like Lind's:
take two patients with the same disease, mesmerize one of them
and not the other, make all other conditions equal, and see who
fared better. But Franklin, the wily rationalist, understood that
there was a bigger obstacle to the truth: self-interest, especially
that of the doctor and his patient. He designed a test to eliminate
human subjectivity from the experiment.

Franklin's proposal also eliminated Mesmer, who, upon hearing
of it, withdrew from the proceedings. He sent another mesmerist

to Franklin's house on the appointed night, willing patients in tow. In front of the commission, he focused the magnetism on parts of their bodies. Asked to locate where he was directing the energy, the patients—all women—responded accurately. Then they were blind-folded. When the mesmerist repeated the procedure, the women located the sensations, according to the commission's report to the king, "at hazard, and in parts very distant from those which were the object of magnetism." Other variations of the blindfold test yielded the same results. "It was natural to conclude," Franklin told the king, "that these sensations, real or pretended, were determined by the imagination."

Lind claimed to have *proved* that citrus cured scurvy, but Franklin seems to have understood that this was more than an experiment could say, and that there is an inexhaustible supply of variables, known unknowns and unknown unknowns alike, that might have been at work in mesmerism. He controlled for the one he deemed most likely—imagination—and when he did so, there was a difference in outcome. Or, to put it another way, he started with the idea that there would be no difference between a blindfolded and a non-blindfolded treatment—and *disproved* it. He didn't *prove* anything; his conclusion from the proceedings might have been natural, but it was also inferential.

This may seem like a distinction without a difference, especially when you consider the different purposes of these experiments: Lind's to ratify and Franklin's to debunk. But what Franklin seems to have understood was that enthusiasm for a treatment wasn't just another variable in the pursuit of scientific knowledge. It was the enemy. Self-interest, hope, the ineffable qualities of the doctor-patient relationship—in short, subjectivity—would always haunt our attempts to understand the world, and the role of the experiment was to rein in its effects, whether by tying on a blindfold or acknowledging the limits of a controlled experiment to establish the truth.

The modern RCT is much more sophisticated than Franklin's

experiment. But it begins with the same recognition of our limited ability to circumscribe credulity and follows the same logic; it starts with a null hypothesis—that the treatment won't work—attempts to confirm it, draws inferences from the results, and then tries to strengthen those inferences by replicating the experiment. In citing Lind, Lasagna seems to have forgotten that an RCT is much more suited to disproving than to proving, that it can only give us probabilities, that its primary purpose is negative: to rain on an experimenter's parade, to put the kibosh on therapeutic enthusiasm rather than to inflame it.

In 1928, twenty-five years before Lasagna touted the virtues of the blinded RCT, and just a year before the stock market crashed, nervous investors wondered if something was wrong in the house of Morgan. Anne Morgan, sister of J.P. and usually no less retiring than the rest of the family, had suddenly turned up as a paid spokesman for Old Gold cigarettes. She reported to newspaper readers that she had "taken the blindfold test, smoked four brands of cigarets [sic], and found that 'the smoothness' of one cigaret [sic] was 'so obvious.'" Miss Morgan urged other smokers to repeat the experiment in the privacy of their own parlors. Advertisers had evidently figured out that by reassuring the consumer that his tastes were not a figment of his imagination, that his own ever-unreliable subjectivity had been neutralized, they could build brand loyalty. They had hit upon a way to use the blindfold test to stoke enthusiasm rather than to curb it. (Miss Morgan's finances, as it turned out, were sound; she apparently donated her thousand-dollar honorarium to charity.)

A decade later, Cornell scientist Harry Gold (no relation to Old) was trying to figure out whether or not xanthines, a group of stimulants that included caffeine, really deserved their reputation as a remedy for angina. He realized that he couldn't trust the data provided by doctors studying the question. They asked leading questions, assigned patients nonrandomly to get the drug or placebo, and interpreted ambiguous results in a way that favored the drugs. It wasn't enough, Gold concluded, to control for patients' credulity; doctors also had to be placed in the dark. Experiments had to be double-blind, Gold said. He cited the Old Gold campaign as the inspiration for the method's name.

Gold was using the blindfold test as Franklin had intended—to impose restraint. He would probably be discomfited to see the ease with which his method has been used to create certainty rather than to limit uncertainty. But perhaps no one would be more upset at the way that RCTs have become one of the drug industry's greatest marketing tools than the British geneticist who invented the mathematical language in which RCTs are reported—a language intended, like the experiments themselves, to rule out rather than to rule in.

Ronald Aylmer Fisher, who was so blind that he had to do his calculations in his head, developed modern statistics while working for an agricultural research institute just after the First World War. Fisher was trying to sort out fact from opinion when it came to crop yields. A farmer could plant two different varieties of grain and at harvesttime reap a much bigger crop from one of them, but most scientists understood that this didn't necessarily mean one strain was more vigorous than the other. Maybe the soil varied from plot to plot, or the exposure to sunlight, or the population of varmints. Without isolating those variables, there was no way to know if the change was due to what the farmer did on purpose or to what just happened to occur.

Fisher's solution was to divide the field into strips, randomly assign different strains (or different fertilizers or tilling practices or some other *independent variable*) to different strips, and then measure the differences in yield (or color or time to maturity or some other *dependent variable*). This approach may seem obvious, and indeed farmers had been doing something like it for centuries, but Fisher was the first to figure out how to systematize the procedure and to express the results in numbers.

To do this, however, Fisher first had to change his colleagues' minds about what they were doing. While most scientists thought that ignorance was their greatest enemy, that with their experiments they were establishing the bedrock facts about how nature works, Fisher, like Franklin before him, saw it differently. To him, the adversary of knowledge was chance. Nature's randomness was a wily foe, one that could never be fully accounted for. Indeed, as Fisher's biographer put it, the best a scientist could do was

> to equalize the chance that any treatment shall fall on any plot by determining it by chance himself. Then if all the plots with a particular treatment have the higher yields, it may still be due to the devil's arrangement, but then and only then will the experimenter know how often his chance arrangement will coincide with the devil's.

The reason to plant one kind of wheat in one row and another in the next is not so much to prove that one variety is superior. Rather, it is to minimize the possibility that the differences between them were the result of random events, which in turn increases the possibility that what the humans did, rather than what just happened to occur, is responsible for the outcome.

Randomization may have put nature's perversity into the hands of the scientist, but only at a cost. You could crunch the numbers all day long—and Fisher developed a series of formulas for just this purpose—but all they were going to tell you was the probability that you had merely caught the devil at play. That's why statistics-guided studies are more like Franklin's experiment than Lind's. They are not designed to confirm a positive hypothesis—that, say, basmati rice grows faster than long grain brown—but rather to disconfirm the hypothesis that any differences between the strains are due to random chance, which is to say that there is no real difference. Repeatedly disprove the null hypothesis and you will gain some certainty that the result most likely wasn't random. In a fallen world, inference was the best that a scientist could do, and the job of the numbers was to specify how confident he could be about those conclusions.

But when Lasagna and other champions of statistics-guided RCTs adopted Fisher's methods, they did not also adopt his humility. Researchers using statistics-guided double-blind methods placed credulity in the devil's camp—as part of the background noise from which the drug's signal could be discerned. They claimed that the method would "free a researcher from the accusation that his beliefs had affected a study's execution" and that therefore the study itself had uncovered the plain facts. It also meant that Miss Morgan's confidence could now be expressed in numbers—a language that, like the method itself, seems beyond the manipulations of self-interest and the vagaries of imagination. Drug scientists didn't have to settle for an alliance with the devil, statistics that could only measure the uncertainty turned up by an experiment designed to dampen enthusiasm. Instead, they could defeat the devil, claim certainty about their results, and use them to assert that their drugs really work.

Some doctors saw immediately how easily the combination of statistics and the RCT could be used to make experiments say more than they really could. Some of them were even bothered by this. At a drug evaluation conference in 1958, one participant said:

> I have an intense prejudice against the mass of statistics that accompanies the introduction of new drugs. It seems to me that statistics . . . are too often misleading. Many people believe that they eliminate chance when in fact they merely give an idea as to the probability of the results being due to chance.

The difference between "eliminat[ing] chance" and "the probability of the results being due to chance"—or between proving efficacy and disproving a null hypothesis—may seem minor. And sometimes it is. Virtually any diabetic given the right amount of insulin is going to survive, just as almost anyone with bacterial pneumonia will get better with an antibiotic. You don't need an RCT to prove this. If you ran one, you wouldn't need statistics to interpret the results. And if you did use statistics, there would be no meaningful difference between saying you had disproved the null hypothesis and that you had confirmed that the drugs worked. The restored health of nearly all the patients would speak for itself.

But if you had a drug-approval scheme that allowed regulators to ignore the preponderance of evidence and a drug that was no more effective than an inert substance at fixing the problem it claimed to fix, that achieved these meager results only after the best minds had concocted the conditions under which the drug was most likely to succeed, and that impaired a patient's sex life and caused him to gain weight and sometimes even made him suicidal—if, in short, you had antidepressants, then you would really want to pay attention to that distinction. You would want to bear in mind the warning of one early statistician:

> Let the experimenter who is driven to use statistical methods not forget this, that the very fact that he is compelled to use statistical methods is a reflection of his experimental work. It shows that he has failed to attain the very object of his experiment and exclude disturbing causes.

You would want to remember that when it comes to disturbing causes there's nothing more disturbing, or harder to exclude, than consciousness, especially when it comes to drugs that change it, and that if those statistics mean anything, if they have any significance at all, it is not that your drugs work. It is, at best, that sometimes they don't not work, and that even when that happens it may not be the result of the drug inside the pill.

But that kind of caution is not going to move product. It is better to cite the numbers as proof that the drugs cure depression and leave out all those picky details—not to mention all those other numbers from all those studies that don't count, the ones that say that the drugs don't work. Then the RCT becomes the opposite of what it was intended to be—not a device to debunk the claims of parties with a stake in the outcome, but a way to harness the power of science to justify their enthusiasm by putting it into the persuasive language of numbers. As it had for Kraepelin, and as it had for the advocates of the catecholamine hypothesis, the rhetoric of science, if not science itself, was going to lend authority to the pronouncements of the drug experts. It would give them the appearance of objectivity, allow them to claim that they were not expressing mere opinions but rather revealing what was really in nature.

And that's not all. What the FDA meant by efficacy was proof that a drug was effective *with a particular disease.* Suddenly, it was more important than ever to find an *indication,* the specific illness on which the drug could be tested and for which it could be approved (and advertised) as a treatment. The fate of a new drug was now tied to the skill with which doctors identified and isolated a disease for it to treat. For a disease like depression, whose definition had proved changeable and whose very existence had been questioned, those government-sanctioned numbers now had a hidden function: they could give the public confidence not only that a drug worked, but that its target was a genuine disease. It was only a matter of time before an official, statistics-friendly definition of depression emerged, one that indicated that vast numbers of people harbored that disease in their brains and should turn their discontents over to their doctors.

CHAPTER 11

DIAGNOSING
FOR DOLLARS

S ometime during my first session with a new patient, usually toward the end, the touchiest subject in therapy comes up. Which is not what you might think it is. People will pour out, sometimes unbidden and in more detail than you may want, their illicit romps, their most ignoble or kinky wishes, their peccadilloes and deceptions, and other carnal secrets long dammed by shame and other family traditions. But ask these same people how much money they earn or have in the bank and they will seize up like a frozen pipe. Sometimes they even tell me that it's none of my business, and, as a rule, the more affluent they are, the more likely they are to feel that way. It is as if they are putting me on notice that there is a nakedness that even I'm not allowed to see.

So when the m-word arises in that first session, it's not because I'm asking. I've learned that much in twenty-five years. It's because despite the intimacy of the encounter—which, by the way, is real; nothing draws people closer than simple, honest talk—we are doing business. I'm renting myself out by the hour, so as the clock ticks down, just as they might wonder if a first date will end with a handshake or a kiss or a wave good-bye, patients find themselves thinking about the transaction that will bring our time to a close. And more often than not, the way they let me know this is to say, "Do you take my insurance?"

This often ends up being a much more interesting conversation than patients bargain on. I tell them that I would be glad to submit the paperwork to their insurance company so that they can get reimbursed for what they have paid me. If they can convince me that their cash flow can't handle that approach, I'll even wait for the insurance company to pay up, although, I remind them, the responsibility to pay me is ultimately theirs, and they need to bear in mind that insurance companies make their profits by paying for as little health care as the law and, occasionally, common decency will allow. And then I tell them what I must do to have a chance of opening up the insurance coffers. "I'll have to tell your insurance company that you have a mental illness."

You wouldn't think that would be a big surprise; they don't call it *medical* insurance for nothing. But very often it is. In general, the people who come to see me for therapy don't think that they are mentally ill. Mostly, they think they are worried or unhappy or reeling from setbacks. So they are often nonplussed when I point out the obvious—that they will soon be officially sick in the head, that this fact will be part of their permanent medical record, that should they seek life insurance or a security clearance or high political office, should the nosy manager in the human resources office happen to get a look at their file, they might come to regret having received that diagnosis.

This discussion sometimes ends with a decision to skip the insurance, which spares me paperwork and which I may reward with a reduction in fee. But in many cases, people feel like they must rely on their benefits, so the discussion next turns to the question of which mental illness they have. Sometimes I tell a patient what I think. If he has come in unshaven and rumpled and tells me it's the first time he's been out of bed in a couple of weeks and that he's been feeling sad and guilty and apathetic, that he doesn't eat and is considering suicide, then that's an easy call: if we're using his insurance, he's got to have major depressive disorder. But other times, I take out the DSM and give a patient some choices.

Thanks to the authors' foresight—some of them were private practice psychiatrists, after all—we can usually find a disorder that fits the facts without sounding too terrible. Adjustment disorder—which is what the DSM calls "a psychological response to an identifiable stressor or stressors that results in the development of clinically significant emotional or behavioral symptoms," and which you can have in various emotional flavors, including *with depression, with anxiety,* or *unspecified*—is an obvious favorite; there is no shortage of identifiable stressors in everyone's life, and simply coming to my office indicates clinical significance. So is generalized anxiety disorder, in which a patient has "excessive anxiety and worry, occurring more days than not for a period of at least six months"; Dr. Dording has already provided an excellent example of how slippery "excessive" can be in capable hands, and on the days you are short of things to worry about, there's always global warming and the coming exhaustion of the oil supply. The only problem with these diagnoses, I explain, is that they are not on the relatively short list of disorders that insurance companies are obliged by the law in my state to pay for in the same fashion that they pay for any other illness. So the benign diagnosis might lead to reduced benefits.

Occasionally, this discussion ends our relationship—something I usually find out when the patient calls back to cancel or just doesn't show up for the next appointment. A few of these ex-patients have been generous enough to tell me, when I asked, that my behavior seemed, well, a little unusual to them. They may have expected me to make mountains, but not out of that molehill. My doing so made them uneasy—not in a let's-talk-about-this way, but in a get-me-the-hell-away-from-this-nut way. I can't blame them for this. They were undoubtedly correct in their assessment that they couldn't work well with me.

That's why I don't try to explain to these people that if the whole point of therapy is not to take anything for granted and to make explicit what would prefer to stay invisible, then it doesn't really make sense to hide the truth of what we—my patient and me—are doing when it comes to the insurance companies. Why make bad faith central to a relationship that is supposed to be the antidote to that?

That used to be my main justification for bringing the business of diagnosis to the foreground of therapy, if only briefly, and to exercise some due diligence when it comes to giving and receiving labels. And I still believe it, but my clinical trial experience taught me another reason to do it: being told that you are mentally ill can really mess with your head.

That may seem like a reason *not* to have the diagnosis discussion, to just let the medical-industrial complex chug along unremarked and get to the real work at hand. But even if such a distinction can be drawn, the truth is bound to emerge. Maybe the insurance company will ask for details about what is wrong with the patient, or will refuse to pay for further treatment unless the patient seeks medication, or, as happened to me the very first time I was in therapy, a patient will catch a glimpse of his statement and realize that there is a diagnosis code on it. If you are in therapy, sooner or later, implicitly or explicitly, it is inevitable that you will figure out that you have become a mental patient. And you should be prepared for what that might mean.

I certainly should have been prepared. By the time I showed up at Mass General, I'd been doling out the labels for a long time—and not thoughtlessly. Like many people who went to college in the 1970s, I had read Erving Goffman and Michel Foucault, sociologists who had pointed out the inescapable ideological dimension of diagnostic labeling. Identifying people as depressives or schizophrenics, they suggested, was a way to exert power over them, to stifle dissent and impose conformity to a particular way of being human. Sometimes, as in the case of a person shut up in a mental hospital or forced to undergo a lobotomy, the exercise of power was obvious and brutal.

But more often, especially according to Foucault, the label doesn't function to put a person in his place so much as it gives him the language to put himself there. People are, in this view, socially constructed, and the power of a diagnosis is that it changes the way people understand and identify and treat themselves. For theorists like Foucault, this is a crucial, and perhaps the central, problem of modern life: that the power to tell us what kind of life we ought to live, and what kind of people we ought to be, could be wielded not directly but diffusely, not through force but through culture, not by jackbooted thugs but by experts. Tell people what they ought to want, help them to think that they are freely choosing, and you've gotten around any resistance they might have to being told what to do. Power exercised this way is invisible and in some ways even more dangerous than the kind that is obvious. The power that hides in the plain light of day can fashion people in its own image without their even knowing it.

Those experts include doctors. For a psychiatrist to say that you have the disease of depression is to tell you not only about your health, but also about who you are, what is wrong with your life and how it should be set right, and who you would be if only you were healthy. In making these pronouncements, the doctor draws on the authority of science, which presumably has no stake in the outcome. He couches his judgments in the language of sickness and health rather than sin and virtue, which means that he is cloaking his morality, even from himself, in science.

It's easy enough to see the way this happens in a case like Sheila's. She was a woman in her midthirties, married with three children, who came to see me in the midst of a crisis. She'd fallen in love with a man who wasn't her husband. He lived far away, but they were in constant contact by phone and e-mail and had arranged a tryst. She was, she said, standing on a precipice, as terrified as she was excited, but there was no doubt she was going to leap.

Which she did, with the predictable results: a blissful weekend, a redoubled attraction, more e-mail and phone contact, a longer trail of incriminating evidence for a husband made suspicious by her sudden preoccupation, the inevitable discovery and confrontation, fifteen years of marriage reduced in a moment of cyber carelessness to a smoking heap of anguish and recrimination. Within a few weeks, she was beside herself with anxiety and guilt, sleepless and without appetite, unable to conceive of a way forward that didn't destroy either her family or herself, flirting with thoughts of suicide.

Sheila had been on antidepressants before—when, a few years before her affair, she had found herself interested in another man, this one a local friend. The drugs had been helpful, she told me. While on them, she had refrained from acting on her attraction and her discontent with her husband dissipated. She was considering taking them again, but she wondered if that was a good idea. Perhaps, she told me, her time on the drugs had only postponed the inevitable. Her husband was the kind of man who inspected the home every day when he returned from work and presented her with a checklist of the flaws in her homemaking, a right he claimed as part of his role as head of a Christian household. While taking antidepressants, she had found it easier to go along with this regime, but since she'd stopped—just before she'd struck up her affair—she had come to see her unhappiness differently.

Her compliance wasn't strictly a pharmacological effect, she said. "When you think you're sick—you know, *really* sick," she told me, "you don't have to pay attention to those feelings. They're not real, they're just part of the sickness, and that was a good thing then. It calmed me down and it gave me a reason to stay. There's a part of me that wants to go back there, but there's a part of me that's afraid that that's exactly what will happen. I'll end up back there, where I can be the good little Christian wife."

I know Sheila's psychiatrist. He's a secular Jew who has no interest in promoting fundamentalist Christianity or preserving a bad marriage for the sake of the institution or otherwise duping a patient into thinking that her unhappiness is the symptom of an illness rather than the indication that something is wrong with her life. He just wants to reduce his patients' suffering, to prevent suicide, to treat their illness. He'd be the first person to say that her circumstances bear exploration, and he would no doubt urge a patient to use her drug-induced relief to do just that. I am also sure he didn't give Sheila the depression-is-a-disease-like-diabetes talk. He didn't have to. To receive the prescription from a doctor and then to take the drug every day and then feel the difference—that's enough to give a person pause, to make her wonder if everything she thought about herself and the world she occupied was wrong, just another symptom of the illness that the drug is evidently treating.

That, at any rate, is what happened to me during my clinical trial. Even before the doctors started to tell me I was getting better, the idea that depression was a malfunction in my brain was working on me like a worm. When I woke up in the middle of the night in a cold sweat or craved my afternoon nap, when I found myself frustrated by my shortcomings or deflated by the seeming impossibility of getting done what I wanted to get done, when I felt sorry for myself and for all of us for having to live in such a broken world, I wondered if indeed I'd been suffering from an illness all along. I began to notice how *physical* the experience was, how it made my legs feel shaky and my chest feel tight, as if a snake were wrapping itself around me and choking the life out of me. I also began to see that the items on the tests, stupid as they were, corresponded to my experience; I may not have accepted that these were symptoms, but I couldn't deny that those sensations, the sleepiness and the lack of appetite and the sadness and the guilt and the indecision, all occurred together. I began, in short, to wonder if depression had colonized my consciousness in exactly the way that Peter Kramer described, fooling me into thinking that it wasn't a disease. I wondered if all those unacknowledged (and untreated) legislators of mankind had manufactured me as the kind of person who would mistake neurochemical noise for existential signal, if I should let myself be remanufactured—not by the drugs, but by the idea behind the drugs, by the label. And I wondered if I shouldn't just bag the fish oil and go to my doctor and get the real thing.

I hated to admit that I was thinking this way. I didn't want the depression doctors to be right. I didn't want to be mentally ill, but even more I didn't want their authority over who we are, over what it means to be human, to be legitimate. That's the problem with science. When it's right, it's right. It's a big comedown to discover, for instance, that the earth is not at the center of the universe. Or, for that matter, to see a CT scan of your abdomen and realize that your inflamed appendix doesn't really care what you were planning to do today. This is probably another reason why some patients run for the hills when I suggest that we collaborate on their diagnosis. They know that science can't be a democracy.

And this is where my resistance to the idea of my own mental illness gains some purchase and my native orneriness finds some comfort. Because it turns out that when it comes to the depression and the other diagnoses into which psychiatrists carve up the landscape of psychic suffering, science (or perhaps I should say "science") is much more of a democracy than you might think. Underneath its placid surface, its dispassionate description of diseases and their symptoms, is a history of heated battles—"fevered polemical discussions," in the words of one of DSM's main authors—that nearly tore the profession apart and were settled in the most humble of democratic ways: at the ballot box.

I may have given you the impression that psychiatry's march toward scientific respectability in the 1950s and 1960s was sure and steady. But even as the drugs were flooding the market and the statisticians were figuring out how to please the regulators and the doctors were claiming that they were hot on the trail of the twisted molecules that caused depression, most psychiatrists, glad as they were to prescribe the drugs their neuroscience-minded colleagues were inventing, were still tending to people's psyches, treating conditions like depressive neurosis and helping their patients understand how their personal histories, their past traumas had shaped their current suffering—or, as psychiatrist Karl Menninger once put it, "what [was] behind the symptom."

Karl Menninger and his brother William, a psychiatrist who had served as a general in World War II, were at the forefront of a postwar psychiatry that was confronted with an unavoidable question. Soldiers who had gone off to war as healthy, well-adjusted people had often returned in a devastated condition. "We must attempt to explain how the observed maladjustment came about and what the meaning of this sudden eccentricity or desperate or aggressive outburst is," Karl Menninger wrote. The answer, said his brother, was not to be found in the internal disease process implied by a classification scheme like Kraepelin's, but rather in "the force of factors in the environment which supported or disrupted the individual." This view dovetailed with Adolf Meyer's contention that psychological problems were reactions to life stresses and with the Freudian notion that the psyche was a dynamic organ whose balance could be upset by trauma. This synthesis was enshrined in the first DSM, which came out in 1952, and in which nearly all mental illnesses were named as reactions: schizophrenic reaction, depressive reaction, anxiety reaction, and so on.

The biopsychosocial model, as this something-for-everyone approach came to be known, seems as obvious and unobjectionable as any other ecumenical view. It's even lucrative; as we have seen, if mental illness is the result of collisions between the complex and demanding modern world and the psyche, then virtually anyone could be insane. But it required taking an approach to the question of what constituted mental illness that was hard to reconcile with scientific medicine and its carving up of the world into disease entities. It created a natural tension between therapy-oriented psychiatrists like the Menningers and biological psychiatrists like Roland Kuhn.

"Instead of putting so much emphasis on different kinds . . . of illness," Karl Menninger wrote in 1963, "we propose to think of all forms of mental illness as being essentially the same in quality and differing quantitatively." There was no discrete boundary between mental illness and mental health, let alone among mental illnesses. To say that the whole world was insane was only to say that we were all vulnerable to psychic suffering, each of us in different ways and measures, and that what was important was not what disease we had but what troubles our lives had visited upon us and how we had reacted.

The clinician's focus, according to this model, should not be only on the individual patient. The Group for the Advancement of Psychiatry, an influential organization founded in 1946 by William Menninger, among others, sought to practice prevention psychiatry, in which the social conditions that could make people mentally ill were as much a focus as the individual patient. GAP advocated for an education that taught psychiatrists to take "an objective critical attitude on a broad perspective of Man in transaction with his universe." Psychiatrists could then weigh in on the psychological dimensions of the important political arguments of the day, give a scientific answer to the question of whether one policy or another was better or worse for our mental health.

Unless, that is, they didn't even know what mental health and illness were.

Even before the Rosenhan study about being sane in insane places was published in 1973, psychiatrists and, increasingly, the public had begun to wonder about that. The troubles began in 1949, when psychologist Philip Ash showed that psychiatrists presented with exactly the same information about the same patient agreed on a diagnosis only about 20 percent of the time. By 1962, published studies indicated that at their best psychiatrists reached the same diagnosis in a dismal 42 percent of cases.

Some defenders of the industry pointed out that the studies used transcripts or summaries of cases, and that the problem might be in the relative distance of the doctors from the subject. So in 1964, a team led by Martin Katz, a National Institutes of Health psychiatrist, showed films of diagnostic interviews to panels of psychiatrists and asked them to make a diagnosis. Even then, they couldn't agree on the broadest of distinctions—whether or not a given patient was psychotic or neurotic—let alone on what variety of mental illness he or she was suffering with. No matter how disturbed the patient seemed, no matter who was conducting the interview, no matter how experienced or august the psychiatrists on the panels, the doctors came up evenly divided.

Katz even tried manipulating how much information he gave the doctors. He had a film of a "clearly depressed" woman in her fifties who, because she had had a manic episode, seemed to have an obvious case of manic-depressive psychosis. He showed one group of psychiatrists a version of the movie in which her history of mania was edited out. Those doctors came up evenly divided about whether she was psychotic or neurotically or involutionally depressed. When he showed the whole movie to the other group, despite what he thought would be a slam-dunk for a manic-depression diagnosis, the vote was again split; even when they were telegraphed the desired answer, only half the psychiatrists got it right.

The bad news was compounded when Katz showed the interview to a group of British doctors. Only one thought the patient was psychotic, seven that she was neurotic, nineteen went with personality disorder, and four split their tickets, voting for *mixed diagnosis*. This study, along with others that specifically investigated schizophrenia and manic-depressive illness, helped to explain a mystery brewing since the late 1950s, when epidemiological studies showed that manic depression was much more common in Great Britain than schizophrenia, while the reverse was true in the United States. It turned out that the diagnostic problem wasn't a result of, say, the differing genetic stocks of the two countries or their different approaches to childrearing. It wasn't in the patients at all, but in the doctors. Something in their educations, their training, perhaps even their countries' differing cultures made transatlantic psychiatry a profession divided by a common language.

The difficulty that doctors had agreeing on the meaning of the words they were using not only worried them; it made them feel inadequate. "There is a terrible sense of shame among psychiatrists," one of them wrote, "always wanting to show that our diagnoses are as good as the scientific ones used in real medicine." They might have kept their shame in the closet, however, had it not been for some of their colleagues who were tired of remaining in a closet of their own. In 1970, gay psychiatrists began to object loudly to the fact that their sexual orientation was considered a mental illness. Emboldened by the activism (and by the sexual revolution) of the sixties, they disrupted APA meetings around the country, picketing and protesting, sitting in to demand that homosexuality be deleted from the DSM, then in its second edition. At one annual meeting, a masked man appeared at a panel discussion and announced that he was both gay and a psychiatrist and that there were at least two hundred others like him at the convention. Donnybrooks like this could not help but attract widespread media attention.

After bruising and embarrassingly public bureaucratic battles, the protestors got what they wanted. In April 1973, an APA committee recommended deleting homosexuality from the DSM, then in its second edition. As part of the decision, the APA nominated a new disease, ego-dystonic homosexuality, which occurs when a gay person is distressed by his or her sexual orientation. Later that year, the APA's board of trustees voted in these changes to the DSM-II. In 1974, after a rearguard action had forced a referendum, a majority (58 percent) of the voting membership ratified the decision. This may have been the first time in history that a disease was eradicated at the ballot box.

Some psychiatrists, mostly those who opposed deletion, pointed out the obvious problem raised by this solution. "Referenda on matters of science makes [sic] no sense," said one dissenter. "If groups of people march and raise enough hell," another observed, "they can change anything in time . . . Will schizophrenia be next?" These doctors understood that they were dealing with a much different problem from the one uncovered by Katz. It was one thing to say that psychiatrists couldn't agree on which illness a given patient had, that, in the parlance of experimental science, diagnosis was not *reliable*. But it was another matter entirely to say that even when psychiatrists achieved reliability, the diagnosis they rendered was not *valid* because the condition in question was not a disease. The lack of reliability may have been shameful, but it was correctable; all that was needed, most thought, was a tightening of standards, better education, more research. But the validity problem was a downright disaster. What kind of doctor doesn't know the difference between sickness and health?

The answer, it seemed, was psychiatrists. And as humiliating as they might have found their lack of reliability, the possibility that the best they could do was to reliably diagnose illnesses that didn't really exist was even worse.

Some psychiatrists had noticed the validity problem long before gay people started raising hell—notably Britain's R. D. Laing and Thomas Szasz, an American. Laing focused on schizophrenia, which he argued was the result of people finding themselves in a social environment that didn't make sense to them; the insane place in which they were sane was not an asylum but a world filled with nuclear weapons, economic exploitation, ecological degradation. Szasz, less explicitly political, had a different idea, one that was less fanciful but that struck closer to the heart of the validity question. Psychiatric problems are not medical problems at all, he argued in *The Myth of Mental Illness*, but "problems of living." This didn't mean that people shouldn't seek therapy or that therapists did not provide a valuable service. But that service was not, properly speaking, a medical one, the illnesses psychiatrists claimed to treat were not valid, and most of their patients were not, strictly speaking, sick.

Psychiatrists may have hoped that deleting homosexuality from the DSM would strengthen their validity case—the fact that they had read an impostor out of the kingdom overshadowing the fact that he'd slipped past the gatekeepers in the first place—and that better training would solve the reliability problem, but their important patrons saw it differently. A 1978 presidential commission with influence over federal funding decisions warned that "documenting the total number of people who have mental health problems . . . is difficult not only because opinions vary on how mental health and mental illness should be defined, but also because the available data are often inadequate or misleading." In 1975, a Blue Cross executive told *Psychiatric News* that his industry was reducing mental health treatment benefits because "compared to other types of services, there is less clarity and uniformity of terminology concerning mental diagnoses," and added that because "only the therapist and the patient have direct knowledge of what services were provided and why," the insurers couldn't be sure they were even paying for the treatment of an illness.

To make matters even worse for psychiatrists, all kinds of non-medical professionals—social workers, psychologists, counselors, even nurses—were claiming (and getting) the right to deliver psychotherapy services. This was an indication that Freud had been correct about lay analysis, and that whatever else therapy was, it wasn't strictly speaking medicine. The diagnosis was obvious, the prognosis grave: as the president of the APA put it in 1976, the biopsychosocial model, "carrying psychiatrists on a mission to change the world, had brought the profession to the edge of extinction."

The second edition of the DSM appeared in 1968. The spiral-bound 132-page manual tried to be user-friendly. It offered a handy listing, in numerical order, of all 158 official mental illnesses, a set of sample tables for clinicians who might wish to keep track of their diagnostic habits, and even a postage-paid card on which users could send "criticisms and recommendations" back to the APA. The book also included, as its last chapter, "A Guide to the New Nomenclature," which explained, among other things, how this edition differed from the earlier one. Of particular note was the elimination of the word *reaction* from the diagnostic labels. *Schizophrenic reaction* had become *schizophrenia*, *manic-depressive reaction* was now *manic-depressive illness*, *depressive reaction* had been rechristened *depressive neurosis*, and so on. This explanation came with a reassurance. "Some individuals may interpret this change as a return to a Kraepelinian way of thinking, which views mental disorders as fixed disease entities," the authors wrote. "Actually, this was not the intent of the APA Committee on Nomenclature and Statistics."

It's possible that Robert Spitzer, lead author of the chapter, really meant that Kraepelin was the farthest thing from his mind and only subsequently came to see the old German's wisdom. Or that he was splitting hairs by writing about the committee's intent while remaining silent on his own; he, after all, was only a consultant to the committee. Or that his denial was, as any psychoanalyst would suspect, an unconscious affirmation of his wishes, his protestations of peaceful intent really a warning of impending hostility. But this much is certain: when the smoke cleared twelve years later to reveal Spitzer's magnum opus, the DSM-III, with its more than 225 diagnoses, its symptom lists, and its differential diagnoses (and no comment card), psychiatry had indeed returned to a Kraepelinian way of thinking. It had also, not coincidentally, been plucked from the precipice and restored to respectability—although some argued, and continue to argue, that Spitzer had destroyed the profession in order to save it.

Spitzer's affinity for Kraepelin might well have been personal. They shared a distrust of inner life, saw it as too raw and unruly to be of much use to doctors. The reasons for their antipathy differed, however. While Kraepelin thought the patients were untrustworthy, for Spitzer the doctor was the culprit. "I was uncomfortable with not knowing what to do with their messiness," he once said. "I don't think I was uncomfortable listening and empathizing—I just didn't know what the hell to do."

Spitzer did know what to do about diagnostic messiness. The problem was obvious. A manual that defined, say, depressive neurosis, as "an excessive reaction of depression due to an internal conflict or to an identifiable event such as the loss of a love object" was bound to get its users into trouble. What's a neurosis? What does *depression* mean? Isn't that the term the diagnosis is supposed to define? How much of it is excessive and how should that be measured? And what is the guarantee that one clinician's internal conflict or identifiable event will match another's, or that the same patient will provide the same story to two different therapists? Too much depended on the rendering of inner life into language, the ineffability of the one compounded by the approximations of the other, and both entirely dependent on a prior theoretical understanding of how the mind worked.

The solution was also obvious. As Kraepelin had discovered, there's no need to go on a Nantucket sleighride while the patient—who is, after all, sick in the head—sounds his woes. A doctor is much better off with pure description of what he sees and hears, which is presumably what any other person with trained eyes and ears will see and hear. If you want reliability, in other words, you have to stick with observation; a mental illness is no more or less than the group of symptoms that a careful observer has noted—perhaps by sorting index cards—to occur together.

That's not the kind of psychiatry Spitzer had learned while training at the Columbia Psychoanalytic Institute. If the Kraepelinian approach had come up at all, it would have been as a cautionary case, an example of how *not* to eliminate messiness in service of an account of suffering that, while tidy, does not get at the real nature of the problem. Psychiatry had, or so its practitioners thought, long ago left Kraepelin's therapeutic nihilism and dry categorizing behind in favor of the Freudian/Meyerian synthesis, its promise that with a little hard work and introspection, and a great deal of money, suffering could be treated at its source in the mind.

Which is why Spitzer, in perhaps his first act of what he would come to call "nosological diplomacy," had to issue his denials in the first place. He knew that *reaction* was central to the way that psychiatrists viewed and assessed mental illness—as the various, highly individualized ways that mental disorder manifests itself when a dynamic psyche is exposed to a traumatic environment. To eliminate that word was to eliminate an entire view of suffering—and, by extension, of human nature. But, Spitzer protested, nothing could have been farther from the committee's mind. To the contrary, it had only purged *reaction* as part of an overall attempt "to avoid terms which carry with them *implications* regarding either the nature of a disorder or its causes" (emphasis in original). The problem with *reaction* was that it implied both nature and causes: that mental disorder was the result of "Man in transaction with his universe," as Karl Menninger had put it, that there was a healthy way of conducting that transaction, and that psychiatrists knew what it was. These notions didn't pass scientific muster. If psychiatry wanted to remain in the scientific camp, *reaction* and all its attendant metaphysics would have to go.

Editing a single word out of the DSM proved much easier than eradicating the idea that our discontents are the result of the interaction between psyche and world, and the committee's reassurances that they meant no harm did not buy off professional resistance forever. But by the time the battle was joined, Spitzer had a huge advantage. Up until the mid-1970s diagnostics was a sleepy backwater of psychiatry, a subject of interest to most doctors only as the key to the insurance treasury, its professional discussions relegated (according to Spitzer) to poorly attended late afternoon sessions at professional meetings. While his colleagues were getting a head start on cocktail hour, however, Spitzer and a small group of researchers were busy at work, creating a reliable nosology. And when it came time to write a third edition of the DSM, in the late 1970s, these doctors were ready with their diagnostic criteria—lists of observable symptoms, cleansed, presumably, of any ontological implications, that would define mental illnesses in a way that was reliable.

Rank-and-file psychiatrists did eventually figure out that a big change was afoot, and much of the ensuing consternation focused on another word slated for elimination—*neurosis,* which, the new Committee on Nomenclature argued, should not appear in DSM-III because it "assumed . . . an underlying process of intrapsychic conflict resulting in symptom formation." In other words, you needed a theory about mental illness, about what caused it and where it came from, to diagnose a neurosis, and theory—especially the Freudian theory in which neurosis played a central role—was exactly what had gotten the profession into its reliability troubles in the first place. It had to go.

Eradicating *neurosis* was not as easy as getting rid of *reaction* had been. Neurosis, and especially depressive neurosis, was the psychiatrist's stock in trade, the general label for the everyday discontents that Meyer had long ago said were a proper indication for outpatient psychotherapy. The proposal to erase the word crystallized opposition to the remaking of DSM-III. Some doctors took a historical approach, pointing out that neurosis had been first described not by Sigmund Freud, but by Scottish physician William Cullen in 1769, and thus had earned a place in medicine. But others got right to the point. "DSM-III gets rid of the castles of neurosis and replaces it with a diagnostic Levittown," one psychiatrist said; most psychiatrists knew where they would rather live. Another colleague dispensed with metaphors and appealed directly to doctors' self-interest. Without neuroses of various kinds, he wrote, "many patients who are not in prolonged therapy will be said to have no disorder." The whole world could not be insane, at least not reimbursably insane, if the net were cast so much more narrowly.

But that was exactly the committee's intent—to prune the taxonomic tree of its less reliable branches, of which neurosis, weighed down with the Freudian idea of a dynamic inner world, was perhaps the most rotten. So when a psychiatrist lamented that proposed changes would turn the DSM into "a straitjacket and a powerful weapon in the hands of people whose ideas are very clear . . . and the guns are pointed at us," he wasn't as ready for his own strait-jacket as he sounded. Indeed, Donald Klein, a pharmacologist and prominent defender of DSM-III, only confirmed those suspicions when he proclaimed that opponents of DSM-III "wish [neurosis] reinserted because they wish a covert affirmation of their psycho-genic hypotheses." Taking such malcontents seriously could only spell disaster for "scientists attempting to advance our field via clas-sification and reliable definition."

Spitzer did take seriously the "pro-neurosis forces"—in the infe-licitous term used by one of them—at least seriously enough to offer any number of compromises. He suggested allowing psychoana-lytic-minded doctors to insert an *N* after the diagnostic code, indi-cating that the clinician thought the problem had something to do with conflict in the psyche. He floated the idea of adding *neurotic* as a descriptor after certain labels, but only in parentheses. He promised to allow the pro-neurotics a large role in "Project Flower," which would produce a companion volume to DSM-III that would allow theoreticians to fill in the diagnostic picture beyond the criteria lists— and whose name, Spitzer said, was inspired by Mao's aphorism, "Let a thousand flowers bloom." He even invited psychoanalysts to add some of their Project Flower material to the DSM's introduction.

In due course, however, the *N* modifier disappeared, the intro-duction idea was dropped as "extremely embarrassing and extremely divisive," and Project Flower somehow failed to bloom. After their trip to the diagnosis wars, all that the pro-neurotics ended up with were lousy parentheses: anxiety disorder became *anxiety disorder* (or *anxiety neurosis*) and depressive neurosis became *dysthymic disorder* (or *neurotic depression*). And in April 1979, after five years of diplo-matic nosology, after the Talbott Plan and the Offenkrantz Com-plaint and Washington Challenge and the Modified Talbott Plan, after the APA's assembly elected to approve the DSM-III, the APA's board of trustees once again voted on the existence of diseases.

This time, the stroke of their pen didn't eliminate a single illness but rather a whole class of them, even as it created some fifty more that hadn't previously existed. But these were new and improved diseases, the kind that could be reliably diagnosed without recourse to theoretical notions about how the mind works.

The DSM-III was a huge hit. Purged of theory, of any pretense to saving the world, and of any claim to know how the mind worked or what caused mental illnesses, the book was invaluable to psychiatrists' attempt to secure their place in "real medicine." Thanks to the descriptive approach, there would no longer be any question about who was schizophrenic and who was manic-depressive, or, for that matter, who had major depressive disorder (MDD), as it was now called, and who was merely unhappy. Nine out of ten doctors using the criteria agreed on diagnoses, a spectacular improvement over the old days of theory-laden nosology.

The DSM criteria for MDD were straightforward: take one from column A ("dysphoric mood or loss of interest or pleasure in all or almost all usual activities and pastimes . . . sad, blue, hopeless, low, down in the dumps, irritable"), four from column B ("poor appetite or significant weight loss . . . or increased appetite or significant weight gain . . . insomnia or hypersomnia . . . psychomotor agitation or retardation . . . decrease in sexual drive . . . fatigue . . . feelings of worthlessness, self-reproach, or excessive or inappropriate guilt . . . diminished ability to think or concentrate . . . recurrent thoughts of death, suicidal ideation, wishes to be dead or suicide attempt"), and rule out the symptoms in column C ("mood-incongruent delusion . . . [or] bizarre behavior," which are indications of other disorders), and you've got your diagnosis. A similar process could lead to dysthymic disorder or adjustment disorder with depressed mood. "Clerks rather than experts can make this kind of classification," one psychiatrist grumbled. But of course that was exactly the point.

These criteria weren't original to the DSM. In fact, Spitzer and his committee had lifted them, sometimes word for word, from the Feighner criteria, invented by a group of researchers at Washington University in St. Louis who, in 1972, had developed descriptive diagnostic standards for depression (and fourteen other psychiatric disorders). The Washington team was the first to achieve those excellent reliability numbers, and other researchers were soon scrambling to hitch their wagons to the Feighner star, using the criteria to back their own studies or as a model for their own tests. By 1989, the paper introducing the criteria had become the single most commonly cited article in the psychiatric literature.

What the Feighner criteria didn't address was the old Kraepelinian problem, the one about the symptoms constituting the diseases and the diseases comprising the symptoms. All the reliability in the world does not add up to validity. That's an especially glaring omission when you consider that the paper came out at the height of the battle over homosexuality, a condition whose presence doctors could reliably agree upon even without fancy criteria. Not to mention that in the particular case of MDD, the Feighner criteria bore a strong resemblance to the items on the Hamilton Depression Rating Scale, whose own author had long cautioned that his test was not valid for making diagnoses.

The authors tried to gloss over the issue by conflating reliability and validity. "This communication will present a diagnostic classification validated primarily by follow-up and family studies," they wrote, as if those follow-up studies could do more than show that people could be reliably grouped by their symptoms—something Kraepelin had already shown, but that didn't prove anything about whether or not those common qualities constituted a disease.

The team also wrote that their "criteria for establishing diagnostic validity in psychiatric illness have been described elsewhere," as if others had settled the question. But thirty-five years later, scholars Allan Horwitz and Jerome Wakefield examined those other studies and found that their authors never claimed that they had built a bridge from symptom to disease. One researcher warned that he couldn't guarantee that his "depressed" patients' symptoms weren't the result of some other illness, another concluded his paper by noting that defining "clinical entities by symptom pictures" remained a "serious problem in psychiatry," and the final source pointed out that in the absence of some theory about what causes depression, it was impossible to sort out the unhappy from the sick. Taken together, Horwitz and Wakefield concluded, "these sources neither justify nor even address the validity of the specific definition" offered by the Feighner criteria. Indeed, the only reason to believe that descriptive psychiatry had solved the validity problem was wishing it was so—which the industry had plenty of incentive to do.

Adopting the Feighner criteria, the DSM-III committee also adopted this wishful thinking. But reality soon intruded. A psychiatrist had discovered that many people who had recently been bereaved met all the Feighner criteria. For all their atheoretical purity, those standards, now incorporated into the DSM, couldn't even distinguish between the diseased and the merely bereaved. And this psychiatrist, Paula Clayton, wasn't just some pro-neurotic dead-ender. She was on the faculty of Washington University and a member of the DSM-III Task Force on Affective Disorders.

So when the DSM-III committee were reminded that, according to Clayton, grief was indistinguishable from depression, when, in other words, the validity problem emerged from the avalanche of reliability statistics under which it had been buried, neither she nor the committee should have been terribly surprised. Neither could they simply ignore it, even if they wanted to. A diagnostic manual that turned a person in the throes of grief into a mental patient was a scientific nightmare and a potential public relations disaster. It threatened Spitzer's strategy of rescuing psychiatry through a return to Kraepelin, to make all the professional blood spilled in the name of reliability a vain sacrifice.

The committee's response was to solve the public relations problem, if not the scientific one, by establishing a loophole in the definition of MDD—the bereavement exclusion. "A full depressive syndrome," the DSM-III eventually said, "is a normal reaction to the death of a loved one," so a recently bereaved person does not have major depression, even if he is depressed. Instead, he is suffering from uncomplicated bereavement, which doctors could still treat if they liked, but, because it was listed in the section of the DSM-III devoted to "conditions not attributable to a mental disorder," they were unlikely to get reimbursed for. There was good news on this front, however. The exemption was time limited. After two months, uncomplicated bereavement could become major depressive disorder.

It's not clear why the exclusion expires after two months. There is some statistical evidence that grief begins to wane, on average, about ten weeks or so after a loss, but while this says something about the usual course of mourning, it hardly proves that an unusual course is a disease. Even more important, it's not clear why bereavement is the only exempt condition, why, for instance, misfortunes like betrayal by a lover or severe financial loss or political upheaval or serious illness—or for that matter a noncatastrophe, the slow accretion of life's difficulties or a loss of faith in one's government or simple existential despair kindled by an awareness of mortality—do not also spare people from the rolls of the diseased. If the whole point of the DSM-III was to eliminate considerations of the nature and causes of a condition in favor of pure description, if indeed depression was no more or less than its symptoms, then why was it suddenly, and only in this one case, acceptable to talk about nature and causes?

The scientific answer is that there is no reason. The bereavement exclusion is like the epicycles that Ptolemaic astronomers added to their models of planetary motion—little loops within the orbit of planets that allegedly explained why they showed up in places where Ptolemaic astronomy, with its insistence that heavenly bodies moved in perfect circles, said they shouldn't be. Epicycles worked on paper, sort of, but they did a much better job at keeping astronomers respectable and their models intact than at describing the actual movements of heavenly bodies; they have come to be known as the epitome of bad science.

Doctors couldn't ignore Clayton's findings any more than ancient astronomers could ignore the actual orbits of the planets, but if they had responded by trying to figure out which other setbacks kindled responses indistinguishable from depression, if they generated a list of exemptions, then their diagnoses would have gotten awfully unwieldy and unscientific sounding. Not only that, but people might well have started marching and raising hell about why their particular cause for grief should (or should not) be on the no-diagnosis list. If, on the other hand, the committee had left the winnowing of diseased sadness from healthy sorrow to the judgment of doctors, then the profession would have been back to the bad old days when the circumstances of an individual's life mattered, when one doctor's "normal reaction" could be another doctor's disease. The bereavement exclusion—a single, time-limited exception based in common sense, as if it was just an unfortunate coincidence that bereavement mimicked depression—steered a middle course between these hazards.

Even as a one-time deal, however, the bereavement exclusion doesn't really work to keep theory, opinion, and judgment out of the clinical picture. It doesn't, for instance, specify the degree of relationship a patient must have to the deceased in order to qualify. Should a person really be allowed as many days to get over the death of, say, Michael Jackson as that of his own mother? Wouldn't the diagnosis still require a clinician to make a judgment about just how important those people ought to be to the patient?

The answer, of course, is that the therapist is supposed to do no such thing. If the symptoms are present, and if they are causing trouble for the patient, then it shouldn't matter if the loved one is a close relative, a celebrity, or even a fictional character whose television show has ended its run. In real life, however, it is hard to imagine that a clinician is going to count the days or consult the family tree as he decides who among the mourning is sick and who is only walking wounded. I haven't polled my colleagues on this, but I think they do take actual circumstances into consideration when trying to understand and diagnose someone else's misery. It seems nearly inhuman not to.

.I confess that I do.

Not that that makes things any less confusing. Consider my patient Eliza. A couple of years ago, a sheriff called to tell her that her mother had been found in her apartment, dead from a drug overdose. That wasn't the only momentous news Eliza received that day. Her mother had once, in a druggy fog, mentioned that it was possible that the man whom Eliza had always known as her father, the parent who remained when her mother was on her benders, who had terrified her and cared for her in equal measure, was very likely not her biological father. Eliza's mother mentioned the other man's name. Eliza tracked him down on the Internet, and he agreed to a DNA test, which, an e-mail on the day of her mother's death informed her, was positive.

Eliza showed me a photo of her biological father, taken when she went to meet him. His daughters—four of them now, not three as they had once thought—descended a staircase in front of his large and lovely home. He stood on the landing, above this quartet of pretty blond women, strong and confident and protective. He told Eliza that day that he was terribly sorry that her mother had never informed him of Eliza's existence, that he would of course have taken care of her, had he only known.

When she talks about her week with her new father, when she recalls her terror of her old father, when she describes her mother's chaotic life and squalid death, and above all when she remembers the day that her entire history was rewritten, Eliza is often overwhelmed with grief and regret—and with a self-pity that she finds as contemptible as it is inescapable. She has, in other words, at least the

requisite five of the current symptoms (the DSM-IV has rearranged
the DSM-III's diagnostic menu a little), they cause her distress, and
they have persisted well beyond two months from her bereavement.
Nine out of ten doctors (or clerks) would agree that Eliza qualifies.

But is Eliza's depression really best understood as an illness? Is
this a valid way to think of her suffering?

These are exactly the questions that the DSM-III and its succes-
sors are designed to avoid. Looking to the symptom itself, and not
to what lies behind it, means that it doesn't matter how Eliza came
by her depression any more than it matters if a person with lung
cancer smoked three packs a day or never touched a cigarette in his
life. There is no difference, except perhaps in degree, between my
major depression and Eliza's and that of someone whose suffering
seems entirely unrelated to circumstances. They're all the same disease, ready to be coded on the insurance form.

Here's another confession: that is exactly what I did with Eliza—
so that I could get paid for seeing her. I didn't anguish over this
decision, and I'll bet her psychiatrist didn't either when he billed
the insurance company for her visits (which in turn meant that the
costs would be covered). Certainly Eliza benefited from her diagnosis, but that's probably more of a comment about how we define
and fund health care than about her actual suffering. She could get
the benefit of therapy without a diagnosis, but it's not clear that her
psychiatrist and I could get our benefits that way.

Which is the whole point of turning psychic suffering into mental illness and diagnosis into a bureaucratic function in the first place:
to take these questions out of the therapists' hands and so to eliminate the possibility of the professional embarrassments wrought
by Rosenhan or Katz or gay people marching and demanding to be
struck from the sick rolls. Erasing reaction, deleting neurosis, over-
looking nature and cause, the DSM version of depression realizes
its major goal: enhancing the reputation of psychiatry, consolidat-
ing its power, turning it into real medicine. Inner life—personal and
political—remains important, if it is important at all, only as symp-

tom, only as the evidence that the diagnostic criteria are met, as the raw material for a disease the mental health industry has become expert at churning out.

This may be the most brilliant achievement of the DSM. By adopting and deploying a scientific rhetoric, it has not narrowed the patient pool at all. Instead, it has given increased authority to the pronouncements of people like me—so much so that state and federal governments have determined that insurers must pay for the treatment of depression in the same way they pay for any other illness—and at the same time have given us opportunity to apply the diagnostic criteria as broadly as possible, to turn everyday suffering into a disease. We may have given Blue Cross the "clarity and uniformity of terminology concerning mental diagnoses" that they wanted, but those diagnoses, and especially the diagnosis of depression, have been rendered by doctors for their own use. So everyone can be insane, only now insanity is a real illness, deserving of health care dollars.

This creates a perverse incentive to render diagnoses, which may have something to do with the ever-burgeoning statistics on the prevalence of depression. This in turn has led therapists to do what my Mass General doctors did with me—to tell more patients than they once would have that they are depressed, and to tell them why: because they have this symptom and that, and because those symptoms have lasted for more than two weeks. Which, of course, the patients are primed to hear because a friend has confided that they have been diagnosed in this fashion, or they saw a discussion of depression on *Oprah,* or because they overheard a conversation at the 7-Eleven. When people are more likely to think of themselves as depressed, to pay attention to certain feelings as symptoms, it's only a matter of time before those statistics start to rocket skyward and before a doctor says, in a prominent, peer-reviewed medical journal, that "depression in the Western world will affect half the population during their lifetime."

There are some other perversities that follow on the DSM revolution—notably, that insurance companies are increasingly insisting that treatment go a certain way, that if a patient is to have any therapy at all, it should be cognitive-behavioral therapy, which I'll tell you about in chapter 13. But the biggest problem with the DSM-III approach to mental illness is that while it hinges on the elimination of theory—which is to say on the elimination of metaphysics, of any kind of preconceived notion of where our suffering comes from or what it means to who we are—it has silently substituted its own metaphysics.

Because there is a theory behind the DSM's atheoretical approach. If your mental illness isn't a function of history or culture or geography, if it doesn't matter whether you got your five symptoms because you were abused and abandoned and then one day bereaved of everything that was familiar or because you show up one day at a clinical trial with a melancholy cultivated through fifty years of absorbing life's quotidian blows, if it's not a reaction or a neurosis, if there is nothing behind its symptoms and nothing of psychological or spiritual significance in them, if depression is not, in short, about your transactions with the universe, but only about whether or not you have the signs of the illness, then there is only one thing left for it to be: an internal dysfunction, as stupid and brutal and meaningless as diabetes or cancer. It's inside you—not in your thoughts and aspirations and dreams, in your fulfillments and frustrations, in your terror or despair or uncomprehending recognition of the irreducible gap between your little life and the infinity of time, not in a flaw in your soul, but in the vast and complex apparatus that gives rise to those thoughts and feelings and apprehensions, and that could just as easily, so it would seem, give rise to others. It's in your molecules. What matters, when it comes to depression, is matter. The rest is for the poets to worry about.

CHAPTER 12

MAD MEN ON DRUGS

Toward the end of my second visit to Mass General, just before I got my pills, George Papakostas asked me how long it had been since I had felt good for any appreciable time.

"Good?" I asked him.

"Symptom free," he said.

"For how long?" I asked.

"Thirty days," he said. "Or more."

I wanted to remind him that I was a writer, that I counted myself lucky to feel good from the beginning of a sentence to the period. I wanted to ask him if he had ever heard of betrayal, of disappointment, of mortality.

But after having spent nearly two hours cooperating with him, helping him to transmute my messy words into precise data, my inner world into bits as smooth and featureless as Chicken McNuggets, I somehow didn't feel free to remind him that we hadn't really agreed that I had symptoms. I'd submitted to his alchemy. I couldn't just turn myself back into lead.

"I'm sorry," I said. "But I have no idea what a month of feeling good would feel like."

I'm sure this only confirmed his diagnosis.

But "thirty days" was ringing in my ears as I left his office with my brown bag full of pills. And much as I wanted to dismiss the

253

very possibility of that symptom-free month, chalk up the idea to a laughably circumscribed view of humankind, much as I wanted to cite the research about depressive realism and to point to Aristotle and Lincoln and the James brothers and other important sad sacks as evidence against the neurochemical reductionism that lay behind this whole enterprise, I had to admit something: thirty days of unbroken contentment, of peace of mind, of resilience and, yes, even of optimism, a month of bright light unfiltered by a black veil—that sounds pretty good. If that was what I'd been missing, if that's what happens if you take the cure or have been lucky enough to elude the scourge in the first place, if health is happiness in month-long blocks, then suddenly the idea that unhappiness is a curable disease didn't seem like such a bad one.

I ducked inside a restaurant. I wasn't hungry, but I ordered a sandwich anyway. And a glass of water. I gulped down my six golden pills. I waited for my month to begin.

Was George Papakostas thinking of the placebo effect when he asked me that question? I don't think so. But maybe he should have been.

Placebos trouble doctors. There's too much magic in them and not enough science. They highlight a subject that most physicians would prefer to avoid: that they may not entirely deserve the power that they wield. The word itself, which is Latin for *I will please,* contains more than a hint of condescension, as if the doctor is merely tossing a pill at whiners, and as if the reason that the placebo effect persists is that people are too credulous (or perhaps too dumb) to get well by virtue of science alone.

So if it weren't for methodological necessity, researchers would probably forget about the placebo effect. And the game is set up so they can almost do that. They certainly don't really have to explore or explain or try to harness it, at least not in the laboratory.

In the clinic, however, the story is a little different. Practicing doctors use placebo effects all the time, on purpose. They generally don't admit this—at least not to patients; that might spoil the ruse. But when a team of their colleagues at the National Institutes of Health asked 1,200 American internists and rheumatologists if they prescribed placebos, more than half 'fessed up. Usually, they said, they didn't use sugar pills, but rather drugs that were pharmacologically active—over-the-counter analgesics and vitamins, and prescription antibiotics and sedatives—but had no reason to work for the patient's complaint. They told the NIH researchers that they made the prescription simply for the sake of providing a treatment, any treatment. To patients, however, they merely explained that the drug was "not typically used for your condition, but might benefit you."

Doctors could always furnish a sugar pill along with that explanation and thus avoid risks to patients (or, in the case of antibiotics, to the public health), but they are caught in a dilemma. Deception is necessary to the placebo effect (or so we think; no one has actually done that research) but unethical, and honesty is ethical but potentially ineffective. By prescribing a real drug, doctors steer between the rock and the hard place. When they tell their patients that a pharmacologically active drug might be of benefit, they are leaving out the part about how there is no particular reason why it should be, letting the implication of cause and effect hang in the air. They are also preserving their own plausible deniability—not to mention the dignity of all concerned. No one, it seems, wants to believe that something as unscientific (and as cheap) as hope could actually cure us.

The doctors in the NIH study said that they didn't tell their patients just exactly how the Tylenol or vitamin B might make them feel. They probably didn't have to—most of the patients treated this way are complaining of pain or fatigue or some other hard-to-specify condition, and they will know better than the doctor what kind of relief they are looking for. But it is possible to specify how a placebo will affect its taker. Especially when it comes to psychoactive drugs, you can fool many of the people a great deal of the time. You can, for instance, give people fake alcohol (without, of course, telling them that it only tastes like the real thing) and put them in a crowd of socializing drinkers and watch them get "drunk." You can give them fake morphine and hear them sigh in relief as their pain goes away. You can tell one group of subjects that caffeine will impede their coordination, another group that it will improve it, give both groups decaf, and observe as both groups behave accordingly.

Psychologists have had a great deal of fun investigating this phenomenon, and while they disagree about much of it, they have been able to arrive at the not-so-startling conclusion that when it comes to mind-altering drugs, expectation shapes response. People given fake alcohol already know how to be drunk, knowledge they have acquired from experience, perhaps, or from observing others, or from their doctors or teachers or after-school specials—in short, from what Norman Zinberg, a drug researcher of the 1960s and 1970s, called the "setting" of drug use. Setting, Zinberg argued, was only one of three interacting factors that determined the nature of drug experience. The other two were the biochemistry of the drug and the mindset of the user—his psychological makeup, his expectations and desires and motivations for taking the drug in the first place. The effects of drugs, Zinberg said, had to be understood in all three dimensions—drug, set, and setting.

Zinberg cited all kinds of evidence for the importance of this trinity. He found, for instance, that the effects of marijuana had changed over time. Research in the 1950s and 1960s showed that people had to smoke pot two or three times before they could get high on the drug—an effect thought to be the result of some kind of neurochemical process. But by the late 1970s, this was no longer the case. "As a result of accumulated knowledge about the effects of marihuana use," he wrote, "even first-time users are prepared to experience the high and therefore many have done so." Pot use had become so widespread that virtually no one, not even a pot virgin, could be naïve to its effects—and this change in setting changed the effect of the drug.

Even with more powerful drugs, Zinberg found, circumstances matter. In the mid-1960s, psychiatric hospitals around the country reported that fully one-third of their admissions were related to the use of LSD and other hallucinogens. A social scientist allied with Zinberg, Howard Becker, argued that while some people were undoubtedly vulnerable to psychotic reactions to the drugs, a much more important factor was the secondary anxiety caused by the unfamiliarity of the drugs' effects, compounded by their sensationalist presentation in mass media, which was in turn reinforced by the dire statistics. An LSD user, especially a novice, confronted with a disturbing hallucination or a raw apprehension of the sublime, might think of Art Linkletter's famously defenestrated daughter or the tripping students who were said to stare into the sun until they burned out their retinas or the people whose trip never ended. Encountering images like those while on acid can easily lead to the fulfillment of the hysterical prophecy. Becker forecast that as more people used the drugs and didn't meet tragic ends, as fewer trips ended up in the emergency room, and as this knowledge spread, the actual experience of taking them would change. And the numbers bore him out. By the end of the decade, the bad trips epidemic was over, even though use of the drugs was on the increase. Culture, and not just biochemistry, determined the effects of drug use.

In the course of his career, Zinberg wrote about groups whose drug use seemed to prove his theory. He wrote about people who used heroin for years and never got addicted, about rituals, religious and otherwise, that kept drug use from getting out of hand in some communities, about artists who used drugs to enhance their creativity but otherwise avoided them. He concluded that people could be taught to use mind-altering drugs safely and wisely, and that this was a worthwhile thing to learn. But, as he found out when one of Harvard's lawyers objected to his research on the grounds that a study proving that marijuana was *not* deadly would remove a deterrent to illicit drug use, his research was on a subject about which Americans weren't exactly rational. The climate of opinion that served as the setting for drug use wasn't going to change just because it didn't make any sense.

Zinberg, who died in 1989, never wrote about antidepressants. But there is an obvious connection between his theory and the magic-bullet ideas behind them. After all, if the effects of cannabis and LSD and heroin can change with circumstance, if people can get drunk on the suggestion that they are drinking alcohol, if pain goes away when people think they are taking morphine, then it doesn't make much sense to talk about psychiatric drugs as compounds that merely straighten out the twisted molecules that give rise to psychic suffering—at least not without giving due consideration to the expectations the doctor hands the patient along with the prescription.

It also doesn't make sense to think of the placebo effect as a nuisance, an unwanted artifact of credulity that interferes with the hard facts of neurochemistry, or as some vague and general tendency to feel better when a doctor provides a treatment. And it really doesn't make sense to pretend that what happens between doctor and patient doesn't matter, that when he asks you about your sleep and your appetite and your sex drive, or about that elusive thirty days, he is only assessing your symptoms, and that when he tells you you are getting better he is only reporting the facts. He's also loading the dice, helping his drug give you a particular experience by telling you what to look for.

There's no real scandal here. Doctors have always worked on expectation. And modern doctors have always had help from the drug companies in channeling their patients' hope into relief: industry-sponsored conferences where logo-emblazoned swag is handed out like party favors, multipage ads in medical journals, and lunches and golf games in which charming detailers tout the virtues of the latest remedy. Since 1997, when the FDA permitted direct-to-consumer advertising of prescription drugs, however, Pharma has been able to cut out the middleman. Patients, as you'll see in a moment, now come to doctors with their fires already stoked.

Neither is there anything new, or even disgraceful, in promising the moon to depressed people. Sometimes it doesn't do any good. Sometimes it just makes them feel more freakish or hopeless. But if you stay realistic and measured, you can often help a person buck up under the onslaught of his own self-reproach. What is new, however, and what has always been changing, is just what moon they can expect to arrive at and which drug promises to take them there.

The first drug touted as an antidepressant was amphetamine. An ad that ran in a 1945 issue of the *American Journal of Psychiatry* featured a photograph of a man in a business suit, hands on hips, smile on his face, eyes on the horizon as if he is glimpsing the good fortune that awaits him there. Looming behind him, barely distinguished from the background, is a close-up of his face in a different mood—brow furrowed, eyes downcast, mouth curling into a frown. "If the individual is depressed or anhedonic, you can change his attitude by physical means," the ad copy reads. Doctors, it continues, have known this for at least twenty years, but only in the last decade has the "agent of cure" been available: Benzedrine, "a therapeutic weapon capable of alleviating depression."

Smith Kline and French, maker of Benzedrine, wasn't suggesting that doctors give amphetamine to delusionally guilty endogenous depressives or to psychotic manic-depressives—people who would be hospitalized and for whom the cure of choice was still the shock therapies—but to outpatients with what were then thought of as reactive or neurotic depressions. And amphetamine often pulled such patients out of their funks. Its problems—chiefly that it was addictive and its effects unstable—were soon obvious, however, and it fell into disfavor (until it was resurrected in the 1960s as a cure for attention deficit disorder, an indication that was worth $1.5 billion in sales of various stimulants in 2008).

In 1955, Wallace Laboratories, an arm of the company that made Carter's Little Liver Pills, came up with an alternate treatment for neurotic depression: meprobamate, which the company named Miltown. Full-page ads in medical journals told doctors of the "outstanding effectiveness . . . with which Miltown relieves . . . anxious depression." Detailers detailed its virtues. And, perhaps most important, patients loved it. Within a few years of its introduction, people were "miltowning": turning on with a "Miltown cocktail"— a pill washed down with a Bloody Mary—and then tuning in to "Miltown" Berle in such large numbers that drugstores often had to hang out "No Miltown today" signs.

By 1965, Wallace had sold 14 billion of its little brain pills to 100 million satisfied customers. The only limit on Miltown's sales was another group of minor tranquilizers—the *benzodiazepines*, which included Valium and Librium, both invented in the early 1960s by Hoffman–La Roche. The industry pushed the minor tranquilizers hard—not only to psychiatrists, but also to general practitioners, the doctors most likely to see the neurotically depressed. In addition to eight-page ads in medical journals, Roche sent out forty different mailings to doctors, along with phonograph records on which doctors testified to the virtues of Librium. Librium's primary target was anxiety, but it also "could be safely administered in the presence of depression," the ads said. (And it was, one ad said, perfect for "when the patient rambles"—although it is not clear whether the doctor or the patient had to take it for this effect.) Valium eventually took up more medical journal advertising pages than any other pharmaceutical drug, and by 1972, it was the most commonly prescribed drug in the world—a position it occupied until the end of the decade. And it wasn't just Valium. Doctors— mostly family doctors—were writing 90 million minor tranquilizer prescriptions a year.

The minor tranquilizers' success wasn't all hype. In 1972, David Wheatley, one of the earliest antidepressant researchers in the United Kingdom, reported on a series of trials testing antidepressants against minor tranquilizers and concluded that the latter were better at treating neurotic depression—a finding echoed in studies that appeared in the *New England Journal of Medicine* and the *Journal of the American Medical Association.*

But the real boon to the drug industry was not so much the drugs themselves as the emergence of a vast new market: people whose suffering wasn't bad enough to warrant a visit to a psychiatrist's office but who would confess it to their family doctor and then gladly take Miltown or Valium. Miltown, according to medical historians Christopher Callahan and German Berrios, was the first "product of the pharmaceutical industry (rather than academia) [that] responded to consumer demand," and the success of the minor tranquilizers capitalized on this response. It's impossible to know how much patients' newfound willingness to talk about their discontents was due to their knowledge that it might be rewarded with a Miltown buzz, but industry executives didn't need to consider that. What they knew was that patients were now convinced that the whole world, including them, could be insane, that the insanity could be treated with a minor tranquilizer, and that family doctors, and not psychiatrists, held the keys to the Valium kingdom.

Take some Valium or Miltown (which is still available in a slightly modified formulation called Soma; one can only imagine how Aldous Huxley would feel about that) and, if you're like most people, you'll immediately see why they more or less sell themselves: they make you feel pretty darned good. Take some imipramine, on the other hand, and you most likely won't feel any immediate effects, except maybe some jitteriness or dry mouth. So it's no wonder that while Valium sales were soaring to the stratosphere, amitriptyline (Elavil), Merck's entry into the tricyclic antidepressant market, was down in the dumps—a mere 14 million prescriptions in 1972, only a 40 percent increase from 1964. Too many doctors, evidently, thought their patients were merely anxious and unhappy, rather than sick with pre-DSM-III depression, the kind that was still considered a psychosis. They were all too willing to prescribe "penicillin for the blues," as one doctor called Valium.

To a marketing executive, the problem was straightforward: doctors weren't making the connection between the problem and the solution because the problem had not yet been properly named. "It's hard to appreciate the difficulty of getting [across] . . . the message that we're all depressed," Dan Fellowes once told Emily Martin, a medical anthropologist. (Fellowes is the pseudonym Martin gave to the man who oversaw Merck's Elavil marketing efforts in the 1960s.) But that was his job: to bust antidepressants out of the psychiatric ghetto, he would have to convince family doctors that they didn't realize how many of their patients didn't just have the blues, but rather depression, a specific disease for which the specific treatment was not Valium but Elavil.

Fellowes was nothing if not ingenious. One of his first moves was to recruit Frank Ayd, the Baltimore psychiatrist who spearheaded the clinical trials for Elavil. In 1961, Fellowes got Ayd to write *Recognizing the Depressed Patient: With Essentials of Management and Treatment,* a book in which he explained to general practitioners the implications of the "chemical revolution in psychiatry" for their practice. Thanks to modern psychopharmacology, Ayd wrote, a referral of a mentally ill patient to a psychiatrist armed with the new drugs was no longer a grim sentence. But even more important, mental illness was increasingly a condition that nonpsychiatrists could treat—if only they recognized it.

Chief among these illnesses was depression. In fact, Ayd wrote, "depressions are among the most common illnesses encountered by the general practitioner." The problem is that the doctor often didn't know it, didn't realize that unexplained physical symptoms like constipation and hot flashes, hard-to-pin-down emotional symptoms like irritability and lack of confidence, and disturbing psychic symptoms like impaired memory and indecisiveness may well add up to depression. Ayd's symptom list wasn't as long as George Beard's was for neurasthenia, but the idea was the same: certain commonly seen, but not obviously related, complaints were the signs of a widespread, although largely unknown, disease.

After correcting this lack of knowledge for 114 pages, Ayd spent his last twenty pages reassuring doctors that they could treat depression. "Not all depressed people require the services of a specialist," he wrote. "Many melancholics can be cared for by the family doctor." Because "no advantage is gained by an intellectual understanding of the psychological aspects of the illness," a doctor did not need to know how to conduct psychotherapy. Instead, he could just order rest and relaxation, provide support to the family, hospitalize the patient (in a general, not a psychiatric hospital, except in the most severe cases) if necessary, and, of course, prescribe the right drugs—which, according to Ayd, did not include the minor tranquilizers. And when the depression "cleared," the patient had to be educated about the likelihood of relapse, and reassured that "treatment can be just as effective as it was for this attack and that the earlier the treatment is sought . . . the better." That's why he should come in for regular checkups and "at the slightest sign of relapse, treatment should be reinstituted."

But the doctor's first duty, Ayd emphasized, was "to explain to the patient the nature of his illness in understandable terms." This was also the tricky part. "Depressed people are very suggestible," he wrote, "and an inept comment can do irreparable harm." To prevent this, Ayd provided a script for the fledgling doctor to use in breaking the news, one that uses the patient's suggestibility for better ends:

> You have an illness called a depression. It is very common. Everyone who has it feels just as you do. What is happening is real. It does not mean that you have a serious physical dis-

ease or that you are losing your mind. Your symptoms have
a physical basis.

Doctors who give their version of this spiel today probably have
never heard of Frank Ayd. Nor have the millions of patients who
have listened to it. But his impact, at least to judge from the swollen
ranks of the depressed, has been immeasurable.

Neither have most doctors or patients heard of Dan Fellowes,
not even by his real name. But it was his idea to send *Recognizing
the Depressed Patient* to fifty thousand general practitioners. And, as
if playing Gideon with Ayd's bible of depression wasn't enough, he
also got into the popular music business. "I found a musicologist,
Leonard Feather, who compiled an album of blues songs," he said.
"It was the most beautiful expression of how life and the problems
of life create depression."

Symposium in Blues, which was released in 1966, had all the trap-
pings of a commercial entertainment—a major label (RCA), star
performers (Louis Armstrong, Ethel Waters, Leadbelly), songs like
"I Been Treated Wrong" and "I'm on My Last Go-Round," and, in
case the point wasn't clear, Feather's symptom-targeted liner notes
("songs that reflected anxiety and agitation, guilt, hunger or loss of
appetite . . ."). The album wasn't available in stores, however. Fellowes explained to Martin:

> We gave this to doctors. "Here's a phonograph record for
> you." They took it home . . . The music didn't say "Elavil" at
> all. And there was no intent to give them a lecture. The liner
> notes talked about the fact that the songs came out during
> the Depression . . . It was talking about the psyche of the
> American public.

The package insert for Elavil was slipped into the dust jacket of
Symposium in Blues, right next to the record sleeve, and when a doctor opened the album, he was greeted with the Elavil logo, along

with a list of the symptoms of depression. But the album was more than a way to put doctors in the right mood while they read *Recognizing the Depressed Patient.* It was also one of the earliest instances of viral marketing, the kind in which the advertiser lets loose a message in a dim corner of the culture, where it can circumvent whatever immunity we have to the pitchman's manipulations, attach itself to reputable sources of knowledge, and replicate.

What better host for this kind of virus than the doctor? What better culture than one that has not yet begun to suspect the motives of drug companies and doctors or to worry about their power? Patients surely didn't know about the record or the book, let alone the phalanx of detailers that eventually included one sales rep for every eight doctors in America. They never saw the medical journals, chock full of ads using the latest Madison Avenue techniques to urge doctors to dispense the latest drugs, or attended their industry-funded seminars. A doctor didn't mention to a patient, as he used his Merck pen to write a prescription for a Merck drug, that he'd learned about the disease at a conference at a luxury hotel in the tropics that was sponsored by Merck, or that the free sample he was handing to you had been given to him by the detailer who took him to lunch the other day. He himself may not have known that the journal article the detailer gave him touting the drug's virtues was written by a researcher getting paid by its manufacturer, who had been forbidden to publish his results if they didn't meet the company's approval.

Despite all these advantages, and despite the best efforts of men like Dan Fellowes, the minor tranquilizers were still beating the antidepressants by a mile. But even before Robert Spitzer and his committee reinvented depression, Elavil and Tofranil and all the rest caught a huge break. In 1968, not long after he arrived in office Richard Nixon made pharmacological Calvinism a matter of national policy. He declared war on drugs, or to be more accurate, he declared war on the users of some drugs—largely LSD and mari-juana, which evidently threatened America's moral fiber as surely as Communism had when Nixon was a senator and vice president.

As the drug war machine cranked into action, minor tranquiliz-ers got caught in the crossfire. Vice President Spiro Agnew warned publicly about the dangers of "mood drugs" in 1970, on which, he said, "over one-half million citizens are now dependent." A promi-nent doctor, in a comparison that probably left drug makers reaching for the Valium, said that benzodiazepines were not much different from "other products which also affect the mind but which we do not label medicine . . . for example, alcohol, tobacco, coffee, tea, and marijuana." The National Institute on Drug Abuse reported in 1976 that Valium was one of "the most frequently abused drugs in the United States." In 1979, Edward Kennedy, chairman of the Senate Subcommittee on Health and Scientific Research, and a man who knew a bit about excess, opened a hearing on the benzodiazepines by declaring that "these drugs have produced a nightmare of depen-dence and addiction . . . thousands of Americans are hooked and do not know it."

Whether people were really hooked on Valium remained a matter of debate, but that was never the real problem in the first place. Psychotropic hedonism was. *Addiction* was just a code word for "drugs that you should be afraid of because they make you feel so good." (Indeed, given the fact that Librium and Valium alone were what *Fortune* called "the greatest commercial success in the history of prescription drugs," it is a safe bet that at least some of the people excoriating them knew this firsthand.) It was not difficult for drug warriors to convince the Drug Enforcement Administra-tion to put minor tranquilizers on its list of controlled substances. Doctors prescribing them would now be under the jurisdiction of the DEA, with all the paperwork that entailed—and all the Dr. Feel-good implications. Patients would have to worry about the stigma of taking a drug that professional scolds were lumping in with mari-juana and alcohol. To judge from the falling sales figures—not as sudden or severe as the decline of the "psychic energizers" in the early 1960s had been, but noticeable by the early 1980s—patients and doctors alike were thinking twice about the benzodiazepines, their market dominance a victim of America's confusion about drug-induced pleasure.

The drug warriors' revival of pharmacological Calvinism hit its stride at about the same time that the American Psychiatric Association was resurrecting Kraepelin's categorical disease entities, which in turn coincided with the FDA's effort to implement Kefauver-Harris by approving only drugs that were effective with specific diseases. Antidepressants—drugs that didn't get you high like Valium did, that treated the newly fashioned major depressive disorder, and that allegedly did so by attacking depression at its biochemical source—had hit the trifecta.

Still, their sales continued to languish. In 1970, they had constituted only 2 percent of the prescriptions written for psychotropic drugs, and even when the new DSM came out and minor tranquilizers were falling into disfavor, they remained a minor player. One more element was needed, and, unknown to even the most savvy marketers, it was already on the way—and had been since 1968.

That was the year that Arvid Carlsson, the Swedish scientist who had discovered the role of dopamine in Parkinson's disease and then proved the principle of chemical neurotransmission to the world, began work in earnest on an idea he'd been hatching for a few years. While he believed that Joseph Schildkraut was correct in asserting that depression was a neurotransmitter problem, he thought that Schildkraut had made a mistake by singling out dopamine and norepinephrine as the culprits. Those may have been the chemicals most affected by the tricyclic antidepressants, but, Carlsson pointed out, all the members of that family also blocked serotonin reuptake. And a close look at the statistics showed that the strongly serotonergic drugs were the ones that improved mood—as opposed to the other items on the HAM-D—most dramatically. A drug that targeted serotonin, Carlsson thought, might be the key to a better antidepressant.

Carlsson's suggestion that psychopharmacology return to its pioneering discovery—and to the theory that serotonin was the key to mental health and illness—fell on deaf ears until he finally convinced the Swedish firm Astra Pharmaceuticals to finance his research. By 1971, he and his team had come up with zimelidine (Zelmid), the first drug designed specifically to inhibit serotonin reuptake. A year later, a team at Eli Lilly synthesized its own serotonin-specific drug: 3-(p-trifluoromethylphenoxy)-N-methyl-3-phenylpropylamine.

When the paper announcing Lilly's drug was published, the company was still calling it LY-110140; it hadn't bothered to name the compound yet because no one was quite sure of its commercial value. Serotonin, after all, is distributed throughout the body and plays a role in digestion, appetite, sleep, and blood pressure, among other functions, so 110140 might have had many uses. It might even have been used only for research purposes, to assay or manipulate or otherwise investigate serotonin.

Much as Abbott had done with serotonin itself in the early 1950s, the company solicited opinions from leading researchers about the new drug's possible uses. Arvid Carlsson, who was at one of the meetings, told me that when one scientist suggested that Lilly might try the drug out as an antidepressant, the company representative replied that this wasn't in the company's game plan.

But in the meantime, Astra, spurred on by Carlsson, was taking zimelidine through clinical trials in Sweden. In 1981, the company received approval to market Zelmid as an antidepressant throughout Europe. The drug did well enough that Merck licensed it from Astra and applied to the FDA for approval in the United States. But in 1983, just as the process was getting under way, trouble arose in Europe: zimelidine syndrome, a flu-like condition, and, more ominously, an outbreak of Guillain-Barré syndrome, a sometimes fatal neurological disorder, among users. Astra withdrew the drug in September 1983, and Merck abandoned its application with the FDA soon after. According to Carlsson, neither company thought the population of depressed people would ever be big enough to justify the expensive research necessary to investigate the link between the drugs and the illnesses.

But the brief success of Zelmid had caught Lilly's eye. By 1984, the company was finally interested enough in 110140's antidepressant qualities to give it a name—*fluoxetine*—and to conduct and publish research showing that the drug relieved depression as well as the tricyclics, and with fewer side effects. By the end of 1987, the FDA approved the compound, now known as Prozac, as an antidepressant. Zoloft, Pfizer's entry into the new market, was approved in 1992, and SmithKline Beecham (successor to Smith Kline and French) introduced Paxil a year later.

SmithKline also introduced the abbreviation *SSRI* to the marketplace, underlining the new drugs' major claim to superiority: that they were selective, targeted precisely at serotonin. Marketers had already jumped on this distinction. The first Prozac ad in *JAMA*, in 1988, touted it as "the first highly specific, highly potent blocker of serotonin uptake," and throughout the 1990s, the industry put selectivity at the heart of its campaign. Antidepressants, the journal ads proclaimed, were clean drugs, strong and effective drugs, high-tech, can't-fail magic bullets that destroyed depression without making patients high, without addicting them—and indeed, without causing any collateral damage.

With a come-on like this, it hardly mattered that none of this is exactly true. Not only do the drugs perform poorly in trials, but while they do bind to serotonin receptors at higher rates than they bind to other receptors, and at higher concentrations than the tricyclics do, they by no means bind *only* to serotonin sites. They are active all over the brain, so while they may not cause as many side effects as the tricyclics, they still cause so much discomfort that there is a cottage industry devoted to reducing nonadherence among SSRI takers. Patients, researchers have found, were reluctant to take psychiatric drugs in the first place, and when they start feeling jittery and agitated, or when they can't sleep and have upsetting dreams when they do, or when they get constipated or nauseated, or when they hear about the reports linking antidepressants to suicide and violence, and above all else, when they find that they suddenly can't reach orgasm or don't want sex at all, they often just stop. Indeed, nearly 70 percent of people stop taking antidepressants within the first month.

None of this—the lousy sexual performance, the lousy clinical performance, the lousy side-effect profile—is a secret anymore, if it ever was. A search of PubMed, the NIH database of published research, turns up 744 papers on the subject of sexual side effects of antidepressants alone. The data used by the FDA to approve the drugs, including the ones in which the drugs didn't work, are in the public domain. Even the head of the agency's division of neuro-pharmacological drugs wondered, when it came time to approve Celexa in 1998, if its clinical trial results showed any "clinical value," only to be told by a colleague that because "similar findings for . . . other recently approved antidepressants have been considered sufficient," they had no choice but to go ahead. The agency also knew that reports linking SSRIs to the increased risk of suicide and violent behavior had begun to surface within a year of Prozac's emergence on the market. Still, by 2006, antidepressants had become the most commonly prescribed class of drugs in the United States, at an annual cost of $13.5 billion.

This dramatic success depends on the old tricks—downplaying side effects and overstating efficacy in marketing campaigns directed at prescribers. But it also hinges, at least sometimes, on outright lies. Psychologist Glen Spielmans and his team analyzed a group of ads from leading psychiatric and general medical journals. They discovered that in more than one third of the cases, the sources cited in the ads failed to verify the claim they were supposed to support. And that's when the companies bothered to mention a source. Fully half the time, they didn't even do that— or they cited a source that couldn't be obtained. When Spielmans asked Wyeth for the data cited in an Effexor ad, the company responded, "Unfortunately, our internal policies do not allow for distribution of unpublished data." As Spielmans pointed out, this is ironic given the tag line of the ad: "See depression, see the data, see a difference."

Doctors haven't been keen on investigating how gullible their colleagues are, so it is hard to know just exactly how much of a role these lies play in doctors' prescribing habits. We do know that 80 percent of "high prescribers" rated journals as "an important source of information" and that two-thirds of doctors overall are exposed to drug ads on a weekly basis. And we know that an investment of one dollar in professional advertising yields a five-dollar return in sales. Doctors, in other words, seem to respond to ads directed at them by writing more prescriptions.

But even the shrewdest ad will not work if the market doesn't exist. Before a doctor writes a prescription, especially for a psychotropic drug, he has to believe that the patient is suffering from an illness that can be treated. Which is where the ad men, with a little bit of inadvertent help from the APA and the FDA and the drug czar, found their sweet spot. Because after the long haul through medical school, where doctors learned to think of diseases as targets and drugs as magic bullets, after practicing in a society that expects and even demands results in return for its deference to doctors, after reading in the *Journal of the American Medical Association* that depression is the second leading cause of disability in the world, but that only 10 percent of the 15 or 20 million citizens who will be depressed in any one year are getting treatment, and that this undertreatment is costing society about $43 billion per year (and that was in 1996, when a billion dollars was real money), after learning, in short, that depression is a public health emergency and that "safe, effective, and economical treatments are available," treatments that work as precisely as antibiotics or insulin, that are being talked about by all their colleagues, and taken by their friends and their patients and maybe even themselves—after all that, it's hard to understand why doctors *wouldn't* render the diagnosis and reach for the prescription pad. And it's easy to see why they were four times more likely in 2005 to prescribe drugs for the treatment of depression as they had been in 1987. Not because they were in cahoots with the drug companies, not for any other reason than that they wanted their patients to feel better and they really believed they finally had something to offer.

Even without blocking the uptake of a single molecule of serotonin, the drugs begin to work their magic. Doctors—primed by the ads, the detailers, the enthusiastic (if industry-manipulated) articles by their peers—were ready to give their patients Frank Ayd's pitch: that they had a disease, that it was no different from any other disease (except perhaps for how widespread it is), and that the cure was waiting for them at the pharmacy. They were ready, in other words, to change the setting in which antidepressant use takes place, to name their patients' pain and create expectation for its cure, to mobilize, whether or not they meant to, the placebo effect.

And it wasn't only the doctors. The drug industry also had help from writers like Peter Kramer, from a news media more than willing to report breathlessly on the new wonder drug (within a couple of years of its introduction, Prozac was featured on the covers of both *Newsweek* and *Time*), from an FDA hamstrung by its own lax standards, and, perhaps above all else, from millions of satisfied customers.

For four glorious years, Lilly had a monopoly on this emerging market, and the company made the most of it—the drug's $1.5 billion in sales in 1992 accounted for one-quarter of Lilly's revenue that year. SSRI competition finally emerged, first with Pfizer's

Zoloft, introduced in 1992, and then, the next year, with Paxil. Some of the competition was fierce. Pfizer, for instance, undercut Prozac's price by 20 percent and deployed its detailers on 660,000 annual Zoloft visits, with marching orders to harp on Prozac's side effects—especially anxiety, which Pfizer claimed was less pronounced in Zoloft patients. Ads for Zoloft in professional journals turned *anxiety* into a code word for suicide, reminding doctors of the lawsuits that Lilly had been fending off for years. The success of the whispering campaign quickly showed up on the bottom line: by 1995, Pfizer had captured nearly one-third of the overall antidepressant market, ringing up $900 million in sales. That same year, SmithKline raked in nearly a half-billion dollars from Paxil. Lilly was still on top of the heap at $1.47 billion (capturing nearly 41 percent of the 1995 market), but, as the *Wall Street Journal* reported in mid-1996, "the king of antidepressants [was] under attack."

The news wasn't really that bad, however. Even if Lilly was losing market share, its sales were still up by 18 percent—less than Zoloft's 38 percent increase and Paxil's 45, but still enough to account for nearly 30 percent of Lilly's revenues. Success was breeding success, the market expanding by 33 percent every year. And when the *Journal of the American Medical Association* reported in early 1997 that still only 10 percent of the depressed were getting treatment, it seemed that the rising tide was capable of floating as many boats as the drug companies could launch.

Supply was creating demand. It was enough to inspire exuberance in even the most rational executive. But a company selling a product whose patent expires in a few years doesn't settle for divvying up the market, at least not if it can avoid it—and especially not if the National Depressive and Manic-Depressive Association declares a treatment gap that will only be closed when "patients [become] informed consumers and advocates." Who better than the drug companies to inform them, and what better medium than advertising? Which is why in mid-1997, Eli Lilly hired the Leo Burnett Company, one of America's leading advertising agencies, to launch an ad campaign for Prozac that would skip right over *JAMA* and the *New England Journal of Medicine* and head for *Cosmopolitan* and *Reader's Digest*. After all, as Willie Sutton might have said, that's where the consumers were.

According to a Burnett vice president, "This is one of the most serious assignments we've ever had," and its mission was clear: to inform readers that, as Mike Grossman, Burnett's director of public relations, put it, depression "isn't just feeling down. It's a real illness with real causes." Lilly spent $22 million in the last six months of 1997—nearly two-thirds of its entire advertising budget for the year—"assisting people in their depressed stupor," as Grossman put it, "to raise their hand for help."

The first ad was a three-page spread: a drawing of a rain cloud over the caption "Depression Hurts," a sun shining on the slogan "Prozac Can Help" (both images were crude, as if drawn by a third-grader), and, on the last page, the fine print about side effects. It turned out that you didn't need to be in a depressed stupor at all, but merely under the weather, to have the "real illness," that "doctors believe" may be caused by "an imbalance of serotonin in your body." The copy under the cloud suggested, "You may have trouble sleeping. Feel unusually sad or irritable. Find it hard to concentrate. Lose your appetite. Lack energy. Or have trouble feeling pleasure." And when people feel this way, the reader discovered, just before moving to the sunny side, "the medicine doctors now prescribe most often is Prozac."

One year and $47 million after launching the print campaign, Lilly and Burnett took Prozac to television with three sixty-second spots, including "Checklist." Filmed in black and white, the ad is a montage of suffering and isolation—stills of a woman on a couch, another woman in bed, a man on a pier, another sitting in a restaurant, a woman at her desk, another showering, and finally just a sorrowful face, each person looking as empty and sad and hopeless as Lutherans in an Ingmar Bergman movie. Violins repeat a minor chord while a woman narrator, sounding not a little depressed herself, ties each image to a target symptom:

> Have you stopped doing the things you enjoy? Are you sleeping too much? Are you sleeping too little? Have you noticed a change in your appetite [at this point the man in the restaurant inexplicably vanishes, as if transfigured]? Is it hard to concentrate? Do you feel sad almost every day? Do you sometimes feel like life [pregnant pause] is not worth living? These can be signs of clinical depression, a real illness with real causes. But there is hope. You *can* get your life back. Treatment that's worked for millions is available from your doctor.

The commercial provides a toll-free number that you can call for a personal symptoms checklist that will make it "easier to talk with a doctor about how you're feeling," and ends with Lilly's logo and the "Welcome Back" tagline.

Other companies gave Lilly's campaign the same flattery they had given Prozac: they imitated it. Zoloft and Paxil soon had ads and websites of their own, and by 2000 companies were spending $128.5 million to advertise their antidepressants to consumers. Laboring under FDA guidelines, which, among other restrictions, required them to disclose side effects and not make claims unsupported by research, the campaigns were barely distinguishable from one another when it came to the information presented. But advertising is much more about presentation than information, and the most creative presentation of the same old facts had to be Zoloft's.

In 2001, Pfizer introduced a cartoon character, a Bizarro version of the smiley face—a round blob that bounces along, sighing and moaning and even ignoring a butterfly as a narrator (male, more chipper than Prozac's narrator) goes through the usual pitch. The blob ads even featured an animated version of a serotonin synapse; when the Zoloft logo appears at the top of the screen, the synapse becomes a much more lively place, and the narrator says, "When you know more about what's wrong, you can help make it right." He adds that only your doctor can know for sure if you have depression, but he suggests strongly that now that you're educated, you go and ask.

You have to admire the economy of ads like these. In a one-minute cartoon, they distill a century and a half of medical history into the simple message that if you are suffering, you may very well be sick, that your sickness is internal and biological, that it can be cured with a precision-targeted medicine, and above all else, that anyone can be depressed, that indeed the whole world can be insane. "Depression strikes one in eight," says the Prozac TV ad. The number is 17 million in the print version, 20 million in the Zoloft ad, and 20 percent of women on Paxil's website—a whole lot of fellow citizens, in other words, whose suffering was presumably no worse than yours, but who had finally come to understand that their unhappiness, no less than their infections and their high cholesterol, was just another disease and got on the Prozac bandwagon.

Some critics worried that it just wasn't fair to deploy the techniques of consumer advertising—which, as the *British Medical Journal* put it in an editorial decrying the practice, is "the science of arresting the human intelligence long enough to get money from it"—on vulnerable people. The FDA stepped in from time to time—for instance, in 2004, when Wyeth ran a radio spot for Effexor that started out:

Hey you, listening to the radio . . . How're you feeling these days? Okay? Not bad? Come on, is that where you want to be? When was the last time you did something you once looked forward to doing? You know, symptoms of depression could be holding you back.

The FDA faxed a note to Wyeth, warning the company that "by failing to draw a clear distinction between major depressive disorder and normal periodic feelings of low interest or low energy, the advertisement broadens the indication for Effexor XR." Which was no doubt what Wyeth had in mind. They just weren't artful enough about it to avoid the FDA's displeasure. Mostly, however, the drug companies stayed on the right side of the law—not too hard to do when all it took was the careful use of the subjunctive, as in "depression *may* be related to an imbalance of naturally occurring chemicals between nerve cells in the brain" (emphasis added). When a couple of researchers pointed out to the FDA that, according to *Essential Psychopharmacology*, a standard medical textbook, "there is no clear and convincing evidence that monoamine deficiency accounts for depression," the FDA wrote back to say that this was an "interesting issue," but that "these statements are used in an attempt to describe the putative mechanisms of neurotransmitter action(s) to the fraction of the public that functions at no higher than a 6th grade reading level." The alleged stupidity of the citizenry, in other words, justified the drug companies' lying to them.

And every indication was that the ads, unlike the drugs, really worked. Ad industry research indicated that every dollar spent on consumer advertising yielded $1.37 in drug sales. Adam Block, an independent researcher at Harvard estimated in 2007 that more than a half million doctors' office visits were inspired every year by consumer advertising of antidepressants. Using epidemiological data, he estimated that only one in fifteen of those patients was likely to be depressed, but using statistics derived from other studies, he determined that more than half of them would get a prescription, which meant, he said, that only "six percent of the increase in antidepressant use due to [direct-to-consumer] advertising is by people who are clinically depressed."

Block concluded, however, that this wasn't necessarily a bad thing. Even if the majority of the money spent on the drugs was for nondepressed people, he argued, the cost of untreated depression was so great that "treating everyone in the country with an SSRI would . . . provide a net benefit." The drug industry hasn't yet proposed this as official policy, but maybe they don't have to. People aren't just turning up at doctors' offices with their personal symptoms checklists and asking their physicians to complete the examination. They're coming in self-diagnosed—and already asking for the drug they're sure will help. And doctors are more than happy to oblige.

At least that's what a group of researchers found out when they pulled a Rosenhan on 152 family doctors in 2003 and 2004.

The team, led by Richard Kravitz, a University of California researcher, developed a method kinder and gentler than Rosenhan's for sneaking in and seeing what doctors do when they don't think anyone other than the patient is watching. They deployed standardized patients (SPs)—people, often actors, trained to present the symptoms of a particular disease and generally used to sharpen the diagnostic skills of medical students. (Doctors who agreed to be in the study knew that two SPs would visit them in the next year, but not what their complaint would be or what the researchers were studying.) Kravitz taught female SPs how to simulate one of two DSM diagnoses: major depressive disorder and adjustment disorder with depressed mood. The depressed patients complained of wrist pain and of the requisite five DSM symptoms: feeling "kind of down" for one month, worse in the past two weeks, of fatigue, sleep troubles, loss of appetite, and sensitivity to criticism. The adjustment disorder SPs had just been laid off from a job and were suffering from back pain, fatigue, "feeling stressed," and sleep troubles.

Kravitz sent his SPs on 298 visits to doctor's offices, equally divided between the depressed and the adjustment-disordered patients. In addition to their symptoms, the SPs were armed with two scripts, both of which described something they'd seen on television. On about one-third of their visits, SPs didn't deliver either script. In another third, they talked about an ad they saw for Paxil. "Some things about the ad really struck me," they told the doctor. "I was wondering if you thought Paxil might help." In the remaining third, they said that they'd seen a show about depression. "It really got me thinking. I was wondering if you thought a medicine might help me." (Kravitz chose Paxil more or less at random; he received no drug company money for the study.)

Fifteen of the forty-eight depressives who didn't ask for drugs received prescriptions anyway. Twenty-seven of the fifty-one SPs requesting Paxil got an antidepressant, and fourteen of them got the brand they asked for. And thirty-eight of the fifty SPs who asked for "a medicine" got one. Those posing as adjustment disordered got similar results: 10 percent got drugs when they didn't ask, nearly 40 percent when they did, and 55 percent when they requested Paxil (of whom two-thirds got their brand choice).

These results, while less embarrassing to the profession than Rosenhan's, still don't put doctors in a favorable light. Only half the depressed patients got the minimal indicated treatment, and half of the patients who didn't qualify for the treatment received it. Doctors failed to spot depression in 20 percent of the cases; they diagnosed it in nearly 40 percent of nondepressed people. And diagnosis rates increased significantly—from 65 to 88 percent in the depressed SPs and from 18 to 50 percent in the maladjusted—when a patient asked for drugs, a request that is not a known symptom of depression or of any other disease except substance abuse disorder.

This increase was probably not due only to doctors covering their asses by justifying their prescriptions with a diagnosis. More likely, it occurred because talk of the cure put them in mind of the disease. The marketing effort, in other words, may create a collusion that neither doctor nor patient needs to be aware of. Indeed, it may be best if it works its magic entirely in the shadows.

GlaxoSmithKline might wish that the ads were effective enough to dictate doctors' brand preferences. But the company couldn't miss the fact that although SPs rarely received Paxil if they didn't ask for it, a simple request goosed sales noticeably. This finding is consistent with research indicating that while consumer advertising increases sales of a class of medications, it is old-fashioned detailing that determines the success of specific drugs. The consumer ads soften up the market; the detailers move in for the kill.

And that's the real triumph here, at least for the pharmaceutical industry—and for their ad agencies. For nearly fifty years, they've been on a campaign to convince Americans—doctors and consumers alike—that they suffer in enormous numbers from a disease called depression. This has not been some idle public health effort, but an attempt to link that disease to a particular cure, and it turns out that if you ask your doctor for the cure, your chances of getting the diagnosis go way up. In fact, in people who *don't* have the symptoms, it nearly triples.

Kravitz and his team didn't do more than note the fact that talk of drugs increases diagnosis. They're much more worried about the clinical appropriateness of the prescriptions, and about whether or not the ads lead to overtreatment. But then again, the official diagnosis may not even matter. Doctors may well be prescribing antidepressants to patients who ask for them for the same reason that they prescribe antibiotics to patients who, in their opinion, are suffering from a virus: because they are in the business of relieving suffering, and the patient is signaling that a pill will make her feel better, because, that is, they want to please their patients. And given their performance in clinical trials, what better drug to prescribe as a placebo than antidepressants?

What does matter is that your *doctor* has ratified your request. If you go to the office with your Personal Symptom Checklist and you leave with a prescription for antidepressants, is it really important for him to run the numbers and tell you out loud that you have depression? Does it even matter if he thinks you do? Does a doctor have to say "bacterial infection" to make you think that this is what your antibiotics are for? After a half-century of being carpet-bombed by this message, it is virtually impossible to suffer prolonged sadness without considering the possibility that you have depression. Frank Ayd's spiel is obsolete. Or, more precisely, it has become an essential part of the climate of opinion in which we experience our unhappiness.

As Kravitz's study inadvertently proves, you can teach people how to be depressed. He went to a lot of trouble to teach them well, but I think advertising and all the other channels through which the depression message is broadcast are also good teachers. And even if you have some doubts about whether the ad or the television program is really describing you, when your doctor hands you the pill, he's confirming the diagnosis whether he means to or not.

But what matters above all else about Kravitz's study is that he has actually out-Rosenhanned Rosenhan. He's pulled a prank he didn't even mean to pull. Because in real life, none of those SPs was actually depressed, at least they weren't when they were screened for the job. Yet 60 percent of them got a diagnosis, and nearly 45 percent of them got drugs. Try faking a case of diabetes. I don't care how good an actor you are or how well informed. Unless you brought a real diabetic's urine with you, or your doctor is criminally incompetent, you are not going to go home with a prescription for insulin.

Okay, so maybe that's not entirely fair. But it wasn't my idea to compare depression to diabetes in the first place. That was the drug companies' brainchild, as in "Depression doesn't mean you have something wrong with your character. It doesn't mean you aren't strong enough emotionally. It is a real medical condition, like diabetes or arthritis"—which is what you learn when you go to the Myths and Facts page on Pfizer's zoloft.com website. Or prozac.com's version: "Like other illnesses such as diabetes . . . depression is a real illness with real causes."

It's easy to see why the depression doctors want to make that comparison. Diabetes provides a classic magic-bullet scenario: your pancreas stops producing insulin (or, in the case of type 2 diabetes, your cells lose their ability to absorb insulin), and the deficiency is treated with regular medication. No one would be ignorant or insensitive enough to suggest that your illness is related to your character or your emotional strength. No one would blame the victim or imply that a diabetic is weak for taking his medicine. A depressed person who thinks of himself this way, in other words, is a loyal patient for life.

But doctors don't have to convince their diabetic patients that they have a "real illness." The symptoms generally speak for themselves. A diabetes doctor doesn't have to worry about the clinical appropriateness of treatment. He doesn't have to wait for a new definition of diabetes to be hashed out in committees of his brethren and then learn the new diagnostic criteria. He doesn't have to worry about whether someone is going to show up at the office claiming to be diabetic, or perhaps hiding diabetes, and then embarrass him when he misses the diagnosis; all he has to do is to take a urine or blood sample. He doesn't have to talk about chemical imbalances that he knows aren't really the problem or contend with package inserts that say, in plain black and white, that the drug makers have no idea why their drug works.

And above all else, the diabetes doctor doesn't have to tell the patient that he is getting better. Which is what they kept telling me at Mass General. At the end of my fourth visit, George Papakostas finished jotting in his notebook and told me that my Hamilton score had dropped to fourteen, from my baseline of eighteen. This was the week after he had asked me about the thirty days of symptom-free living that I'd apparently been missing out on because of my disease. Had I heard him right? I asked. How long did he say I should be feeling good?

"For at least a month," he said.

Then I asked him why he wanted to know.

"People, when they're depressed," he answered, "they get a sort of recall bias. They tend to feel that their past is all depressed."

Which meant, I wanted to point out, that depression is more like an ideology than an illness, more false consciousness than disease, and that telling me I was getting better was like dispatching propaganda from a new regime.

But this wasn't the only way in which Papakostas was telling me what my disease consisted of or what health would be like. He also did it through the tests. They asked me about my sleep and appetite; they asked me if I felt guilty; they asked me if I thought my life had been a continuous process of learning, changing, and growth. They gave me zero points for seeing myself "as equally worthwhile and deserving as other people" and three for "thinking almost constantly about major and minor defects in myself." You don't have to be a weatherman to know which way that wind is blowing.

In this respect, the tests aren't much different from the advertising—only the ads can be smarter than the tests. "Prozac isn't a 'happy pill,'" Lilly's first ad reassured. "It isn't a tranquilizer," nor would it "turn you into a different person." It would just have you "feeling sunny again." "Your life is waiting," Paxil reminded people, and "once they got back to themselves," as the Zoloft ad put it, "they would appreciate life even more." "Welcome back," was the Prozac slogan— to yourself, it seems, to the person you were supposed to be all along.

When Papakostas added up my Hamilton numbers and concluded that I was getting better, he didn't have to say in what way that was true. It was already in the air. And when he asked me, "Are you content with the amount of happiness that you get doing things that you like or being with people that you like?" he didn't have to tell me outright that this was the whole point: that to be healthy, to be back to yourself, to occupy the life that's been waiting for you all along, was to be content. Which is a deep philosophical statement, and one that seems at odds with a consumer society and an economy that depends on our never being content, at least not too content to think that there is always some other happiness you could be pursuing at the mall. But he didn't make this claim as a philosopher. He made it as a doctor. So we didn't have to talk about any of that.

And Christina Dording didn't even have to mention contentment on my last visit. She just had to look in the binder, riffle the pages, and say, chipper as always, "Give me one second here." She paused and then smiled. "Look at your scores. Nice response."

I wasn't sure whom she was congratulating, but there wasn't any question who—or what—was responsible for my improved mental health. Or so I found out when she started talking about my next visit.

"Next visit?" I asked. "I thought this was the last."

"You're not coming in for the follow-up?" She seemed surprised and hurt and a little incredulous, as if no one with such a nice response would pass up the opportunity to get even better. I asked her if the follow-up would be any different from what we'd been doing. It wouldn't, she said. So I declined.

But she wasn't done with the subject. By then we'd adjourned to an examination room, where she was performing a cursory physical. "I think you've done very well," she said as she looked into my eyes with a scope. "You're much improved."

But if the treatment made me better, I wanted to know, then why did I need any more follow-up than buying some fish oil at the Whole Foods conveniently located next door to her office? And for that matter, how did she know that it was the fish oil at all? How did she know I wasn't on the placebo?

I asked her if I'd been on placebo or drug. "I don't think we unblind the study," she told me, looking again through my binder. "No, not in this one. No unblinding."

I protested. "I don't get to find out?" Had no one ever asked this before? I wondered. And was it possible that being much improved could have no other meaning than that the drug had worked its magic? Wasn't that what the study was supposed to find out?

"No," she said. "But you had a good response."

I didn't see the point in arguing, but a few months later, I called the doctor in charge of the fish oil study. I asked him why Dording had offered to keep me on the fish oil when she didn't know if I'd been on it in the first place, and why neither of us was allowed to find out. He explained that clinical trials remain blinded so that researchers don't get tipped off by associating certain patterns of response with certain outcomes and thus start behaving differently toward patients whose condition they have deduced. But, he told me, seemingly unaware that he was contradicting himself, it is common practice for the doctors to "take their best guess" and offer follow-up accordingly.

I wasn't going to let this mystery stand. I didn't know if I was really better. Some days I thought so. I wasn't feeling content exactly. But sometimes, on some days, there was some ineffable feeling, a flicker of belief, a floor beneath me that kept me from plunging into darkness, where I could stand and catch and hold love and goodness, dwell with it and feel, if just for a moment, that life wasn't only cruel and stupid. Eventually, I'd find myself paging through the familiar catalog of discontent, thick now with age and experience, but even then somehow less certain that this was the life that was waiting for me, this was the self I'd always come back to, this was the darkness that welcomed me. And maybe it was the fish oil that was making me feel that way.

Or maybe not.

I had some extra capsules. I sent them off to a commercial lab. The report came back a couple of weeks later. There wasn't a drop of fish oil in them. I'd been on the placebo.

CHAPTER 13

EMBRACING THE MODEL: COGNITIVE THERAPY

Of course, Papakostas and Dording didn't mean to horn-swoggle me. They were convinced that the drugs work, and that their conviction is a matter of fact and not faith. And to the extent that they were aware that they were using placebo effects, it was undoubtedly to bolster their patients, to give them hope. They weren't really trying to sell anything. That's why I have had to resort to these literary tricks—deconstruction and interpretation of what they say and of the ad campaigns that shape the meaning of their words—to shed light on their invention.

That's one of the great advantages of being a psychiatrist. George Papakostas may think that he and his colleagues love language, but ever since they decided that psychoanalysis is bunkum, or at least not the science for them, they don't really have to take language seriously. They don't have to articulate what exactly their notion of health is, or what philosophy lies behind it, beyond vague bromides about resilience or leading questions about contentment. There is, however, a group of depression doctors that trades in language, and they leave nothing to the imagination when it comes to spelling out what they mean by healthy. They'll tell you exactly what the good life is and how and why depression has robbed you of it.

I'm hearing all about it right now from Dr. Judith Beck. We're role-playing, and I'm doing my version of Ann, the patient I told you about earlier. She's the woman whose depression was worst when she was doing the best, when she was shining at her job or saving lives on an ambulance crew or sorting out a bookkeeping disaster at her church. Ann might join a club that would have her as a member, but not without telling its membership director that he'd made a terrible mistake and that the invitation would ruin a perfectly good organization. And she would definitely—I have first-hand knowledge of this—tell her therapist that while he seemed a good judge of many things, and a decent guy to boot, his apparent affection for her diminished his stock severely. After that, she'd prob-ably stop at Burger King and grab a Whopper and fries, go home and eat some ice cream and then sit back and wait for her gallblad-der to start to pound in her belly—an orgy of eating and pain that she would tell me about in excruciating detail the next time I saw her. And as she did, as it became clear that my attempt to find a nug-get of gold glinting in the foulness of her emotional life had driven her to this bout of self-abuse, I would have to resist the sharp temp-tation to tell her what Freud said about how the self-reproaching melancholic is undoubtedly correct in her self-assessment. I would find myself agreeing with her that it was a huge blunder to try to find something to love in her.

Ann, in short, was a therapist's nightmare, a nightmare that right now I am inflicting on Beck. I'm discovering that it is much more fun to be Ann than it is to be her therapist. You can really tor-ture someone this way.

Judy Beck has short dark hair and hooded eyes that look right at you without staring. She's not having anywhere near as bad a time as I would in the real-life version of this conversation. Partly that's because it is hard for me to fully summon the awfulness of Ann; while I may know depression from the inside, and am always up for some self-laceration, I have never hated myself the way she does, so I can't quite exude her toxic misery. But it's also because Beck is familiar with people like Ann, and here on her home turf—the Beck Institute for Cognitive Therapy and Research, where she and her staff are teaching the basics of cognitive therapy to me and twenty-nine other therapists from around the world—she is calm and confident as she demonstrates why she is one of the world's leading experts in the treatment of depression and why cognitive therapy is one of the very few nondrug treatments for depression that have been validated in clinical trials.

The non-profit institute is named for psychiatrist Aaron Beck, Judy's father, whom everyone here, including his daughter, calls Dr. Beck. We met him a few days ago, watching as he conducted a therapy session. Afterward, he came into the room in which we've spent this week to answer some questions.* He's almost ninety years old. He probably once was taller and straighter than he is now, but he still wears a jaunty bow tie and fashionably oversized glasses and delivers his nimble answers with wit and charm in a strong, clear voice.

Aaron Beck developed cognitive therapy in the early 1960s. A psychiatrist trained in psychoanalysis, he was, as he put it, "caught up in the contagion of the times"—which included efforts by the National Institute of Mental Health and the Group for the Advancement of Psychiatry to implement "systematic clinical and biological research"—and "prompted to start something of my own." He dabbled in the diagnostic reliability field, but his big opportunity came unexpectedly when he undertook research into the dreams of depressed patients. Freud's theory of melancholia predicted that repressed anger would turn up in dreams, but Beck found something different: "that the dreams . . . contained themes of loss, defeat, rejection, and abandonment, and the dreamer was represented as defective or diseased"—an exaggeration, he pointed out, of the themes of their conscious life. And unlike Ann, most depressed patients responded well when they succeeded and were praised, at

---

* I'd give you more details of this session, but all participants agreed to keep confidential the clinical work and role playing we witnessed.

least under experimental conditions. Perhaps, Beck concluded, it had been a mistake to look past the patient's manifest self-reproach and toward a latent hostility directed toward others. Instead, he suggested, the problem was the negative thoughts themselves, which in turn kindled the recalcitrant unhappiness of the depressed person.

Something had gone wrong with the patient's inner life, Beck thought, but it wasn't related to the relentless discord among ego, id, and superego, to dark forces and incestuous longings, to the clash of Eros and Thanatos or of civilization and instinct. Instead, it was to be found in the schemas, the dysfunctional beliefs that structured a patient's experience and shaped his distorted cognitions. Where Freud saw a self groping around in a dark and treacherous inner landscape, lit only by the ego's dim light (and, in treatment, by the slightly brighter lamp of a talented analyst), Beck saw a self with the potential to process information accurately, to map the inner and outer world and navigate successfully through their obstacles— and whose pathology could be discovered and corrected through a straightforward technique that used Socratic questioning instead of Freudian probing.

Beck based this therapy in part on behavior therapy and in part on the cognitive science that was then emerging at the intersection of linguistics, philosophy, and computer science. In cognitive therapy, he explained,

> therapist and patient work together to identify the patient's distorted cognitions, which are derived from his dysfunctional beliefs. These cognitions and beliefs are subjected to empirical testing. In addition, through the assignment of behavioral tasks, the patient learns to master problems and situations which he previously considered insuperable, and consequently, he learns to realign his thinking with reality.

Beck's theory didn't ignore the past—indeed, the troublesome schemas are often laid down in childhood, by trauma and deprivation and all the other varieties of parental failure. But if your mom and dad will fuck you up, once you see how, there is no point in dwelling on the particulars. In his talk to us, Beck recounted the story of Lot's wife as a cautionary tale about encouraging patients to explore their pasts. The point of recollection in cognitive therapy is not to delve into all the possible meanings of a memory, or the way that personal history reveals the confusion of forces that makes us human, but to identify the original distortions, correct glitches at their source, and restore the patient's operating system to tip-top shape.

Judy Beck is demonstrating just how this works by drawing a picture on a whiteboard. She's trying to help Ann with the crushing self-loathing she felt after an ambulance call in which she had revived a man in cardiac arrest. As Ann, I've just told Beck that I hadn't really done anything to deserve praise, that I probably broke some of his ribs, and it was all just dumb luck, nothing to do with me, not evidence of my competence, and certainly not praiseworthy—which means that if I allow my crew to compliment me, I'm just fooling them, which really makes me a terrible person.

"It's almost like there's a part of your mind that's shaped like this," Beck says, indicating the diagram on the board. "Like a Pac-Man, but on its side. Inside this part of your mind is this idea, 'I'm incompetent.'" The Pac-Man's mouth is rectangular, not a jagged ellipse as in the original.

Beck explains that neuroscientists have not yet found a place in your brain that looks like the Pac-Man; the drawing "just helps us understand it better." Then she asks Ann to tell her about something incompetent that she has done recently. I oblige with a story about the next emergency call after she revived the man, and how she parked the ambulance on the wrong side of the yellow line at the hospital loading dock.

"And when you did this, did you ask yourself, What does this mean that I parked on the wrong side of the line? Does it mean I'm competent, that I'm incompetent . . ."

"I just thought it meant I'm stupid."

"So, automatically. You didn't even think about it. So here we have some information—you parked on the wrong side. It's almost as if that information was in a Negative Rectangle." She draws a rectangle. "You see how the Negative Rectangle can fit right into this part of your mind? It's like every time a Negative Rectangle goes in, it makes this idea 'I'm incompetent' a little bit stronger." She draws an arrow. "Almost anything that happens that could possibly mean you're incompetent, I think that information goes straight into your mind and immediately, automatically, you start to feel incompetent. You don't even think about it. Do you think I could be right about this?"

I nod.

Beck turns now to Ann's lifesaving. "Did you think, Does this mean I'm incompetent?"

"I thought that everyone who was telling me what a good job I did just didn't know the truth."

"And what do you see as the truth?"

"The truth is that most of the time I don't bring the people back and I think I just got lucky."

"So isn't this interesting? Here we have some positive data." She draws a triangle next to the Pac-Man. "The positive data is in a triangular shape, you see how this can't get into a rectangular opening? In fact, in order to get in, it's got to change its shape. It's got to change into a Negative Rectangle, so now it can fit in."

Beck draws a triangle on the board. She asks Ann how many times she was late for work last week.

"None."

"So every day when you got to work on time, what did you say to yourself?"

"Here I am."

"So you didn't say this is a sign of competence?"

"People are supposed to get to work on time."

"So here we have five potential Positive Triangles, where you got to work on time, because not everybody actually gets to work on time, but you didn't take that as a sign of competence. This positive data just bounces off. This way of processing information isn't your fault, just your automatic way, which I can help you learn to override."

Beck is going by the book here. Which makes sense, because she wrote it. Although *Cognitive Therapy: Basics and Beyond* was assigned for this seminar, I can't say I've read it recently, but I do remember it well from when I was a college professor teaching a course that surveyed the various schools of psychotherapy. *Cognitive Therapy* was a hit with my students. After the maddening uncertainties of psychoanalysis, the quasi-fascism of behavior modification, and the touchy-feely vagueness of existential-humanistic therapy, they really appreciated Beck's bullet lists, her step-by-step instructions and verbatim scripts and you-can-do-this-too optimism. And above all, they liked her rational approach, her implicit reassurance that we were equipped to make sense of our lives.

But even if I weren't familiar with the book, I'd have gotten most of its contents from psychologist Leslie Sokol, a Beck Institute staffer who has done the majority of the teaching this week. She's taught us how to help a patient identify Automatic Thoughts and record them on a Dysfunctional Thought Record, how to use the Downward Arrow to point to the Core Negative Beliefs that lie beneath them, how to chart the mental journey from belief to thought to emotion and behavior on a Cognitive Conceptualization Diagram, how to fashion an Alternative Response or other Compensatory Strategy that will, repeated and reinforced, lead to Cognitive Restructuring. She's told us about the Negative Triad—regret about the past, unhappiness in the present, despair about the future— about Task Interfering Cognitions and Task Oriented Cognitions ("Stop TICking and start TOCking," she says), about Key Cognitions and Situation-Specific Thoughts and Affect Shifts, about the importance of Activity Schedules and Coping Cards. She's shown us video of other therapists in action, passing papers back and forth across the table at which they sit with their patients, following the steps of a cognitive therapy session—the Mood (and Medication) Check, the Bridge from the Previous Session, Setting the Agenda, the Review of the Homework, the Discussion of Agenda Items, Assignment of New Homework, the Final Summary, and the Feedback. She's reassured us that this isn't just some cult of positive thinking. "The key is not to teach people to be more positive. The key is to have an accurate perspective on what's going on." She's argued that the problem for depressed people is that "nothing in life is a neutral situation," that their Negative Bias needs to be replaced with neutrality, their inadequate coping skills with new strategies, that when this is accomplished, when they have a good Cognitive Conceptualization of themselves, when they are able to process information as accurately as their mental apparatus is designed to do, they will become resilient, and that this is the goal because "the resilient person is the person who is going to make it."

Sokol seems like someone who has made it. She's around fifty, blond and bright-eyed and trim, her skin scrubbed to a glow. She wears well-tailored suits, talks fast, leavens her PowerPoint with Family Circus cartoons and her patter with fresh and funny anecdotes from her private practice, and delivers it all with panache. She could probably sell iceboxes to Eskimos, and she makes no bones about the sales part of her job. She doesn't call it that, of course, but she tells us repeatedly that we have to "socialize the patient to the model," and it's clear that this is what she's doing with us. "If we're asking them to embrace the model," she says, "we have to already understand and believe in the model. I'm a believer and I'm here to make you a believer."

There's something refreshing about this ideological candor, even if I'm finding its particulars hard to embrace. Every therapist is selling a view of the world, or at least of suffering and its relief, but few articulate it so clearly, and with such certainty. This clarity and single-mindedness is in fact part of the program. "Self-doubt is contagious," she said, striking a theme she will sound all week: that ambivalence—about the model, about our cognitive abilities, about our prospects of making it—is the enemy that stops depressed people from getting reality just exactly right and therapists from staying on the empirically validated, doctor-tested cognitive therapy path. And if we stay on the straight and narrow, we will be amply rewarded for our faith. "My hope is that over the week you're going to see that if you understand the model, if you understand the problem though the model, you're going to know what to do to treat anybody that walks through the door."

That is a very powerful pitch. Therapeutic outcomes are dependent in part on allegiance effects, on the extent to which a therapist believes in what he is doing and conveys that confidence to his patient. So a claim to be in possession of a universal method is good for a therapist's business. Commerce, in fact, has been on Leslie Sokol's mind ever since she started the week by saying, "Let's get right down to business." Not the business of the Beck Institute, which is certainly successful, but the business of living, of making it in a world that constantly throws up obstacles. "We're not here to cure you. We're here to teach you how to cope," is what she suggested we tell our patients as we socialize them to cognitive therapy. "We're here to help you more effectively navigate life. We're here to say that when bad things happen, you're going to be equipped to deal with them so they don't get the best of you." Therapy is not only the venue in which these management skills are taught; it's where they are embodied. "You need empathy, but it's not enough. You get to work," she reminds us. "We're here to accomplish work. And every time we meet here, we're here to get something done."

This is also a powerful idea, and while I'm not exactly embracing the model, I am painfully aware of the fact that I've never talked to Ann or any patient about Automatic Thoughts or Core Beliefs, never swapped papers with them or assigned them Coping Cards or diagrammed their Cognitive Conceptualizations. All week, in fact, I've been wondering if this is connected to the futility of those dismal hours I spent with Ann, to my inability to say whether I accomplished work with Ann. Now that I see what it means to set an agenda and pursue it, I'm not even sure I actually set out to do that in the first place with Ann—or, for that matter, with anyone else.

And if I did, it isn't the kind of work Beck and Sokol have been talking about here. Ann once told me about a dream she'd had in which five red tractors were planted vertically in the ground, and she single-handedly uprooted and knocked them down. We had one of our few fruitful discussions that day, about this dream's multiple resonances with her emotional life. Most obviously, the dream referred to her resentment over the fact that her father refused to allow her to drive the tractor on her family farm (men's work, he said), but there was something much more important. Ann's family (three siblings, two parents, five tractors) had been uprooted when Ann's youngest sister was institutionalized and they moved from the farm to the city to be closer to her. The sister had multiple birth defects, the result of a case of rubella that Ann's mother had contracted while pregnant—and which she had caught from Ann. You can imagine the conflicts this would create, the shame a child could feel over her role in such a catastrophe, the sense of power it would bestow upon her, the rage at being displaced and the self-reproach for causing it and the confusion between them. The dream was such a near-perfect distillation of these themes that even Ann could see it. And when she did, it certainly felt like work, but I have no idea how it would come out on these diagrams. More to the point, it would be hard for me to say right now exactly what we were accomplishing in that conversation.

I thought I knew at the time. Just as I did when Eliza—the woman who learned on the same day that her drug-addicted mother had died and that her father was not the man she had thought, and whose grieving had outlasted the DSM's dispensation for bereavement—told me a story about strawberries. We had been talking about how when Eliza was fourteen, her mother, Mary Ann, had taken up with a new man, a crystal meth addict and dealer, announced that she was finished being Eliza's mother, and ordered her out of the house immediately. Eliza knocked around for a while, actively resisting a life like her mother's (Mary Ann had become pregnant with Eliza when she was sixteen and lived mostly from man to man and bar to bar), until she was taken in by someone she knew from her church. She got a GED, a degree from a technical college, and, eventually, a job in her profession.

One day, Eliza told me about a time when Mary Ann came to visit her for a weekend, almost ten years after she had kicked her daughter out. The visit ended abruptly, with both of them in tears. "It always ended badly," she said, "because her visits brought up a lot of issues for me. About what a pathetic excuse for a mother she was and how I needed her anyway, things like that. A couple of days later I was having friends over, and I guess the buildup from the weekend hit me all of a sudden. I was slicing some strawberries. I have no idea why this triggered it. But all of a sudden, it hit me that my mother was lost to me and I was never going to get her back, and I just collapsed in tears."

Eliza began to cry, just a little. "I just don't want to cry about this anymore," she said. "I didn't want to cry about it then. It just puts a knife in your heart."

"Maybe that's why it happened when you were slicing strawberries," I said.

"With the knife and the heart-shaped strawberries? That's funny." She laughed. "Are you going to go all Freudian on me?"

"This is a pretty remarkable thing," I answered, dodging her challenge. "That you were talking about cutting strawberries and you came up with this metaphor of a knife jammed in your heart." (And why did I say "jammed"? Was the strawberry image contagious?)

"Yeah," she said, sarcastic. "I'm brilliant that way. Maybe as I was cutting the strawberries I was thinking, My mother's put a knife in my heart."

"I don't think you were thinking that. It's just the way it was for you."

She was silent—reconsidering, I think. "Yeah, maybe, maybe. God, I never thought about that. But it's funny that the visual memory of me cutting those strawberries is so emblazoned into me."

"But you know, I wonder if it's also her heart," I said.

"What do you mean?"

"That you want to cut. What is the response to a mother who is a pathetic excuse for a mother?"

I'm not sure exactly why I said that. It was something about inverted hostility, about how the once absent, now dead, and still longed-for mother is nearly impossible to be angry with, so there is no other place to put the knife besides into her own heart. Maybe I was just socializing Eliza to my model. Maybe I just wanted her to get angry with her mother because it fits into my theory about how the world works. That's the thing about being the kind of therapist I am. I'm always catching a case of self-doubt.

And right now I have a whopper. Against all of Judy Beck's polished technique, all her sharp distinctions, all her careful plotting and planning, I'm wondering if I've failed my patients and myself, if I've frittered away twenty-five years of my life and millions of their dollars by focusing on the tractors and the strawberries and all their possible meanings, by the inescapable and sometimes intentional inefficiencies of this method, by my nearly willful avoidance of anything resembling accomplishing work, by my possibly blind and certainly unscientific belief that the best we can do is to integrate all that we can of ourselves into a good story, even the thoughts that don't make sense and the desires that are horrifying or the feelings that shock. Because right now all of that murkiness doesn't seem to stack up to Beck's clarity of purpose and her method for getting there.

Nor can whatever I've been doing all this time, and whatever has come of it, compare to one fact, documented in clinical trials and endorsed by the mental health industry and government alike: that when it comes to depression, cognitive therapy gets results. Empirically validated results, results that give psychologists a place at the depression feeding trough, that both capitalize on and strengthen depression's status as a bona fide disease, and that warrant cognitive therapy's inclusion in the American Psychiatric Association's standards of care—which means that by not practicing it with Ann (or anyone else who is depressed), by worrying about the strawberries and the tractors, I may be guilty of malpractice. So who am I to argue?

I'm not entirely exaggerating about the potential legal consequences of not practicing therapy by the book, at least when it comes to depression. Just ask the doctors at Chestnut Lodge, a psychiatric hospital near Washington, D.C., with a long and venerable history. In 1979, Rafael Osheroff, a forty-two-year-old doctor was admitted to the lodge, complaining of anxiety and depression. He'd been under treatment for the past couple of years; his psychiatrist was none other than Nate Kline, who thought Osheroff was getting better. But Osheroff didn't agree, at least not enough to keep taking his tricyclic antidepressants. He became disabled and ended up in Chestnut Lodge, where the house brand was psychoanalysis. When he didn't improve after seven months of that, his family moved him to a Connecticut hospital where he was promptly diagnosed with a psychotic depression and put on antipsychotics as well as tricyclics. His condition improved noticeably within three weeks, and three months later he was discharged.

By then, however, Osheroff had been in the hospital for nearly a year and his life was in ruins. His second wife had left him, his first wife had gotten custody of their two children, his hospital had yanked his accreditation, and his partners had kicked him out of their practice. In 1982 he sued Chestnut Lodge, claiming that their failure to put Osheroff on drugs was negligence. Five years of arbitration and appeals and hearings that included testimony from leading psychiatrists like Frank Ayd and Gerald Klerman ended with an out-of-court settlement in Osheroff's favor.

Because it never went to trial, *Osheroff v. Chestnut Lodge* didn't establish any official legal precedents. Its impact on the profession was nonetheless profound. According to Edwin Shorter, "The case left the strong impression that treating major psychiatric illnesses with psychoanalysis alone constituted malpractice . . . Any clinician who henceforth treated patients as Chestnut Lodge had Dr. Osheroff ran the risk of incurring heavy penalties." Not only that, Shorter says, but psychiatrists, chilled by the outcome, began to abandon their notebooks and couches for prescription pads and more traditional office furniture, creating a vacuum that was filled by the psychologists and social workers and other non-physician therapists. Sixty years after they had wrested psychoanalysis from Sigmund Freud, doctors evidently could barely wait to hand it back over to the lay analysts.

But the case had an even more direct impact on the treatment of depression. Osheroff had gone into Chestnut Lodge at the height of psychiatry's thrash over DSM-III and filed his lawsuit just after it was published. In the new psychiatric world, with its diagnostic specificity and its magic-bullet drugs, therapists' difficulty in passing scientific muster was a new kind of problem. As Gerald Klerman, writing in 1990 about the Osheroff case, put it:

> If a pharmaceutical firm makes a claim for the efficacy of one of it products, it must generate enough evidence to satisfy the Food and Drug Administration before it can market the drug . . . No such mandate of responsibility exists for psychotherapy. Anyone can make a claim for the value of a form of psychotherapy . . . with no evidence as to its efficacy.

The moral of the Osheroff story, Klerman said, was that it was time to require of psychotherapies what Kefauver-Harris had required of drugs: proof that they worked.

It's not that no one had tried to do that. In 1936, in fact, a prominent American psychologist, Saul Rosenzweig, published a paper examining therapy outcomes and concluded that all forms of therapy, competently practiced, were equally effective. Rosenzweig lifted the subtitle for his paper—"Everyone Has Won and All Must Have Prizes"—from the dodo bird's verdict on the race in *Alice in Wonderland*. That might have been an unfortunate choice. His conclusion has gone down in history as the *dodo bird effect*—not an embarrassment of riches, that is, but just plain embarrassing.

In 1975, a team led by psychologist Lester Luborsky subjected the dodo bird effect to modern statistical methods. They looked at studies comparing one therapy to another, therapy to no therapy, psychotherapy to drug therapy, and time-limited to interminable therapies and concluded that all indeed must have prizes.

Luborsky also determined that there was nothing specific to a given therapy that accounted for its success. In part this was because the therapists generally chose the outcome measures, but even when the measure was an objective test (like the HAM-D), the dodo bird effect held. Luborsky suggested an explanation: "The different forms of psychotherapy have major common elements—a helping relationship with a therapist . . . along with the other related, nonspecific effects such as suggestion and abreaction [Freudian jargon for emotional catharsis]." These common elements—nonspecific factors—accounted for therapy's success.

Luborsky's work got updated from time to time, using increasingly sophisticated and impenetrable statistical techniques, and the result was virtually always the same. Something like three-quarters of patients are better off with therapy than they were without it. Patients themselves ratified this result, at least they did in a survey that appeared in *Consumer Reports*. There was "convincing evidence that therapy can make an important difference" the magazine reported, adding that the most important factors were "competence [of the therapist] and personal chemistry"—not the particular school the therapist subscribed to or the techniques he employed. The conclusion is inescapable: to the extent that therapy succeeds, it's due not to the particular help that's offered, but rather to the fact that something is offered in the first place, and by a person whom the patient expects, and believes, will help. Therapy, no less than drugs, works by the placebo effect.

This shouldn't be a surprise. To the extent that it is understood, the placebo effect seems to be the result of a patient's entering into a caring relationship with a healer, which is a much more explicit feature of psychotherapy than of general medicine. Nor should this be bad news. It just means that when therapists listen with empathy, when we offer support and understanding, when we help people to pick up their pieces and fashion a story out of them, to make as much sense of their lives as they can and to withstand the uncertainty of whatever is left over, when we provide a space in which they are free to be just as confused and demoralized and ambivalent as they really are—that when we do all that, and when we do it well, it really does help. It would no doubt be better to have a world in which we therapists weren't necessary, where narrative coherence wasn't so hard to come by and people weren't driven into private rooms to plumb the depths of their fears and their hopelessness, but that's not this world, so having those rooms, and the professionals who occupy them, is the next best thing.

Still, psychotherapists, and particularly cognitive therapists, have not been content to take their prizes and go home. To the contrary, when psychopharmacologists like Klerman sounded off about the lack of evidence for therapy's efficacy, or when Donald Klein, another leading antidepressant researcher, complained that "psychotherapies are not doing anything specific," the professional guilds didn't make the obvious point about the pot and the kettle. Nor did they take Klein's comment that therapies are "nonspecifically beneficial to the final common pathway of demoralization" as an unintended compliment and trumpet the value of remoralization and their unique ability to bring it about.

Instead, they panicked. "If clinical psychology is to survive in this heyday of biological psychiatry," a task force of the American Psychological Association warned in 1993, "APA must act to emphasize the strength of what we have to offer—a variety of psychotherapies of proven efficacy." The gauntlet had been thrown down, said the task force, and therapists had to pick it up by meeting drugs on their own ground—in controlled clinical trials that would identify empirically supported therapies (ESTs). It turns out that you can make at least one kind of therapy into something like a drug—a specific treatment that can be given in known doses, whose active ingredient attacks a specific disease, and whose effects can be measured. And the DSM-III provided empirical therapies with a perfect target: depression.

It's not an accident that more than 90 percent of EST trials focus on cognitive therapy. From the beginning, even before the DSM-III's clinical-trial-friendly symptom lists, Aaron Beck had set out to create a therapy whose effects on depression could be validated scientifically. He did this by developing his theory that depression is caused by dysfunctional thoughts and core beliefs—and a treatment targeted directly at those causes, one that could be broken down into specific modules, standardized in a treatment manual, and taught to therapists, whose performance could in turn be evaluated by reviewing tapes of sessions and scoring them on the Cognitive Therapist Rating Scale. Beck also developed a test—the Beck Depression Inventory (BDI)—to measure the outcome. If you think there's a circular logic at work here, not to mention a conflict of interest, you're probably right. But it's no worse than what Max Hamilton did when he fashioned his test to meet the needs of his drug company patrons. Besides, it's easy to overlook such matters when the theory allows cognitive therapists to claim that they are attacking the psychological mechanisms of depression in the same precise way that antidepressants attack neurotransmitter imbalances.

In the mid-1970s, Beck got a chance to put his theory to the gold-standard test—a clinical trial. His team got a government grant to compare cognitive therapy to antidepressant drugs as a treatment for neurotic depression (as defined in DSM-II). The study had a simple design. All forty-one subjects were given tests, including the BDI and the HAM-D, at the beginning of the trial. Half were then given tricyclic antidepressants, the other half cognitive therapy, and at the end of the twelve-week trial they were retested. Cognitive therapy won hands down. Therapy patients' scores on the tests dropped significantly more than those of the subjects on drugs. And, presumably because of the unpleasant side effects of the drugs, many fewer people dropped out of the therapy cohort than the antidepressant cohort.

The trial went on to have "a profound effect on the course of depression outcome research"—not only because of its results, but also because of how they were obtained. Beck and his team had done as much as possible to control for nonspecific factors. They had not only carefully measured the dose of therapy and continuously monitored therapists' adherence to the treatment manual; they had also chosen inexperienced therapists, medical residents and psychology interns who presumably hadn't yet picked up the tricks of the trade, who couldn't command confidence or deploy empathy like thirty-year veterans do, and whose successes could thus be attributed more to what was in the treatment manual than what was in their personality or technique. Beck could then plausibly claim that he had obtained his results with a minimum of placebo effects and a maximum of "active ingredient," that the reason CT outdistanced drugs was that there was something in the manual that was specifically therapeutic.

This impression was only strengthened over the next fifteen years as researchers replicated the finding that CT was as good as or better than drug treatment and added studies testing it against no therapy at all (other than an intake interview and placing the subject on a waiting list), and even against other therapies. As the findings mounted, professional and public opinion followed. In 1996, the *New York Times* reported that cognitive therapy was "the most scientifically tested form of psychotherapy . . . as effective as medication and traditional psychotherapy in helping patients with depression." In 2000, the American Psychiatric Association issued practice guidelines asserting that cognitive therapy was among the therapies with "the best-documented effectiveness in the literature for the specific treatment of major depressive disorder." Gerald Klerman's dream of government regulation of therapy hasn't yet come true, but a therapist not using cognitive therapy for depression would find himself on the margins of his profession. At least according to the *Times,* by 2006, cognitive therapy had become "the most widely practiced approach in America."

Dig into the clinical trials that give cognitive therapy its stranglehold on depression treatment, however, and its claim to the status as the most effective therapy begins to seem less than scientific. It turns out that cognitive therapy resembles antidepressant treatment in a way that Aaron Beck couldn't have intended: like the drugs, it owes its marketplace dominance less to science than to its unique suitability to the particulars of the scientific game, and much more to the placebo effect than anyone wants to admit.

Some of the trouble is built into the idea of validating therapy. It's hard to think of an enterprise less suited to lab testing than psychotherapy. What are the criteria of success and how do you measure them? How do you take all the thousands of words that are exchanged between therapist and patient—and for that matter all the nonverbal exchanges, the averted eyes and the fidgeting, the fleeting smile and brimming tears—and render them into data bits? The solution that researchers have hit upon is to ignore as much of that fuzzy stuff as possible and focus instead on what they can measure. This generally means doing exactly what Beck did: standardizing the treatment in a manual, aiming it at specific targets, such as the symptoms of depression found in the DSM, and then measuring the change in those symptoms after the therapy is implemented.

Critics complain that while this approach may work well in the laboratory, it has precious little relationship to what goes on in the real world. The lab therapist, indeed, does exactly the opposite of what most real-life therapists do: refrains from clinical judgment in favor of the manual and limits his focus to a set of symptoms rather than to the patient as a whole. "Psychotherapy is essentially concerned with people, not conditions or disorders," wrote one dissenting psychiatrist, "and its methods arise out of an intimate relationship . . . that cannot easily be reduced to a set of prescribed techniques." Add to this objection the fact that for both subject and therapist the proceedings are framed as a research project rather than as an encounter whose intention is to ease psychic suffering, and you have to wonder if the therapy studied in clinical trials is merely an artifact, a bell jar version of the real thing.

Cognitive therapists are aware of this disconnect, or at least Leslie Sokol is. On the first day of our workshop, she told us not to sweat the data too hard, at least not the part about people getting better after a prescribed dose of sessions. "Cognitive therapy is thought of as time limited because research demanded it," she said. "We delivered this amount of sessions not because there was a magic number but because we were running trials and we can't run them indefinitely. Time limited," she added, "really means goal limited." (It also evidently means "having it both ways," as in claiming to have a lab-validated treatment model that specifies a certain dose of therapy, but then, when out of the glare of the lab lights, not sticking to it.)

Cognitive therapists don't only claim that their treatment works; they also assert that it is superior to therapies that haven't been tested. This is another advantage of adopting the drug model; according to the logic of clinical trials, absence of evidence is evidence of absence. That's why Steven Hollon, an early collaborator with Aaron Beck and a leader in the field, can get away with writing that the fact that "empirically supported therapies are still not widely practiced . . . [means] that many patients do not have access to adequate treatments"—as if it had already been proved that the only adequate treatments are empirically supported therapies.

That's not the only way that the fix is in. Consider what happens when researchers try to institute placebo controls. In a drug trial, the placebo is a pill, and it is at least arguable that the only difference between the placebo and the drug is whatever is inside the two pills, so long as the patient is otherwise treated the same. Early EST trials used waiting lists as the placebo treatment; people do indeed get better merely by being told that help is on the way. But that procedure does not allow researchers to zero in on the active ingredient—assuming such a thing exists—of a given therapy.

The remedy is to compare two kinds of therapy that differ only in their specific interventions. But most forms of psychotherapy weren't designed to be manualized—not to mention that the people who practice them aren't leading the charge to measure therapy outcomes. It has been left to cognitive therapists to invent their competition, with the predictable results. One study, for instance, pitted cognitive therapy against "supportive counseling"—a therapy made up by the researchers for their trial—as a treatment for rape victims. The subjects in the supportive counseling group were given "unconditional support," taught "a general problem solving technique," but "immediately redirected to focus on current daily problems if discussions of the assault occurred." It's not surprising that the patients who couldn't talk about their assault didn't fare as well as the patients who could (and who were getting cognitive therapy), but that does cast doubt on the conclusion that cognitive therapy should take home the prizes. Proving that a bona fide therapy provided by someone who believes in it, who is inculcated with its values and traditions, works better than an ersatz therapy, implemented by someone who doesn't think it is going to work, may only show, as one critic put it, "that something intended to be effective works better than something intended to be ineffective."

Allegiances do matter, both the therapist's and the client's. Even in the lab, outcomes are consistently much higher when clinicians believe in what they are doing. I may not be entirely certain of why I want to talk with Eliza about her strawberries, I may indeed be flying by the seat of my pants, but I do believe that we're going to land somewhere better than where we were in the first place, and I'm sure I convey that confidence to Eliza. This kind of confidence shows up in the numbers as clearly as Judy Beck's belief that substituting Positive Triangles for Negative Rectangles will help cure depression. Furthermore, clients who don't have some loyalty to their therapists, or who don't believe that whatever is happening between them is going to help them, don't stick around.

This is why critics object to another statistical procedure common to clinical trials: excluding from the bottom line the subjects who don't complete the study, people who presumably didn't feel that confidence or loyalty. Rather than counting them as failures, most studies simply treat dropouts as if they never enrolled in the first place, which, mathematically speaking, makes the treatment look stronger than it would otherwise. And the numbers also exclude those people who were not allowed into the study because their case wasn't diagnostically pure enough—a move that allows researchers to improve their numbers by cherry-picking the patients most likely to benefit from their treatment.

Researchers can study the effect of these and other methodological problems by using meta-analysis, a statistical technique that allows them to determine the mean of means, or, in layman's language, what all the studies lumped together say about a particular factor—even one that the original scientists didn't necessarily intend to examine. So, for instance, two independent groups of researchers have used meta-analysis to factor out the advantages that cognitive therapy has when it goes up against treatments intended to fail. They scoured the literature for studies in which all treatment groups were given bona fide therapies. After crunching the numbers, they came to the conclusion that when the competition was fair, there was no difference in the effectiveness of the treatments.

Two other psychologists—Drew Westen and Kate Morrison— meta-analyzed thirteen leading studies of psychotherapies for depression, eleven of which used some form of cognitive therapy.

Overall, about half the subjects improved—results that put the treatment in the same ballpark as antidepressant drugs. But Westen and Morrison discovered that only one-third of the patients who tried to get into the studies were accepted—presumably because they didn't pass diagnostic muster—which limits the generalizability of the study. And of that select few, so many dropped out before the trial ended that the overall number of subjects who improved was only 36.8 percent. And when they looked at the handful of studies that followed their subjects over the long haul (and this is another way that therapy trials mirror drug trials; the book is closed after eight or ten or twelve weeks, and only rarely does anyone ask if the treatment remained effective), of the 68 percent of completers originally reporting improvement in those trials, only half remained improved after two years.

Westen and Morrison are quick to point out that they aren't saying that the therapies don't work. They help a carefully chosen portion of patients for a short time. That's not trivial, but it is less than the claim that cognitive therapy is the scientifically proven treatment for the disease of depression, and far less than what you would expect of a therapy that has become the standard of care for the AMA or "the most widely practiced approach in America."

Westen and Morrison acknowledge that their work is not exempt from allegiance effects. They think that cognitive therapy's success depends on a redefinition of psychotherapy with which they disagree. Their objection, they warn, may have unconsciously influenced their choice of studies to include or the hypotheses that guided their results. Indeed, all the critics of ESTs seem to be similarly inspired by a disagreement about how therapy ought to be practiced and evaluated and a distaste for cognitive therapy's answer to that question, for the way that therapy, like Heisenberg's subatomic particles, is changed by the very act of measuring it. The dispute, in other words, is not about the effects of therapy but the nature of therapy—and, by extension, the nature of human suffering and its relief. And like all ideological arguments, this one cannot be settled by numbers alone.

But there is one set of numbers that bears particular weight: findings generated by a group of loyal cognitive therapists. The team, led by prominent cognitivists Neil Jacobson and Keith Dobson, set out to investigate Beck's pivotal claim that his therapy has active ingredients that target the psychological cause of depression. Jacobson and Dobson wanted to determine whether some of those ingredients could be effective in isolation from the others—presumably because this might make an even more efficient therapy. They separated patients into three groups—one that received cognitive therapy according to Beck's manual, one that was given only the component in the manual directed toward behavioral activation (using activity schedules and other interventions to get patients into contact with sources of positive reinforcement), and one that got the modules that focused on coping skills, and in particular on assessing and restructuring automatic negative thoughts. The experimenters, all of them seasoned cognitive therapists, had an average of fifteen years' clinical experience, had spent a year training for this study, and were closely supervised by Dobson. And at the end of the twenty-week study, to everyone's surprise, there was no difference between the groups. Everyone benefited equally, just as the dodo bird hypothesis would predict.

Other studies, like one in which two cognitive therapists discovered that most improvement in cognitive therapy occurs in the first few sessions and before the introduction of cognitive restructuring techniques, strengthen the finding that to the extent that cognitive therapy works for depression, it is not because its specific ingredients act on specific pathologies. Instead, according to the meta-analysts, cognitive therapy's success depends largely on the therapeutic alliance, therapist empathy, the allegiance of the therapist to his technique, and the expectations of the patient—the same nonspecific factors that Aaron Beck intended to eliminate in the first place. "*How* therapy is conducted is more important," as one researcher put it, "than *what* therapy is conducted." As it does in drug therapies for depression, the placebo effect deserves most of the prizes.

But in real life, the prizes go to cognitive therapy, especially the prizes doled out by insurance companies. A therapist can't get sued for not practicing cognitive therapy, at least not yet, but there are other, more direct ways, to persuade us. According to the *New York Times*, insurers "often prefer their consumers" to go to cognitive therapists. Only a few health plans—the ones that employ their own counselors—can directly enforce this preference. But they can all require, as most companies do on the treatment reports they make me fill out as a condition of reimbursement, that therapists specify a "definition of successful treatment," with "desired observable outcomes" and deny coverage if those goals—themselves lifted from cognitive therapy manuals—don't address dysfunctional thoughts and core beliefs. They can also limit therapy sessions on the grounds that it has been scientifically proven that depression can clear up in fifteen or twenty visits, and that if this didn't happen then a therapist must not be providing adequate treatment.

I don't mean to sound unreasonable here. Insurance companies have every right to figure out how to spend their money; that's just the price we pay for placing our health care in the invisible hand, and every therapy dollar we coax out of it is a small miracle. You can't blame health care managers for favoring cognitive therapy, not when it's advertised to take half the time of traditional therapies. Nor can you blame the media for reporting breathlessly on this "scientifically proven" therapy. Or universities for hiring cognitive therapy researchers, who can bring in government money for investigating ESTs, over adherents to other schools. Or the professional journals for running article after article touting the virtues of the therapy without paying much attention to the bell jar problem, or to the statistics indicating that caution is in order. Intentionally modeled on drug therapy, fashioned as a kind of verbal bolus targeted at cognitive symptoms, an antidepressant without side effects or pharmacological Calvinist implications, cognitive therapy just fits in too perfectly with medicine's magic-bullet aspirations to be resisted, no matter what the numbers really say.

★   ★   ★

Cognitive therapy resembles antidepressants in one other way that Beck probably didn't intend—and that he surely doesn't mind. By succeeding as a treatment whose active ingredient is targeted at a specific pathology, it provides a backdoor validation of the idea that depression is a disease. It also provides a benefit for the disease model that the drugs don't. Drug manufacturers are vague about exactly what it means for people to "get back to themselves" or just what kind of "life is waiting" for them. It's enough to reassure people that their medication is only going to restore them to "health"—rather than to remake them—and leave it at that. But cognitive therapy is very clear about who we will be when we are cured: smoothly functioning processors of information, resilient navigators of life's ebbs and flows who can "take off those tinted lenses and see the world for what it really is," as Leslie Sokol exhorted us, and get down to business.

After four and a half days in this airless room, I still haven't accepted the idea that the world really is a place that offers up nothing I can't handle, if only I can restructure my negative thoughts and shed my self-doubt, that when I repair the glitches in my software, I will finally be able to make it. Indeed, I'm chafing against Beck's and Sokol's relentless can-do optimism, weary of their talk of coping skills, their agendas and strategies, their paperwork. Their model of life as a series of challenges to be managed efficiently is as bland and disappointing as this suburban office building. It just doesn't do justice to the perversity of our nature or to the seemingly limitless tragedy on which it feeds.

Which is why I'm not going along with Judy Beck right now. That's easy to do in my role as Ann, who was nothing if not perverse. After she tells Ann about her Negative Rectangles and Positive Triangles, Beck asks, "Do you think this is true, that your mind does work this way?"

"I think you're just trying to get me to set the bar lower."

"What does that mean exactly?"

"Either I've got you fooled, or you're trying to fool me. You're trying to argue me out of something I wasn't argued into in the first place."

"Well," Beck says, "let me apologize. Because I wasn't trying to argue with you at all."

I am silent. I have no idea how Ann would respond to that. I know what *I* would say: "HUH?" Because arguing is exactly what I think has been going on here, not just in this role play, but for this whole week. After all, they call their examination of a patient's thinking "Socratic questioning," and didn't Socrates seek to argue his students into the correct position by posing questions to which he already knew the answer?

Still, I can see Beck's point. She's not arguing. In fact, she's taking her own medicine, looking for the problem not in Ann's contrariness (or mine), but in her own cognitive apparatus. She breaks out of character to explain this to the class. "I think what happened in the role play is that I started this process too early with her," she says. "She doesn't buy this yet." But Ann's resistance is probably futile. Beck addresses her again. "We have to be really careful that I not sell you a bill of goods. I don't want you to start seeing reality differently if it's not realistic," she says. "On the other hand, I don't want you to sell yourself a bill of goods and see everything as negative if it's not really that way. Anyway, could we send this piece of paper home with you and you think about it a little bit and we can talk about it more next week?" That's not arguing. It's just waiting for Ann to come to her senses.

There's no need to argue about what is right here in front of our eyes. If Ann just realizes that her way of looking at the world isn't working, if she just thinks about it a bit, if she just listens to this self-assured and calm woman who doesn't seem a bit troubled by her own success, she'll understand that the problem isn't out there, in an economy that doles out reward and punishment in strange and unequal measure, in a nature that doles out a case of rubella at the worst possible time, in a world where fluke often trounces honest toil, in which it is possible for a woman to learn, on the very same day, that her mother is dead and her father is not her father, where anyone can meet the fate of Job at any moment; the problem is *in here,* in the faulty cognitions that prevent us from seeing the world as the benevolent place that it really is and moving on to the next item on our agenda.

This may not sound like positive thinking to Leslie Sokol, but the father of positive thinking would surely have a different opinion. Ten years or so before Aaron Beck began to craft cognitive therapy, Norman Vincent Peale was saying things like, "The basic reason a person fails to live a creative and successful life is because of error within himself. He thinks wrong. He needs to correct the error in his thoughts." And "The world in which you live is not primarily determined by outward conditions and circumstances but by the thoughts that habitually occupy your mind." And "A man's life is what his thoughts make of it." And "An inflow of new thoughts can remake you regardless of every difficulty you may now face." And finally: "If you think in negative terms you will get negative results. If you think in positive terms you will achieve positive results." *The Power of Positive Thinking* first appeared in 1952 and ministered to the demoralized, especially to the demoralized businessman, the one who was feeling downtrodden despite his successes, who had given in to circumstance. Peale wanted that man to understand that he could always pick himself up, shake off his losses, and reinvent himself. He wanted him to remember that in America, if you put your mind to it, anything is possible.

Unless you're depressed. Indeed, this is the essence of depression, why it's a pathology. What the cognitive therapists spell out, and what the psychiatrists only imply, is that depressed people, all depressed people—the melancholic and the neurotic, the endogenous and the reactive, the dysthymic and the majorly depressed— have in common their demoralization, their inability to see a limitless horizon, their despair over the possibility that their longings will never be satisfied. They have stopped pursuing happiness.

What reassurance there is in this idea that a properly function-ing mind is one that is always able to get on with the business of liv-ing! Peale, like the Becks, shifts the burden of demoralization from the *out there* to the *in here* in exactly the same way that Eliphaz did with Job, but then they go him one better. Because the modern person can do more than shrug his shoulders and figure that God knows what he is doing; he can use the principles of science to call himself sick and take the cure. If he takes off those tinted glasses, he won't see anything he's not equipped to see and then meet with resilience. He doesn't have to settle for Job's choice between bit-ter denunciations of the terms of existence and awe in the face of the whirlwind. Because, evidently, God (or is it nature? or just the brain?) meant for us to be happy.

What a felicitous coincidence—to be an organism designed for happiness in a land dedicated to its pursuit! And here is another way that cognitive therapy helps us understand depression's wild suc-cess in the marketplace of ideas about us. Because to be told that depression is a disease is to be reassured that when we are discour-aged, we are not really sick at heart. We are just plain sick. Which means we can get better. We don't have to look back at the fire that once rained down on us or outward to the inhumanity inflicted in the name of our prosperity or forward to the certainty of our own suffering. We don't have to be stunned at the cruelty—or, for that matter, thrilled by the tragedy—of life on earth or worried that pur-suing happiness the way we do is also pursuing destruction. We can be healed. We can get our minds to work the way they are supposed to. And then we can get back to business.

CHAPTER 14

THE NEW
PHRENOLOGISTS

I should have quit while I was ahead. My clinical trial ended well for me, I thought, at least for the me for whom I was rooting all along, the one whose depression has some meaning and whose failure to take antidepressants is not a mistake that puts me and my family at risk. The raw material out of which depression is manufactured—the idea that this pageant of selfhood, the stories we cobble together out of our lives, and the epic of history that all those stories together make is all just a byproduct, an illusion manufactured by our molecules, something to keep us busy while they go about their business of dividing and replicating—is repellent to me. It's demoralizing. It's nearly intolerable. Who wants to be the tail of some electromolecular comet?

So I was happy to have my ontological commitments confirmed by Harvard scientists. The cognitive therapy expedition, on the other hand, was a more measured success. The method didn't prove itself ineffective, but the conditions of its effectiveness, its dependence on our very peculiar social arrangements and on the corporatism that has come to dominate our self-understanding, were unmistakable. I got a glimpse of the finishing room in the depression factory, the place where the last touches are put on the gleaming new self. Much as I could see the appeal of resilience, of being able to get down to business no matter what, I also wasn't ready to "buy in," as Judy Beck put it. I wanted to leave my capital where it was.

There's no double-blind test to verify that my view is right (or wrong). Which is why, I must confess, I was pleased when the lab called and told me that I'd been on the placebo. Score one for the good guys, I thought. And why the relative ease with which cognitive therapy showed itself to be a method of indoctrination into the pieties of American optimism, an ideology as much as a medical treatment, was a relief. And why, as I looked back at where this idea about depression came from in the first place, I was gratified to find that it came from us, that its history is not that of a law of nature slowly revealed but of an invention, of aspiration and compassion and the determination to alleviate suffering, and that the human voice, however muted, can still be discerned behind the clanking machinery. Not only because I proved, to myself anyway, that I wasn't entirely crazy to think that depression was invented and not discovered, but because it meant that the climate of opinion still was not fully formed, that there could be a break in the weather.

Like I said, I should have quit while I was ahead.

Of all the evidence that the depression industry is selling a bill of goods, perhaps the strongest is the problem that has been there all along: the fact that depression is nothing more or less than its symptoms. The American Psychiatric Association's claim to have solved this problem with the DSM-III was only a renewal of the promissory note that Emil Kraepelin had issued eighty years earlier. Even the highest priests of psychiatric orthodoxy will, at least in private company, admit that they haven't resolved this conundrum so much as legislated it out of existence. "The DSM-IV . . . has 100 percent reliability and zero percent validity," Thomas Insel, the director of the National Institute of Mental Health told psychiatrists gathered for the APA's annual meeting in 2005.

Did rank-and-file psychiatrists, upon hearing from America's psychiatrist in chief that their diagnoses were fraudulent, stop doling them out? Did they stop delivering their version of Frank Ayd's speech along with their prescriptions? Did the drug companies stop comparing depression to diabetes? No, they did not. But then again Insel's candor wasn't intended to convince his brethren to repent their zero-validity ways. It was to announce the happy news that redemption from the Kraepelinian purgatory was at hand.

> Brain imaging in clinical practice is the next major advance in psychiatry. Trial-and-error diagnosis will move to an era where we understand the underlying biology of mental disorders. We are going to have to use neuroimaging to begin to identify the systems' pathology that is distributed in each of these disorders and think of imaging as a biomarker for mental illnesses . . . We need to develop biomarkers, including brain imaging, to develop the validity of these disorders. We need to develop treatments that go after the core pathology, understood by imaging. The end game is to get to an era of individualized care.

The holy grail of psychiatry, according to Insel, was in sight—literally. Their vision extended by scanners like MRI and CT and SPECT and PET, doctors would soon be able to peer into the brain and find there "the basic pathophysiology of each of the major mental disorders." Psychiatry, recast as clinical neuroscience, would then be able to claim to have found what lies behind the symptoms, what makes mental disorders diseases. And having identified the targets with precision, psychiatry would be able to "emerge once again as among the most compelling and intellectually challenging medical specialties" and be "integrat[ed] into the mainstream of medicine." Perhaps even more important, it would be on its way to a world in which everyone has his own disease and his own personal magic bullet.

Insel wasn't saying when these halcyon days would arrive, but at least one doctor thinks the future is already here. Daniel Amen has been using single photon emission computed tomography (SPECT) to find biomarkers for mental illnesses for fourteen years. He's a casually dressed man who, despite his baldness, looks younger than his fifty-six years. A Lebanese-American who grew up Roman Catholic, converted to Pentecostalism while he was a soldier in Germany, and got his undergraduate education at an Assembly of God Bible college, Amen was a member of the charter class of the Oral Roberts University School of Medicine. "I was suspicious that they accepted me because of my last name," he once wrote. "The first graduate of their medical school would be Dr. Amen."

Amen has made the transition from psychiatry to clinical neuroscience, and he's practicing the new discipline right now, detecting psychopathology in my brain. He's looking at a set of SPECT-generated photos taken over the past couple of days at his office in Orange County, California. Amen's waiting room is crowded, mostly with adolescents and their parents; three women behind the reception counter are answering phones and clacking away on computers. A display case features Amen's books and DVDs: *Healing the Hardware of the Soul; Change Your Brain, Change Your Life; Magnificent Mind at Any Age.* They're all for sale at the desk.

Everyone here is kind and friendly. The receptionist smiled at me and touched my hand as she took my paperwork and my check—$3,250, which covered two scans and two hours with Amen. The woman who recorded my psychiatric history yesterday seemed genuinely moved when I described my struggles, and Mike the nuclear medicine technician apologized for the pinch when he punctured my arm with an IV line for the radioactive dye that would allow the camera to capture the blood flow in my brain.

Yesterday, Mike injected the dye while I was attending to my concentration task—sitting in front of a computer, hitting the space bar every time a letter flashed onto the monitor, unless it was an X, in which case I was supposed to refrain. I found this nearly impossible—the tedium alternately made me inattentive or impulsive. Today, he hit me up and then told me to sit in the darkened room and do nothing for fifteen minutes—not even fall asleep, which was a little tricky given that I'd been forbidden caffeine (and all other psychoactive drugs) for the past two days. On both days, I lay down afterward on a hard table in a chilly room while a nuclear-sensitive camera circled my skull, snapping 120 photos of cross sections of my brain. The prints Amen is holding are computerized reconstructions of the SPECT scan, and they're giving up my secrets.

He leads with the good news. "Your brain is really healthy," he says. He's showing me how he knows that, pointing to various parts of the picture, talking about temporal lobes and basal ganglia and regional cerebral blood flow. I'm having a hard time following him. I'm distracted by the picture itself, the yellow and pink and purple and green hues melting into one another on the pockmarked surface of my brain. I wasn't expecting my brain to look so much like a tie-dyed moon rock.

But maybe that's not the only reason I'm distracted. The good news told, and the disclaimer given that it would be silly to make a diagnosis based on brain scans alone, he says, "You probably have ADHD," using the abbreviation for attention deficit/hyperactivity disorder. We've moved on to the active view of my brain, in which it shows up as a helmet-shaped latticework woven from threads of deep blue and red. Behind some sections of the grid are white bundles wrapped in more red and blue threads. "When you concentrate your cerebellum drops fairly dramatically," he says, holding two versions of the same view and pointing to the relative lack of red strands in one of them. "Normally, it stays the same." And he can see my depression too. "It's this triad—cingulate gyrus, basal ganglia, thalamus." He's connecting areas of the lattice with his pen. "Your cingulate is up, which usually means your serotonin is low, and the basal ganglia usually goes along with anxiety."

And that's not all. "This right here?" He's connecting some other blotches of color. "That's the diamond-plus pattern. When I see this pattern I ask, Have you ever been traumatized? Sounds like growing up there was plenty of trauma in your family, that your mother was giving it regularly." Which means, he says, that I am a candidate for a technique called EMDR—eye movement desensitization and reprocessing—to go along with the omega-3 fatty acids and L-tryptophan and maybe some SAM-e that he is recommending for my depression and ADHD. He's not averse to pharmaceutical drugs—although he's also skeptical of them and has done all his research independent of drug company money—and if the supplements fail, he thinks Effexor is the "right bullet" ("I guess that's not such a good word for an antidepressant," he adds quickly) because it "works on serotonin and dopamine and if you did just serotonin, your mood would be better and your ADHD would be worse and to do dopamine without serotonin your focus would be better and your worrying would be worse."

Amen says some sensible things. "A diagnosis of depression is like a diagnosis of chest pain," for instance, which strikes me as a concise way to get at the problem created by the DSM's eagerness to turn all depression into a single disease. He thinks that the symptoms point to many diseases, each with its own brain pathology, and that the DSM-VI (an inevitable development, although the APA is only now working on the DSM-V) will be organized accordingly. He's also honest about the economics of his efforts. He's not in the pocket of the drug companies, secretly fueling their marketing efforts with his research. Instead, he asks his patients to sign a consent form allowing their scans to become part of his database, which he then uses to strengthen his case that certain psychological illnesses go along with the brain pathologies that he, more than anyone else, knows how to recognize and treat. These overlapping roles—researcher, clinician, entrepreneur—may create the grounds for all kinds of murky ethical problems. I didn't exactly feel like I could refuse to sign the consent form, and the experience gave me a new appreciation for the team at Mass General; they hadn't made me pay to be their guinea pig. But at least there isn't any mystery about who is benefiting from Amen's research, and how.

Still, however, as Amen goes on in his calm and confident way about empty cerebellums and hypoactive cingulates, too much serotonin and not enough dopamine, trauma diamonds and depression triads, and the bounty that awaits me when he balances my brain— "With a healthy brain," he tells me, "your free will is greater, and so is your ability to have people trust you over time, be able to be engaged in a loving long-term relationship, be a good dad, to be more thoughtful and loving"—it's impossible to shake the feeling that Daniel Amen is a high-tech, nuclear-armed quack and that I've stumbled into a twenty-first-century medicine show.

I'll admit that this sentiment is not only uncharitable—not that he needs my charity, not at $3,250 a pop, not to mention the bestsellers and the lecture tours and the television specials—but it's awfully convenient, self-serving even. I mean, it's not like I didn't go into this meeting with a deep suspicion of the project of rendering human life as a series of biochemical events, and of Thomas Insel's vision of the psychiatry of the future. For that matter, it's not like I didn't go into this book without an idea about what is wrong with the medical industry's invention of a brain disease out of our daily troubles and aspirations. After all, just because he's a little too slick and greedy for my tastes, that doesn't mean he's wrong.

Amen has met the likes of me before—people who don't want to accept the morphing of psychiatry into clinical neuroscience, of the mind into the brain, of discontents into glitches. "When I first started doing this work," he told me, "I got no end of grief from my colleagues." So, he says, he learned to shut up, speaking very little about what he was up to and only to what he thought would be sympathetic audiences. But in 1996, he gave a talk to a professional group, and one audience member, incensed, Amen says, by his claims to have discerned the physical foundations of mental illness, hauled him before the California Medical Board. He complained that Amen was practicing outside the standards of care, using equipment that only neurologists had the training to use—an offense for which a doctor can lose his license. Amen battled the board for more than a year and finally was cleared of the charges—but not before he realized what he was up against. "One is often labeled a heretic for trying to change religious beliefs," he says.

But compared to what the French and Dutch authorities did to Julien Offray de La Mettrie in the eighteenth century for suggesting that consciousness was the product of a machine-like brain, litigation with a medical board is a walk in the park. Similarly, my showing up here with a check, a brain, and my skepticism doesn't seem so bad compared to what Mark Twain did to the American phrenologist Lorenzo Fowler in 1873.

By then, America had proven itself a hospitable home to the brand of phrenology originated by Franz Gall's assistant Johann Spurzheim, who had redrawn the skull map in a way that eliminated the bad bits. Spurzheim had also declared that evil and suffering were not inherent properties of the brain but the result of pathologies that could be corrected. He did well enough in Europe, but Spurzheim's science of self-improvement set off an outright frenzy in the land dedicated to the pursuit of happiness. Spurzheim arrived in New York in August 1832 and lectured in New Haven and Hartford before arriving in Boston to a reception so enthusiastic that he soon found himself lecturing to overflow crowds at the Masonic Temple, to Harvard students and faculty, and at the Boston Medical Society. It was a killer pace—literally. He died at the end of September of exhaustion and fever.

Those two months, however, were long enough to capture the attention of a young Amherst student and aspiring minister named Orson Squire Fowler. By the time he had graduated in 1834, Fowler had examined heads in and around Amherst with such success that he abandoned his plans for the ministry and became a full time phrenologist. With his brother Lorenzo, he married Gall's brain maps and Spurzheim's optimism to American pragmatism and took the resulting self-help science to the streets, offering readings intended to help common people learn "what they *are,* and what they *can* be as well as how to make themselves what they should become." What Gall had eschewed, what Spurzheim had only implied, the Fowlers said right out in their phrenological self-instructor, one of America's first self-help books: the purpose of life is happiness, which is achieved when the organism—specifically the brain—is functioning as it should. "Would you become great mentally," they wrote, "then first become great cerebrally." Or as Daniel Amen would say a century and a half later, change your brain and change your life.

The Fowlers built a therapeutic empire that eventually included not only the brothers but their sister, her husband, Lorenzo's wife (who in 1850 became the second American woman to earn a medical degree), and their daughter. They set up shop in New York City, with satellite offices in Philadelphia, Boston, and even London. President James Garfield, Brigham Young, John Brown, and Oliver Wendell Holmes all offered their skulls to the Fowlers' hands. Like everyone else, these luminaries went home with charts of their heads, on which their strengths were outlined in black and white for all the world to see.

By midcentury, phrenology had reached into most areas of American public life. But it was a particular hit with the literati. Horace Greeley published Combe's lectures and suggested that railroad men be selected on the basis of the shape of their heads. Horace Mann called phrenology "the greatest discovery of the age." Edgar Allan Poe wrote that "Phrenology . . . has assumed the majesty of a science, and, as a science ranks among the most important which can engage the attention of thinking beings," and built a poetics on phrenological tenets. Ralph Waldo Emerson spoke favorably, if grudgingly, about the Fowlers' achievements.

But no literary figure was more effusive than Walt Whitman. "Phrenology, it must be confessed by all men who have open eyes," he wrote, "has at last gained a position, and a firm one, among the sciences." The relationship between Whitman and the Fowlers was a mutual lovefest. He sang their multitudinous praises—inspired, no doubt, by Lorenzo's reading of his skull, which proved him to be well endowed in the areas of Friendship, Sympathy, and Self-esteem, not to mention Voluptuousness and Amativeness—and Fowler & Wells, phrenology's publishing arm, published his poems, including *Leaves of Grass,* which they then allowed him to review (anonymously, of course) in the *American Phrenological Journal.* Whitman gave the book a favorable notice.

We can only imagine how all this piety and respect sounded to a skeptic like Mark Twain, although his telling makes it clear that his motives were no purer than Joseph Wortis's or David Rosenhan's (or, of course, mine):

> I made a small test of phrenology for my better information. I went to [Lorenzo] Fowler under an assumed name. He examined my elevations and depressions, and he gave me a chart . . . He said I possessed amazing courage, an abnormal spirit of daring, a pluck, stern will, a fearlessness that were without limit. I was simply astonished at this, and gratified too; I had not suspected it before . . . However, he found a CAVITY in one place where a bump should have been in anybody else's skull . . . He startled me by saying that that CAVITY represented a total absence of a "Sense of Humor"!

Twain went back to Fowler three months later, this time "bearing both my name and my *nom de guerre.*" This time the news was different:

> The CAVITY was gone, and in its place was a Mount Everest—figuratively speaking—31,000 feet high, the loftiest BUMP OF HUMOR he had ever encountered in his life-long experience!

It wasn't Twain's prank that undid phrenology, however, but rather the advent of scientific medicine, especially the nascent field of neurology. In 1848, for instance, a railroad worker in Vermont, Phineas Gage, was loading a hole with gunpowder when an errant spark set off the charge, sending the tamping iron through Gage's skull, carrying with it a good portion of Gage's frontal cortex. Gage was brought to a local physician, John Harlow, who treated him with the usual purgatives and poisons, amputated the herniated brain tissue, and closed his skull as best he could. Gage floated in and out of consciousness for a couple of weeks, but he was soon up and around and ready to get his job back. But the railroad didn't want him. As Harlow reported, Gage was a changed man. "The equilibrium or balance . . . between his intellectual faculties and animal propensities," he wrote in a medical journal, "seems to have been destroyed." Gage, who was once a gentle and reliable man, had become irascible, unpredictable, and even mean—so much so that his friends, according to Harlow, reported that "Gage was no longer Gage." Gage left Vermont, worked as a coach driver in Chile, and died in San Francisco in 1860.

Harlow was sympathetic to phrenology, but he didn't report the fact that Gage was a changed man until 1868—perhaps because the areas of the brain wiped out by the tamping iron, according to Gall's charts, governed functions like poetical talent and acquisitiveness, and not equanimity or amativeness. But by then, other discoveries had called the phrenological brain map into question. In 1861, for instance, Paul Broca, a doctor in Paris, autopsied an epileptic man who, before he died of gangrene, had lost the power of speech. He found a "softening" in an area at the rear of the left frontal lobe, a lesion that showed up subsequently in other people who could not speak. Broca's area, as that section of the brain came to be known, was soon joined in the brain gazetteer by Wernicke's area—a region discovered by German neurologist Carl Wernicke that, when damaged, causes receptive aphasia, the loss of the ability to understand speech. And in the 1870s David Ferrier, a British doctor experimenting on monkeys, found that frontal cortex damage like Gage suffered could induce profound changes in behavior.

These discoveries didn't kill phrenology, not entirely. Doctors digging around in brains may have discredited the idea that the skull revealed the brain beneath it, but they also confirmed Gall's basic insight (or maybe it was just a lucky guess): that the brain is divided into regions, and that those regions correspond to various functions. Gall was wrong in the particulars, but right that the brain, much as La Mettrie had said, was an elaborate clockworks.

At least that's the story that historians of medicine like to tell: that real scientists managed to distill the gold of neurology from the dross of phrenology, Broca's epistemological modesty triumphing over the Fowlers' hubris, real doctors vanquishing the quacks. But the Fowlers' version lives on, and not only at Daniel Amen's clinic, where anyone can plunk down some cash and come away with a picture of the bumps on the inside of his head, a reading of their significance, and a program for brain improvement. It's in the inescapable images in every magazine and newspaper and television program about health, the full-color scans showing the brain at work and relaying the news that doctors have planted the flag of science in our brains like sixteenth-century explorers, that they have discovered the real basis of empathy or racial prejudice or sexual orientation, claiming for the brain territories once thought to be the province of the mind.

Especially depression. Pick up the October 16, 2008 issue of *Nature*—okay, it's not exactly light reading, but it is about as mainstream science as it gets—and you'll find out just how far into the brain the depression doctors have penetrated. After you hear from two of the leading lights in the field that "the official diagnosis of depression is subjective" and that its cause is "far from being a simple deficiency of monoamines"—once again, doctors can acknowledge to one another that what they tell their patients is not true—you'll learn that depression can be found in the *grey-matter volume* and *glial density* of the *prefrontal cortex* and the *hippocampus,* in the *amygdala* and *subgenual cingulate cortex* and the *nucleus accumbens.* You'll find that the reason SSRIs work, when they work, may be that they cause *secondary neuroplastic changes,* themselves perhaps the result of *upregulation* of the *calcium-binding protein p11* or the *transcription factor CREB,* which, following the logic of psychopharmacology, indicates that those are the causes of depression. Or maybe the culprit is a lack of *brain-derived neurotrophic factor* or *vascular endothelial growth factor* leading to a decrease in *hippocampal neurogenesis,* or too many *cytokines* or *glucocorticoids* floating around the brain or just glitches in the *intracellular interactions between brain macrophages.* You'll also discover that *resilience,* the ability to withstand stress and adversity that the depression doctors define as health, may be due to a good supply of *ΔFOSB,* especially in the *midbrain periaqueductal grey nucleus* and a relative lack of *Substance P,* or even a reduction in the stress-related *increased excitability of VTA dopamine neurons* caused by *upregulating voltage-gated potassium channels.*

And if that doesn't convince you that doctors are hot on depression's trail, take a look at the *New England Journal of Medicine,* in which a team of French doctors reported that a very strange thing happened when they were treating a sixty-five-year-old woman with deep brain stimulation (DBS) for her Parkinson's disease. That procedure involves inserting tiny electrodes into carefully chosen individual neurons and turning on a low-voltage electrical current, which generally relieves the tremors and spasticity associated with Parkinson's. In this case, within a few seconds of receiving the charge, the woman, who had previously shown no signs of depression, started to cry. "I no longer wish to live, to see anything, to feel anything," she said. "I'm fed up with life, I've had enough." And in case her doctors missed the point, she continued, "Everything is useless, always feeling worthless, I'm hopeless, why am I bothering you?" Ninety seconds after the stimulator was shut off, her depression disappeared.

A few weeks later, the woman agreed to try it again. This time, the doctors took pictures, which they published with the article. The first photo shows a relaxed woman with a Mona Lisa smile. Within thirty seconds, her right hand is gripping the lower part of her face, pushing her mouth into a pout; her eyes are downcast, and she looks thoughtful and worried. Four minutes later, she is weeping, her eyes squeezed shut against the feeling that, as she puts it, her "body is being sucked into a black hole." Five minutes after that, she is smiling again, this time broadly, as if in celebration. "The neural networks involved in this particular case have not been clearly identified," the doctors wrote, although it may have involved the *ventral nuclei* of the thalamus or perhaps the activation of the left *pallidum*. Whatever the specifics, they wrote, the important thing is that the stimulation affected only "a few cubic millimeters of neural tissue," which means that "the depression probably resulted from the stimulation of afferent, efferent, or passing fibers within the substantia nigra or from the inhibition of those fibers."

If you don't understand any of this brain talk, don't worry too much. I don't really understand it either, any more than Walt Whitman really understood the Fowlers when they went on about his philoprogenitiveness or his inhabitiveness or than I fully understand what Daniel Amen means when he tells me about my *cingulate-basal ganglia-thalamus* triad. I mean, I know where those regions are, more or less, and I have some idea about what goes on, electrochemically speaking, in them—and so could you if you spent just a little time with a basic neurobiology textbook—but I really don't understand how they bring about my experience of depression. Even more important, I'm not entirely sure that they do, at least not all by themselves. I'm not sure what makes the depression doctors so certain that they do know, but when they write about depression as a disease that is the result of pathophysiological processes, when they say that the circuitry of depression, whatever it is exactly, gets turned on by stress—"kindling," they call it—and then in some unlucky segment of the population stays on even when the stress is over, creating its own stress, which creates its own depression, the brain consuming itself like a serpent eating its tail, depression an exercise in meaningless suffering, they sure sound as if they do.

But maybe we—those of us not equipped with scanners and advanced medical educations and prescription pads—are not supposed to understand. Maybe the point isn't to help us gain self-knowledge, at least not in the sense that that term usually carries, but rather to help us realize that we're way too complicated for ourselves, and so we might as well turn that work over to the only people who are qualified to understand the mumbo-jumbo, to the new phrenologists, who have used the SPECTs and PETs and MRIs to look inside our skulls and figure out what has gone wrong to make us suffer, and what will cure us. Maybe the point is to help us see that there's no point in looking elsewhere.

Because what else explains the fact that the French doctors never even considered in their article the possibility that the patient's sudden sadness isn't merely an artifact, a random by-product of their treatment, but that she is actually very, very sad about something—like maybe the fact that she has a debilitating and degenerative illness that will eventually ruin her—and that they somehow released a cascade of pent-up feeling, that her experience was not just a random electrical storm but real and potentially important and meaningful? Or that they had simply stumbled on one way in which the brain provides intense sadness, activated it in the same way that a tap on the knee activates a kick reflex, thus proving only the highly unremarkable fact that brain events are necessary for conscious experience? For that matter, what else but a conviction that the explanation for our suffering can be found in our molecular biology accounts for the total absence in the *Nature* article of any consideration of the actual lived experience of depression—not even a gesture at the possibility that what they are describing is the *result* and not the *cause* of a person's troubles, the *how* of our suffering and not the *why,* that our lives as we live them actually matter, and that the conviction that this is so is not a mere illusion (itself no doubt arising from some neurochemical trick) but is as real and as significant as any bit of grey matter? To write about depression without ever even mentioning what it feels like, or how it might be related to the actual circumstances of a person's life, seems nearly willful, as if subjectivity is a forbidden topic.

Which makes sense. All religions need taboos, and in the church of clinical neuroscience, subjectivity must be off-limits. To talk about a self as something that exceeds the sum of its parts is to challenge the notion, central to materialism and the scientism behind it, that we are nothing more or less than what our brains do. This idea has always been latent in the attempt to turn magic-bullet medicine toward our psychological suffering, but it has only become manifest recently, as scientists' ability to monitor and measure the brain has grown—and with it their belief that they have found the reality behind the illusion of self.

The conviction that biochemistry is more important than biography is as faith-based today as it was when Julien Offray de La Mettrie first proposed it. Psychoanalyst-turned-neuropsychiatrist (and Nobel laureate) Eric Kandel has no more scientific proof when he claims that "the mind is a set of operations carried out by the brain much as walking is a set of operations carried out by the legs" than La Mettrie did when he declared that the brain "possesses muscles for thinking as the legs do for walking." Kandel does, of course, have much more certainty about what is actually going on in those "muscles," but he doesn't know that that's all there is to it. Daniel Amen might be correct to say that he is seeing my depression in my hypoactive cingulate, but he can't tell me why it feels the way it does, how the ebb and flow of neurotransmitters through my brain turns into a pit in my stomach, how it nails me to the floor of my study. He can tell me everything about my brain and he still will have told me nothing about my self.

Or so I believe. And I may be a majorly depressed abuser of illegal substances, but I do have some sober company here. Carl Woese, for instance. He's one of America's most prominent molecular biologists and has been trying to alert his colleagues to the fact that to reduce mental life to its molecular correlates is one thing, and to claim that it is nothing more than whatever is happening in that chunk of neural tissue is quite another. This is the difference that Woese had in mind when he wrote:

> Empirical reductionism is in essence methodological; it is simply a mode of analysis, the dissection of a biological entity or system into its constituent parts in order to better understand it. Empirical reductionism makes no assumptions about the fundamental nature . . . of living things. Fundamentalist reductionism . . . on the other hand, is in essence metaphysical. It is ipso facto a statement about the nature of the world: living systems (like all else) can be completely understood in terms of the properties of their constituent parts.

Materialism may arise out of the wish to be rid of metaphysics, of something that simply can't be explained by science, of a doubt that can only be resolved by faith, but when it crosses the line into fundamentalism, it turns into a metaphysics of its own.

And when that metaphysics purports to explain our inner lives—as it most surely does when doctors tell us our depression is a disease of the brain—it has profound implications. Philosopher Karl Popper outlined them in 1977:

> With the progress of brain research, the language of the physiologists is likely to penetrate more and more into ordinary language and to change our picture of the universe, including that of common sense. So we shall be talking less and less about experiences, perceptions, thoughts, beliefs, purposes and aims; and more and more about brain processes, about dispositions to behave, and about overt behaviour. In this way, mentalist language will go out of fashion

and be used only in historical reports, or metaphorically, or ironically. When this stage has been reached, mentalism will be stone dead, and the problem of mind and body will have solved itself.

Repeat the fiction about the legs and the brain often enough over two hundred and fifty years, and eventually it comes true. Recite the gospel of brain chemistry, invoke your authority to see inside us and show us the miraculous pictures of our despair, name the secret regions of our brains in a language as incantatory, as mysterious and incomprehensible as a Latinate Mass, and then offer the sacramental pills that will absolve us of our original sin, of the imbalances that are in us but not of us, promise us the salvation of mental health in the form of the ability to meet whatever sorrow we encounter with resilience—illness and loss and the death of loved ones (after two months, of course), our own limitations and failures and those of our leaders, the creeping awareness that we're suffocating the planet, that living high on the hog requires billions of people to be our fodder, troubles both personal and political adding up to a world broken beyond repair: do all that, and we will line up for communion, we will take the sacrament, and we will be transformed into neurochemical selves, reinvented as the people of the pill. Not only because the drug changes our brain chemistry, although it undoubtedly does, but because it changes our idea of who we are. It's the biggest placebo effect of all.

So let's say you're writing a book about the way the medical industry has manufactured depression as a brain disease, how it has assembled the magic-bullet hopes and the yearning for self-improvement in a time of diminishing prospects, the Age of Enlightenment and the march of progress and the pursuit of happiness, the doctors and the ad men and the government regulators into a sleek and gleaming product. And let's say your best evidence that this invention is not all it's cracked up to be is the fact that for all the scientific language and scholarly discourse, for all the doctors' claims that they've found the wellsprings of demoralization, there's still no actual biochemical glitch that lies behind the symptoms, which means that the targets are only targets of opportunity and that the depression doctors are still working off a century-old promissory note. And then let's say you get to the part where the captains of this industry finally acknowledge this, although only by way of saying that they are on the verge of solving the problem, that soon, very soon, they will be able to take snapshots of depression right there in the brain, which will give us not only the true targets but also the final proof that the problem is not in our selves but in our molecules. Whereupon, let's also say, you find a doctor who insists that these days have already arrived, who has a book full of Technicolor brain pictures to prove it, and who shows you one of them (a teenage boy's, he says, who wanted to murder his mother), points out the "huge dent" in his amygdala ("dark evil thoughts here") and the hole in his interior prefrontal cortex ("and there the inability to control them") and adds for good measure that if only the boy "had good frontal lobes, he would just have those horrible thoughts but he'd have a brake; he doesn't have a brake in his head." And then let's say this conversation takes place in Southern California in late 2008, in one of the few office parks with cars in the lots and tenants in the suites, where building after building is festooned with tattered signs bearing futile messages about lease and sale opportunities, where the shuttle driver who ferries you to your deserted hotel from the equally deserted airport used to be a computer programmer, where the waitresses in the half-dark restaurants eye you like lions who haven't seen prey in a month (but where just down the road in Newport Beach, people are buying handmade dog biscuits at The Barkery and snapping canvas covers over the teak decks on their luxury yachts), and the doctor tells you that he can *see* your depression and your ADHD (and your mother too) in your head, and you think that now, more than in your clinical trial or your cognitive therapy workshop, now you're finally seeing what's dark and Satanic about the depression mills, that it's not just the arrogant scientism of the Harvard doctors or the slick corporate shine of the cognitive therapists, but the ambition, the hubris, the greed, from which this effort grows, the nearly desperate desire to reduce us to what can be grasped with the hand or glimpsed with the SPECT-aided eye, to make our suffering into a need that can be satisfied with a product, to *fix* us so that the outside world no longer really matters—even if it means robbing us of our tragedy, denying us our rage, playing Eliphaz to our Job, and in the bargain rendering each of us as an electromolecular stew just a dash short of this or that crucial salt. And then let's say it dawns on you that, when the doctors tell you that the reason you can't sleep in your nearly empty hotel and are worried and sad when you walk in the deserted streets is that there is something wrong with your brain, they are also telling you that you are nothing more than the output of the inputs, a cloaca through which chemicals pass, a mere residue that can be scraped and polished and recast. Which makes you a perfect customer, someone who will buy the illness to get the cure, who will take heart at the news that your brain is mostly healthy and that all it needs is a few grains of this or that, and who is ready to believe that a whole society of resilient people with chemically balanced brains would make for a better world, that this desiccated self is the paragon of health, and that none of this is a matter of conviction, but of fact, not a climate of opinion that arose from the hurly-burly of history and desire, but truth that has descended from the scientific heavens. And, finally, let's say the guy with the brain pictures turns out to be a shamelessly entrepreneurial, dyed-in-the-wool American optimist who is also a fundamentalist Christian graduate of a medical school founded by a faith healer, a man named Amen no less, who is faintly ridiculous and whose ridiculousness gives you an opportunity to compare the brain scanning frenzy to a long-discredited nineteenth-century fad that is a powerful reminder of just how easy it is for doctors to claim that they know more than they know and how willing we are to believe them, an opportunity, that is, to point out that there's *still* no there there, that they really can't claim to have found what lies behind the symptom without a leap of faith about how we get from the depression triad to depression itself, that they have not gone beyond metaphysics, but only replaced one metaphysics with another. Well, if all that happened to you, you could be forgiven for thinking that you'd gotten the perfect slam-dunk, wrench-in-the-works, really long one-paragraph ending for your book, the final Amen.

But you would be wrong.

Because I'm sitting here with these pictures. I'm holding my brain in my hands. My brain on no drugs, not even caffeine, my brain concentrating here and bored there. I'm thinking about Daniel Amen telling me I have ADHD and how I've taken Adderall a few times, and how sharp it made me feel, how it turned my work into metal and my brain into a magnet, and I'm wondering if this means it was really making me well. Maybe my brain knows more than I do.

I'm also thinking about what happened when Arvid Carlsson unveiled his photos of neurotransmitters at work to the scientists who had held out for a decade, hoping against hope that something inhabited us, that the brain only served the mind, and who clung to their notion that we couldn't possibly be just a chemical soup, that the spark of God had to animate us, and who finally, in the face of incontrovertible photographic evidence, had to relent and admit they had been wrong. Those white bundles wrapped in the blue and red tendrils—they aren't ghosts trapped in a machine. They're chemicals, made up of molecules made up of atoms made up of ever tinier particles, and I am nowhere to be found in them, I am not behind them looking out. They are me.

Which, as I said before, is not how I want this to turn out. I have a favorite rejoinder to this conclusion—that to learn about the pipes and wires of the brain is to discover the necessary but not the sufficient conditions for our selves and our suffering, only the infrastructure and not the edifice, that it's like cataloging the pigments of the Mona Lisa and claiming that you've said something about why it is beautiful. But even this response seems inadequate right now. Maybe I'm just tired after a couple of years of swimming upstream, maybe I'm depressed, but even if I know that Daniel Amen has only prettied up pictures of the blood rocketing around in my brain, even if I think that this is a crabbed and even repellent way to look at us, I can't deny what's in my hands and before my eyes. I can't deny that all that stands between me and accepting this idea—that every thought and feeling, not to mention my depression, is just the outcome of biochemical events that could easily, with a change in my cerebral climate, be otherwise—is the simple desire for it not to be so, and the conviction that it isn't, which, for all I know, will someday be found in my brain and become a page in Dr. Amen's book. I can't prove that they are wrong.

All I really have is belief. That's all the manufacturers of depression have too, and much as I wish they would admit this or at least not so ruthlessly exploit their claim to be on the side of the facts and the facts alone, much as I think their failure to do so is just plain bad faith, I can't deny the attractions of their conviction. They are on the side of progress and optimism and I am on the side of . . . what? Of suffering? Of some ancient, outmoded idea about the necessity of storytelling, the inescapability of tragedy, the uniqueness of consciousness, the importance of meaning?

I once talked to Donald Klein, the Columbia psychopharmacologist, about the placebo effect. Or I should say, I tried to talk to him. He wouldn't engage the subject. "For the same reason that I don't debate creationists," he told me.

Maybe I'm just a creationist. Maybe I don't want to accept that the world simply showed up one day and started evolving stuff, which eventually included people who could, if they were unlucky, be depressed. Maybe I don't want to accept that when I am lying on the floor of my study and I can't get up, when my discontents multiply into despair, when I qualify for the diagnosis that I am just experiencing some neurochemical bad weather. Maybe I don't want to live under this climate of opinion, any more than your average geocentrist wanted to live under Copernicus's sky. Maybe I don't want to believe that depression is a disease of the brain because I am a coward, afraid of a not-so-distant future when doctors know more about me than I possibly can, including how to make me better, when the self that suffers finally surrenders to their certainty that it is nothing but a few millimeters of neural tissue, its suffering reduced to renegade molecules that can be brought to ground with well-placed bullets.

CHAPTER 15

THE MAGNIFICENCE
OF NORMAL

I'm not going to stop there either. You've stuck with me this far, so you deserve something more for your trouble than a gloomy forecast of inexorable climate change—some consolation at least, maybe even some advice.

Not that kind of advice. If you or your doctor thinks you are depressed and you want to know what to do about it, you already have many places to go: self-help books, biographies of depressives like Abraham Lincoln, memoirs like Andrew Solomon's or William Styron's (both of which are beautiful and fascinating), Internet support groups, your pharmacist, your therapist, your friends. You won't necessarily get coherent advice from these sources. More likely you'll hear cacophony and contradiction, one voice beckoning you this way and another that way. But you shouldn't be afraid of complexity. We're pretty complicated creatures, no more so than when in the throes of an emotional state that colors all of our experience. And among all those voices, chances are good that sooner or later you will hear something that hits home, reaches down to you and lifts you out of your darkness. That's part of how the placebo effect works: a doctor, someone you trust and look to for help, gives you reason to hope that things will be getting better soon.

But simplicity can be a good thing too, at least once in a while, and here's the simplest thing I'm going to tell you, the closest to advice I'm going to come.

Whatever else you do, don't let the depression doctors make you sick.

This is harder to do than it sounds. Because you have to grant the brilliance, the irresistible narrative power of the story they have manufactured. The depression doctors have found a middle way between Job and Eliphaz, one that can be truly comforting: that your depression is a flaw, in you but not of you, that causes you to see the world as darker and meaner than it really is and that can be corrected with a quick trip to the drugstore. And especially as we move into a time of deep uncertainty, economic and other-wise, and the opportunities to feel worried and disconsolate and even despairing increase, as more and more patients show up in doctors' offices sleepless and upset and agitated and worried and wondering whether life is worth the candle now that their jobs and their 401(k)s have disappeared, or now that they have to tell their kids that they're losing their house or moving from their hometown because there just aren't any jobs left, or now that they realize they may never be able to retire—as, in short, more and more of us meet the criteria, doctors are going to be dispensing the diagnosis even more often than they already are.

If you're going to resist this prescription, which has the signal virtue of leaving the world intact and giving you the opportunity to live in it more comfortably, you will have to remember that the depression doctors exact a price for their diagnosis. They want to tell you who you are. They want you to see yourself as the kind of being whose unhappiness is a sign that you need to buy their ser-vices. And you're going to have to remember that hidden in the diagnosis, and in the vast assumptions about humankind behind it, is a whole history of accident and misunderstanding and overreach-ing, of unwarranted leaps of logic and wishful thinking and the misapplication of scientific rhetoric, of bad faith and greed. They don't teach this history in medical school, and few doctors have the time or patience or inclination to learn it. So now that you know more than they do, you're going to have to consider the possibility that even if you take their drugs and feel better, that doesn't mean you were sick in the first place. And then you are going to need new ways to understand your discontents.

I'm not going to tell you what those new ways should be. You'd want to be pretty sure of yourself before you try to convince people you've never met that you know what causes them to feel a certain way and what they ought to do about it. And the simple fact is I've been working with depressed people and I've been depressed on and off myself for a quarter century, and I've spent the last couple of years writing a book on the subject. And I still don't know the answer to those questions.

Because you could write that book, as I have, and then as your reward you could take yourself to a tiny cabin on an island without either cars or roads (and whose sole concrete sidewalk features speed bumps), and you could be awakened at three in the morning by a sudden shift of the wind, by palm leaves slapping the sides of your cabin and an occasional quarter-sized raindrop plunking its tin roof. You could sit up in bed and look out your window at an orange crescent moon on the eastern horizon, lighting a faint path across the sea, and before you could congratulate yourself for having found such a place, before you could even begin to try to tame this riot of beauty with words, you could become aware of something else: your pounding heart, your stomach hollow with nausea, the sudden dead certainty that something is wrong and will never be put right. This is all so disappointingly familiar, and because you don't want to be depressed on your vacation, you hope that this ill tiding blown across a vast ocean from an unseen continent is bound for some other destination.

The feeling has descended like the weather, unbidden and dismal and blacker than the night, and you can't help but think that this is a random event, a synaptic thunderstorm signifying nothing that you don't already know or need to be reminded of: that the flesh is infinitely vulnerable, that our lives are lived inside the thinnest of biochemical margins. This rude awakening may even be the vestige of some adaptation, the remnant of a capacity developed hundreds of thousands of years ago to respond to some forgotten contingency of prehistoric life—the presence of the sublime triggering the genetic impulse to curl up and wait for danger to pass—but now just an unpleasant paleontological curiosity.

But then again, maybe not. Maybe this is a visitation from your own distant reaches, the solitariness you've been reveling in transformed into loneliness and the memory of an uncomforted childhood, the sudden tempest, reminiscent of the roiling and dangerous household in which you once lived, plunging you into that familiar miasma of fear and self-loathing, into the echoing prison of depression. Maybe, in other words, it is the margins of biography that are narrow. Maybe there is no escape from a story so deeply inscribed.

Or perhaps it is a visitation of a different order, the intrusion of another unwelcome fact: that just beyond the ersatz poverty of your crude wooden shack lies the real thing and just beyond that a skein of injustice and suffering and exploitation in which you have a hand and that seems illimitable. Maybe what has crept over you is only the hopeless truth behind your ecotourist pretenses, the knowledge that you would rather live with the self-devouring discontents of bad faith than chuck your comfort and your familiar life into the sea. Or the fear that whatever is important to you will crumble, that even love will disappear.

Or maybe Hippocrates was right all along. Maybe you really do have out-of-whack humors, but they are in a place where scientists haven't yet thought to look or that their instruments won't find anymore than a magnet can find feathers. Or maybe it's karma or the alignment of the planets or the chemicals in your food. Or, as you will think later, standing behind a beautiful young couple kneeling together under a stained glass window in the transept of a church on the mainland, praying (or so you imagine) for a happy marriage or a pregnancy or the good health of a loved one, devotion radiating like sunlight off their bowed shoulders, maybe none of these stories make much sense. Maybe these children of Job—who believe (so you think) that good and bad fortune reflect a heavenly order—are right, maybe hope and despair, pain and consolation are part of the battle between God and Satan, and maybe you should get down on your knees and pray.

Every one of these accounts of depression—and there are undoubtedly others—can account for this sudden malaise, and each has its uses. I'm not going to tell you which one is right, because I don't know. I'm just going to tell you to be wary of people who tout certainty at the expense of truth, especially when what they are certain about is something so complex and baffling and weighty: the nature and causes of our suffering and what we ought to do about it.

Even some depression doctors are beginning to wonder about the truthfulness of their story. They're beginning in particular to think they might have rendered too much of the world insane. Some of this concern is sparked by business considerations. The widespread distribution of the depression diagnosis, they worry, can result in brand dilution or in other marketing difficulties. Darrel Regier, the American Psychiatric Association's chief of research, put the problem this way:

> Various critics of the current diagnostic system have characterized the expansion of diagnostic categories as a "guild" attempt to justify payment for any condition a psychiatrist might see in practice, or as fabrications of the pharmaceutical industry to justify the sale of their products.

Regier went on to point out that the dire estimates of mental illness in the population—in any one year, using DSM criteria, something like 30 percent of Americans qualify for one diagnosis or another—raise some red flags even without the critics. For instance, he wrote, the mental health treatment system is in no way prepared to treat the 100 million patients forecast to meet the criteria every year. This embarrassment of riches could be a public relations disaster.

Regier is among the psychiatrists who think the ease of diagnosis is mischief wrought by Kraepelin's ghost, that running down a checklist of symptoms and concluding that people are sick if they say yes often enough is bound to lead to overestimates. What's missing, they think, is sufficient consideration of whether and to what extent those symptoms are actually problems for people—"clinical significance," they call it. This standard was supposed to replace intrapsychic conflict as the übercriterion of mental illness when the DSM-III was revamped in the wake of the homosexuality debacle. But Regier believes, and more psychiatrists are coming to agree, that doctors are not any better at agreeing on how to assess significance than they were at standardizing intrapsychic conflict. Nor are they particularly eager to try:

> Despite the prominence of clinical significance in diagnostic criteria, there is currently no consensus as to how it should be defined or operationalized. In large epidemiological surveys, direct clinical judgment is rarely used because of the high cost of clinical time and the large number of subjects.

Doctors, in other words, are too busy diagnosing patients to worry about whether or not they are really sick.

Regier thinks that the data are there to assess clinical significance. Researchers—epidemiologists and clinicians alike—do ask patients how much a given symptom interferes with their lives or whether it ever prompted them to visit a doctor. But the answers aren't necessarily factored into the final results, so, as I discovered at Mass General, a person who gives little or no indication of significant impairment but who has five symptoms of depression is still depressed. The data remain, however, and Regier and his team extracted them to reassess the outcomes of the major epidemiological studies of depression, the ones that lead to those dire estimates. And it turns out that, once people are ruled out who have the symptoms but aren't impaired or distressed by them, the prevalence of depression is cut nearly in half.

This problem has led some psychiatrists to suggest that the current categorical approach to diagnosis should be replaced with a dimensional approach, in which only people at the extreme end of the symptom spectrum, the ones with the most symptoms that rise to the highest levels of clinical significance, would receive diagnoses. Regier's position as vice chair of the DSM revision committee has led some psychiatrists to worry that if this happens, the disorders at the mild end of the spectrum will soon go the way of homosexuality and neurosis. These doctors have already struck back with some statistics of their own, showing that the mildly symptomatic will eventually get worse. We don't wait for people with high cholesterol and blood pressure to have a stroke before we diagnose and treat them, they argue, so why should we wait for the mildly depressed to cross the threshold?

The dimensionalists suggest that such people ought to get treatment and that any unmet need this creates "should be addressed by developing comprehensive triage rules that allocate available resources based on evidence-based assessments of the cost-effectiveness of available treatments." Interestingly (and speaking of a "guild attempt" to increase business) this proposal not only casts the net wide; it also creates an argument for early assessment (which can only increase the numbers of diagnoses) and preemptive treatment—much of which will no doubt be with medication administered by psychiatrists, usually the cheapest of the available resources.

Whichever way this dispute breaks, one thing is certain: you are very unlikely to hear much about the dispute over mild disorders. The squabbles over neurosis and homosexuality taught the American Psychiatric Association a lesson about airing the family linen in public, so it has made people serving on the DSM-V committees sign a confidentiality agreement as a condition of participation. That's too bad, not only because we'll miss a debate bound to be as absurd as it is enlightening, but also because from what I can gather from *A Research Agenda for DSM-V*, a book the APA brought out at the beginning of the planning process for DSM-V in 1999 (a mere five years after DSM-IV was released; you would think that they'd realize how all this dithering looks to the rest of us), the guild is going to make the zero validity problem part of its proceedings as it fashions a new diagnostic manual:

> The major problem for mental disorders as currently defined is that their causes and pathophysiological mechanisms remain largely unknown. It is expected that, at some point in the future (perhaps decades from now), the pathophysiological states predisposing or contributing to major mental disorders will be identified . . . Once it is possible to define a mental disorder based on the identification of its underlying pathology, then it would surely make sense to follow the course of other medical conditions and have the presence of the disorder be based solely on pathology and not on the effect this pathology exerts on the individual's functioning.

It would be nice to hear psychiatrists acknowledge in public that even though they've been telling people for two decades that they know what the underlying pathology of depression is, they really don't. But with the pathophysiologically based classification system that the book says will solve this problem decades away, it's no wonder that the APA wants to keep a tight lid on the proceedings. They don't want us to know that they're still working off that promissory note until they're ready to put paid to it.

*   *   *

There is some evidence that reform-minded doctors should be careful what they wish for. The transformation of psychiatry into clinical neuroscience may hold some unpleasant surprises for them. Consider what psychiatrist Max Fink has been saying recently.

Fink thinks that the APA took a disastrous wrong turn when it resurrected Kraepelin's categorical approach but left buried one of his most important categories: melancholia, a diagnosis that Fink thinks fits the subgroup of the depressed who feel anxious and despondent for no particular reason, who wring their hands and sleep all day, who are delusionally guilty and self-reproachful—patients like the woman I've called Ann. This, Fink points out, is the cluster of symptoms that Hippocrates originally observed, that has been reported by Kraepelin, Freud, Meyer, and Kuhn (and nearly anyone else who has bothered to look). The disease, as Fink sees it, has none of the Chinese menu fuzziness of the DSM's major depressive disorder, but instead has four symptoms, *all* of which must be present for the diagnosis. And, best of all, the fourth criterion is a lab test—either a sleep study that shows irregular brain wave activity or an endocrine panel that reveals abnormalities in cortisol, a hormone whose levels increase with stress. Fink's melancholia, in other words, is a disease in the modern sense—a form of suffering with a specific biochemical signature. To Fink, this means that it would behoove his profession to restore it as a diagnosis in the DSM-V. It could even be the flagship disease in that longed-for pathophysiologically based classification system.

Fink is not the first doctor to propose cortisol tests to verify depression, and they are compatible with current neurochemical theories, which see depression as a stress reaction gone amok. But studies have shown that fewer than 50 percent of the subjects who meet DSM criteria for depression turn up positive on the bioassays. On the other hand, when researchers weed out the nonmelancholics from the subject pool, then that number goes up to 70 percent or higher. Fink also points out that studies showing homogeneous brain chemistry or structure tend to be strongest when the subjects are more severely depressed, which suggests, he says, "that what is now considered the pathophysiology of major depression is best restricted to melancholia." Make the category less heterogeneous, in other words, and it may actually start to have some scientific integrity.*

But not without a cost—market share. Melancholics in Fink's sense make up only a small portion of people who meet the DSM criteria for depression. But that's not the only reason that the depression industry is not beating a path to Max Fink's door. It's also because Fink, who in his late eighties, is one of the world's leading proponents (and practitioners) of electroconvulsive therapy, which is a highly effective treatment for melancholia—as doctors have known since the 1940s. But while doctors continue to provide ECT, very quietly, it's hard to imagine who is going to pay for clinical trials for a device that lost its patent protection long ago, and which has such a terrible reputation.

Nor are the drug companies likely to get behind Fink's proposal anytime soon. They do have a cure—tricyclic antidepressants, whose effectiveness with melancholia approaches that of ECT and far outpaces the SSRIs. But they're also off-patent; there's not much money to be made there either. Some electricity-based therapies less dramatic than ECT—deep brain stimulation, transcranial magnetic stimulation, and vagus nerve stimulation—have shown promise for melancholia, but these are hardly blockbuster treatments, especially if the patient pool is limited to those who fit Fink's diagnostic scheme. Given the fact that the burial of melancholia was essential to the expansion of the depression diagnosis, it is very

---

* This proposal would help the depression doctors make good on what is perhaps their strongest claim: that because melancholia is unresponsive to external circumstance, it must be the result of internal dysfunction. But a problem remains: who gets to decide which depressions are meaningful and which are neurochemical noise?

unlikely that without the lure of large profits the industry is going exhume it anytime soon.

Fink's proposal to restore melancholia to the DSM suggests a reason for the poor performance of SSRIs in clinical trials. Doctors know that the subjects who show the most melancholic symptoms do worse in clinical trials than the other patients. And the HAM-D, the test that measures the performance of SSRIs, was standardized on hospitalized depressives, many of whom would very likely meet Fink's criteria for melancholia. The drugs, in other words, may be good for something, just not what the doctors are looking for.

Actually, the drug companies already know at least one thing SSRIs are good for. They're just not sure they want to make a big deal about it.

Twenty to forty percent of all men suffer at one time or another from premature ejaculation (PE). (These numbers may be inaccurate; in keeping with the DSM's requirement that a disorder be a problem for the patient, the benchmark criterion is ejaculation that occurs "on, or shortly after, penetration and before the person wishes it," so men can have the disease if they think "shortly after" means anytime before they wish it, even if that's, say, after the four hours promised in the Viagra ads, while men who are just in a hurry to get back to the football game don't qualify for the diagnosis at all.) As many as 70 percent of people taking SSRIs suffer from sexual dysfunction—which in men often takes the form of delayed ejaculation. Indeed, repeated studies, most of them conducted in European countries, show that the intravaginal ejaculation latency time of men suffering from PE increases significantly when they take SSRIs.

You don't have to be a marketing whiz to see the lemonade-out-of-lemons opportunity here. Our pharmacological Calvinism isn't quite intense enough to justify an ad campaign promising an antidepressant whose side effects cancel your guilt. ("Makes you feel better, but don't worry. It will ruin your sex life.") But the endless desire for sex and the inescapable feeling that it's never quite good enough offer a chance to turn side effects to advantage by creating a new market for the drug.

But the fact that the SSRIs do reliably treat an illness, even if it's not depression, hasn't yet translated into drug company riches. The reason is obvious. To advertise SSRIs as a cure for PE, the drug companies would have to get an indication from the FDA, and, as one of those European PE researchers pointed out, "focusing on the ejaculation-delaying effects of these drugs would highlight their potent sexual adverse effects and thereby hamper marketing strategies for use of these agents for depressive disorder." Indeed, currently the industry officially denies, and the FDA officially doesn't know about, the sexual side effects. According to the package insert for Paxil, for instance, only 1.6 percent of men taking it experience abnormal ejaculation. (That's probably because clinical trials aren't exactly designed to elicit accurate information about such intimacies. My Mass General doctors sandwiched their questions about my sex life between items about nausea and muscle cramps. It is as if a person's sexual performance were no more difficult to talk about with a stranger than his headaches. "Oh, yes, doctor," we must imagine the subject saying, "I have a sore throat and, now that you mention it, I can't come. Now, about this backache . . .") There's no particular reason to change this. After all, there is nothing to stop the drug makers from having it both ways—whispering to doctors about antidepressants as an off-label treatment for premature ejaculation while shouting from the rooftops that they don't interfere with your sex life.

That may change. Pfizer recently funded a study in which researchers gave Viagra to women suffering from *antidepressant-associated sexual dysfunction* (look for this one in the DSM-V) to see if it undid the orgasm- and libido-inhibiting effects of Zoloft on women. Seventy-two percent of the women reported improvement, a finding that was trumpeted in press releases that resulted in widespread news coverage. Perhaps when Pfizer finally formulates a combination therapy (Vizoft, anyone?) that promises to enhance both happiness and sexual pleasure—pardon me, to cure major depressive disorder and female sexual dysfunction—those package insert numbers will finally change.

But I digress. The real point is that to say SSRIs don't test well is not to say they don't work. Any clinician can tell you that they do, that he or she has been humbled by how quickly and effectively the drugs improve the lives of some of their patients; and the success of the drugs in the marketplace indicates that the customers are happy.

The problem with the clinical trials may simply be that they don't measure what the drugs do, or, to put it another way, that the diagnosis and the treatment aren't well matched. If researchers want the numbers to break their way more often, then all they need to do is to figure out what the people who actually respond to SSRIs are suffering from, find a way to test for it, and then give it a name.

Here's my suggestion: Prozac-deficit disorder.

Okay, that's glib and maybe even unfair. But it's also honest, at least to the extent that it reflects the way doctors actually prescribe the drug. Just ask Richard Kravitz. He's the physician who led the study in which standardized patients faked depression and then asked for Paxil. I called him to ask about, among other things, what it meant that doctors seemed willing to prescribe the drugs without rendering a diagnosis. "The pursuit of a more precise diagnosis often hinges on the relative risks and benefits of the available treatment," he told me. "If the treatment is relatively harmless, then sometimes you give empiric therapy a try, as many of these doctors did." What matters, he added, are results. "People who are unhappy will get better and people with major depression will get better." Or, as the head of psychiatry at Stanford University once told an interviewer, "for the vast majority of . . . the walking wounded, the SSRIs are good drugs." So when someone walks in the doctor's door, as Kravitz's accomplices did, bearing the wounds that the doctor has identified as responsive to the drugs, the patient is likely to leave with a prescription.

This could mean that doctors are lazy or pill-happy. But it could also mean that SSRIs limn a disease, much as Roland Kuhn claimed that imipramine did for vital depression. Is there a homogeneous form of suffering that the drugs relieve? What is the wound that the walking wounded are suffering from?

This is one of the questions that preoccupy Peter Kramer in *Listening to Prozac*. He doesn't approach it head-on, but rather through his careful exploration of the way the drug transformed the lives of his patients. What does it say about the kind of people we are, Kramer wants to know, that a mere pill can do all this?

In a way, it doesn't mean anything that you don't already know, that you can't figure out if you drink a couple of cups of coffee or shots of whisky, if you smoke a cigarette or a joint or just scarf down a donut and get a sugar rush: there is no such thing as a *you* without a body. It would be nice if there were, because then things like cancer and death wouldn't be so scary. (On the other hand, it's hard to imagine what pleasures like sex and eating would be like, or even how you'd manage them, without a body.) Despite what Daniel Amen and other clinical neuroscientists are saying, no one has figured out how your body, particularly your brain, gives you consciousness—and I doubt anyone will (or maybe I just hope they won't)—but it's pretty clear that the old Cartesian idea that the self is like a bird trapped in a cage of bone and blood is untenable. Its modern version—the idea that somewhere in us is an authentic self, waiting to emerge like a sculpture from a slab of marble—is equally unable to withstand the obvious effects of mind-altering drugs, unless you define that self tautologically as whatever is released by the chemical.

This is why Kramer is sanguine, if reluctantly so, about SSRIs: because there is, in his view, no authentic, nonbodily mind to be spoiled by the drugs. The drugs don't create a self so much as create the conditions for us to achieve selfhood, whatever it is. He makes this point most clearly when he tells us about Dr. Yang, the healer in Woody Allen's *Alice*. He gives the title character an herbal concoction that, Kramer writes, "proves to be a sort of instant super-Prozac":

> Yang's herbs allow his patient to experience the world differently—to see husband, lovers, parents, siblings, and children in a new light—and then to bear the possibility of loss inherent in that fresh vision . . . His drugs only potentiate change; ultimately it is Alice's quest that transforms.

The herbs make it possible for Alice to transform herself, but they don't dictate the nature of that transformation. Similarly, in Kramer's view, Prozac is just a neutral technology, a dab of lubricant applied to the neural apparatus that allows it to do what it was meant to do in the first place: to produce a competent, effective self.

Kramer provides plenty of vivid pictures of what happens when you grease the machinery with Prozac. He tells us about Tess, his first Prozac patient, who once was plagued with "low self-worth, competitiveness, jealousy, poor interpersonal skills, shyness, fear of intimacy" and blossomed all at once into someone

> socially capable, no longer a wallflower but a social butterfly. Where once she had focused on obligations to others, now she was vivacious and fun-loving. Before, she had pined after men; now she dated them, enjoyed them, weighed their faults and virtues. Newly confident, Tess had no need to romanticize or indulge men's shortcomings.

Tess joins Sam, the architect who quit his pornography habit; and Julia, the perfectionist who stopped being shrill and short-tempered with her husband and children when she stopped being so demanding of herself; and Lucy, who became less worried about rejection when she stopped looking so deeply (and, Kramer says, accurately) into the nuances of her interactions with her friends, and all the rest of the patients Kramer introduces to illustrate that Prozac's main effect is to release people from whatever rut they have gotten stuck in. It helps people become the selves they think they should have been all along—outgoing and confident, resilient and optimistic—which is probably why he keeps hearing them say, "On this drug I am myself at last."

Being energetic and confident and assertive are surely modern virtues, so it makes sense that people would be delighted to find themselves suddenly in possession of more of them. But there is a virtue beyond those virtues, a peculiarly American aspiration that they also serve: the ability to make ourselves into whatever we want to be, to be limited by no necessity, to seek fulfillment by inventing and reinventing ourselves endlessly. This is the wound that Kramer's patients are walking around with: the inability to secure that freedom, the disappointment that follows, the loss of the American dream. When Kramer tells us that his patients suffer from a "fixed tragic view," it's easy to think the problem is only the pessimism. But it is also, and perhaps more importantly, the "stuckness," as he calls it. You're lying on the floor staring at the ceiling and feeling sorry for yourself and the problem isn't only that you're distraught; it's that you *can't get up.* You can't *move on* from your grief, you can't *get over* your unhappy childhood, you can't *let go* of your past, you can't dust yourself off and pick yourself up by your bootstraps, declare your independence and pursue your happiness to the next horizon.

And it's not just your biographical self that's stuck. It's also your biochemical self, the one whose depression switch is stuck in the on position. There's even a theory that depression occurs when neurogenesis—the formation of new brain cells, which scientists have discovered does continue throughout our lifetimes—is inhibited, and that SSRIs, in some yet undiscovered way, foster the growth of new brain cells. A depressed brain, in other words, isn't free to reinvent itself either, at least not until another chemical rides to the rescue, shooting down those growth-impeding molecular villains with its magic bullets. Then you can be yourself at last.

In this sense, then, mass depression is a disease like neurasthenia. It captures the anxiety and discontents of people living in times of cultural upheaval—in our case, the impending collision between our sense of an ever-opening horizon and the insuperable limits of life on earth, the economic and geopolitical and ecological realities that loom closer every day, the increasing difficulties of pursuing happiness—and turns them into a widespread neurological illness, a whole world gone insane.

But why bother calling it a disease at all? Why not just say that the drugs help us to be the kind of endlessly flexible and resilient self that our culture has long demanded and leave it at that? For that matter, why not get rid of the middleman—in this case the doctor, with his list of diagnoses and his prescription pad—entirely? Why not just make SSRIs available over the counter?

I'm almost serious. You will recall that when the FDA came up with the prescription scheme, it simply meant to limit access to drugs whose safe use was too complex to explain in layman's terms on a label. But, as Richard Kravitz says, most doctors think that the SSRIs are "relatively harmless." That's not entirely true—even though the drug companies deny that SSRIs can increase suicidality, the statistics suggest otherwise, and then there are those other, more common side effects. Even so, it's hard to see how the current model—take this drug and check in with me in a couple of weeks— does much to address whatever dangers the drugs present. More to the point, it's very hard to use the drugs to harm yourself. That's more than you can say for aspirin, which will make you bleed to death if you take too much of it, or acetaminophen, which will ruin your liver in a New York minute, or any of a long list of other dangerous over-the-counter drugs. And doctors generally don't have any particular scientific reason to prescribe you, say, Lexapro rather than Prozac—or, as Kravitz's study makes clear, to prescribe you anything at all; the signal-to-noise ratio in clinical trials is just too low to make confident predictions about which patient is going to respond to which drug. Neither do doctors adjust your dosage or brand based on lab tests or other findings that only they can gather or interpret. They are providing old-fashioned empiric therapy, which is just a fancy way of saying that they work by trial and error, and when it comes to assessing the results they are at a disadvantage in ascertaining the relevant data: how you are feeling. Surely, that is something about which you are more expert than they.

But I don't think for a second that antidepressants are going to go over the counter. To cut doctors out of the loop, which is to say to decouple SSRIs from the disease of depression, would require the drug companies to give up the dispensation that the DSM grants them from the pharmacological Calvinists among us. It would turn the drugs into exactly what the industry doesn't want them to be: enhancement drugs, mood steroids. It would make them much too close for comfort to alcohol or marijuana or any of the other drugs people commonly take to feel better.

It's probably an oversimplification to say that depression as we have come to know it has been manufactured in order to maintain a Maginot line between recreational drugs and antidepressants. On the other hand, you have to marvel at how well the diagnosis protects the pharmaceutical companies from the bad reputations of their illegitimate siblings.

Consider, for instance, the fact that as many as 50 percent of the patients who stop taking antidepressants experience symptoms including headaches, dizziness, fatigue, sweating, tremors, chills, and nausea, not to mention agitation and anxiety, confusion, and memory problems. (It's not clear whether this percentage includes the many patients who report simply not liking how they feel without the drugs, the way that their emotional life becomes once again fraught and unpredictable.) These problems are only temporary—they go away when patients start taking the drugs again. This can lead patients to conclude that their depression is indeed a chronic illness, and to resign themselves to remaining on antidepressants for the rest of their lives. The depression doctors are familiar enough with this phenomenon to give it a name: *adverse events related to antidepressant discontinuation.* But back in your ninth-grade health class, your teacher probably gave a different name to the malaise that occurs when you stop taking a drug and disappears when you start again: *withdrawal syndrome,* which they list, correctly, as a good reason to be very careful about using mind-altering drugs.

Your health teacher, or maybe even the cop who led your D.A.R.E. program, probably also told you about drug tolerance, the need to take increasing amounts of a drug in order to get the same effect. As it happens, many antidepressant patients find that the drug's initially felicitous effects fade after a while, a problem doctors can solve by boosting the dose (or, more often, switching to a different antidepressant). Critics of antidepressants call this "Prozac poop-out," but mainstream doctors have a much fancier name for it: *tachyphylaxis.* Some of them have even suggested that it only occurs in people whose improvement was a placebo response in the first place, as if it has to be the patient's fault when the drug stops working. And none of these doctors make the point that your health teacher would have made about a drug that causes tolerance and withdrawal. But then again, they don't have to. After all, even the D.A.R.E. cop wouldn't say that a diabetic is addicted to insulin.

But as dishonest as this evasion-by-renaming is—and it is really dishonest—it does accomplish one good thing. It is hard to imagine that so many people would avail themselves of whatever relief antidepressants offer if the drugs were officially considered addictive. Neither would regulators long tolerate an addictive drug if it weren't a cure for illness. As long as we live under a pharmacological Calvinist regime, calling depression a disease is perhaps the best way to get drugs into the mouths of the people.

★   ★   ★

Depression is surely not the only disease that lacks a biochemical known cause; indeed, as scientists grasp more of the complexities of our biochemistry, the magic-bullet model becomes increasingly inadequate to explain illness or to generate cures. But depression is probably the only disease for which doctors insist that they have found the pathogen despite the evidence. The positive results of this bad faith, however, cannot be disputed. Indeed, there is at least one way in which thinking of depression as a disease is helpful. I can imagine less mercenary responses to widespread unhappiness, and better drugs than antidepressants to treat it, but at the same time there's no use denying that the depression doctors have succeeded in commanding for the unhappy the resources we generally grant to the sick—money for research and treatment, time from doctors and nurses, and, perhaps most important, the kind of sympathy due a victim. In a society that values science over other forms of knowledge, and materialism over other ideologies, a diagnosis is a ticket to collective assets, and the depression doctors have not hesitated to punch it—to their own benefit, of course, but also to their patients'.

This might be a good working definition of disease: not a condition with a specific biochemical cause, but a form of suffering that a particular society deems worthy of devoting health care resources to relieving. This point has been lost on the depression doctors, who are still scrambling for their place in scientific medicine, but at least one group of depression patients—the Depression and Bipolar Support Alliance—seems to have partially grasped it. DBSA has its own way of laying claim to those assets. The organization lobbies legislatures, pressures insurance companies, and sends out news releases. It also helps patients find and support one another, encouraging them to take responsibility for their illness by becoming educated consumers in the tight mental health marketplace.

Every year, DBSA holds a national conference, part rally, part seminar, part bazaar. In a ballroom, vendors await customers. They're hawking membership in patient advocacy groups, handing out brochures that explain just how real depression is, soliciting subjects for clinical trials. The closest thing to an actual product on display is psychTracker. Sean, a vice-president of the company that developed it, is manning the booth. He shows me how I can use his Web-based program to chart my emotional life. All I have to do is log on, record how many hours I slept last night, and then rate myself from one to ten on a hundred or so items ranging from *happy* and *sad* through *delusions of grandeur* and *inappropriate affect* to *thoughts of death*. If I do this every day (and Sean hopes I do; psychTracker's revenues are ad based, so keyclicks and eyeballs are the key to prosperity) then I will end up with a graph like the one Sean is displaying. It charts my moods across time, like a corporate profit-and-loss poster, so my doctor and I can see the short- and long-term trends. (Sean adds that his company is also trying to amass "a database of the emotional health of the country," but it's not clear what they intend to do with it.)

Kathy Cronkite is here. She's Walter's daughter. In 1995, she put together *On the Edge of Darkness*, a book of celebrities' accounts of their depressions. Joan Rivers talked to her, as did Dick Clark and Rod Steiger. Cronkite's story leads off the book, and she's recounting it now as part of her keynote address. She says she too was once a "young adult, lying on the floor staring at the ceiling for days on end." She tried harder than I did to fix herself, working with "dozens of social workers, psychiatrists, and physicians who failed to recognize my illness," and who wanted to blame it on "my father's position" or something else in her past. Until she met the heroes of her story, a wise psychologist and the doctor he sent her to, who, in a speech that would have made Frank Ayd proud, said:

> Here's what we do about it. Like other illnesses, he said, the earlier it's treated the better. What works best for most people, he said, is a combination of talk therapy and medication,

and he said, you've had years of talk therapy, how about we try some medication?

And, she tells us, that after she got over her worries—"Would it take away the richness of emotional life? Would I still be me?"—and tried medication, in a "miraculously short period of time," she was transformed. It wasn't your usual transformation, the kind where a person leaps into a new and extraordinary role, but nearly the exact opposite. The drugs, she says, made her normal.

> Outside of this room, most people think normal means boring and bland, but we know, I hope we all know, that normal is the most glorious feeling there is. And when you're depressed you can't even imagine the magnificence of normal . . . The medicine didn't take away myself. The medicine gave me back myself.

That's not the only miracle that the medicine performed. The drugs gave her suffering a new meaning—"I am not crazy or bad or lacking in faith. I have a disease. It's called depression." Her deliverance from abnormality was so magnificent that she has devoted her life since then to "spread[ing] the word that depression is . . . a real, treatable, medical condition, with real treatments—and real recovery."

While this is a gospel of hope, she cautions, recovery should not be confused with cure. Indeed, it would be a mistake to think that depression will ever really go away. "I have a chronic illness that I will always be aware of, and always be susceptible to, and probably will always take medication for, much like diabetes," she says. "And much like diabetes, if I take good care of myself . . . it will be no more than a background of my life."

This, it seems, is where the depressed pilgrim's progress ends: in the movable city of recovery, where entry is granted to those born with an affliction. "I've found the reason for depression," Cronkite says. "Bad luck." The "magnificence of normal" is the highest aspiration here, emotion traded like currency and comity achieved by mutual confession in the recovery groups, modeled on Alcoholics Anonymous's twelve-step groups, that are DBSA's stock in trade. One attendee—dressed in houndstooth pants and a striped shirt, grabbing a smoke in the Florida sun—tells me (and anyone else in earshot) about his pilgrimage across the South, driving his RV from twelve-step group to twelve-step group, all that fellowship capped off by his arrival here, on what just happens to be his two thousandth day of sobriety. And as he concludes in his gravelly Tennessee drawl, "They ought to just give us a state!" it occurs to me that this might be a glimpse of the future, in which we sort ourselves according to our diagnoses and where our vital associations will be with other like-minded (or is it like-brained?) people.

Is this what Alexis de Tocqueville had in mind when he said "Nothing . . . is more deserving of our attention than the intellectual and moral associations of America"? Tocqueville went on: "If men are to remain civilized or to become so, the art of associating together must grow and improve in the same ratio in which the quality of conditions is increased." Here in this republic of the afflicted, the citizens are, as Cronkite puts it, not patients or victims or even sufferers, but "consumers of mental health services" whose recovery includes driving a hard bargain with legislators and hospitals and drug companies and universities, demanding more insurance coverage, more research, more treatment.

But at least these consumers are taking care of each other. I can't help but be moved by the way people here accommodate one another's disabilities, their shuffling gaits and off-kilter questions, commiserate about psychiatrists and insurers, and share secrets about medications. They listen to one another's stories—a form of love if there ever was one—even if they are largely stories about the tyranny of their own brains. They haven't taken diagnosis as a call to retreat to their sickrooms and whine to their decreasingly sympathetic support systems. Instead, they are coming together to demand services. It is tempting to think that a society organized around *making the recovery connection* might not be any worse than the society we have.

If you think that's damnation with faint praise, you're probably right. But it is worth pointing out that it's not such a bad thing that the bar to entry is set so low, or as one DBSA speaker put it, that "you have a diagnosis. You just don't know what it is yet." I don't think she was trying to destigmatize mental illness, which would, in that setting, only have been preaching to the converted. I think she was saying that *mental illness* is a valid way to think of our troubles. To say that we're all mentally ill is only to say that we are flawed people living in a broken world. If that's true, then one state might not be enough.

But much depends on what demands a diagnosis leads us to make. To say that we have a chronic illness is to direct our attention to the health care system and not to other social institutions. And there is a danger here: that to be a consumer, whether of health care services or flat screen televisions, is to be essentially passive, to choose only from among the available options. When your choices are only Paxil or Zoloft, it's worth wondering whether you have any real choices at all.

Peter Kramer noticed this danger and considered whether it was built into the pharmacological effects of the drugs, whether antidepressants are a "modern opiate" that "supports social stasis by allowing people to move toward a cultural ideal—the flexible, contented, energetic, pleasure-driven consumer." He was quick to reassure us on this count. We needn't be concerned because Prozac can also "catalyze the vitality and sense of self that allow people to leave abusive relationships or stand up to overbearing bosses." And besides, even if the drug "induces conformity, it is to an ideal of assertiveness, but assertiveness can be in the service of social reform of the sort ordinarily understood as nonconformity or rebellion." That's why he was confident that Prozac will be "on balance a progressive force."

I'm not so sure about this. Prozac doesn't seem to have told any of Kramer's patients to take to the streets. It certainly hasn't done that for mine, and while it does occasionally help to give someone the courage to leave a bad marriage, it just as often helps them to adapt and stay. The two decades since Prozac was introduced have seen enormous increases in injustices like the widening gap between rich and poor, in opportunities to feel hurt and victimized or just plain worried sick about the future, in horror and atrocity in which our own complicity is hard to ignore, but they haven't exactly been a period of nonconformity or rebellion. Indeed, what SSRIs seem to do best, when they work at all, is to help patients live with less anguish in the world as they have found it. That's why the drugs have the reputation as "so-what drugs," why critics like Erik Parens accuse them of "facilitat[ing] better performance in an often cruelly competitive, 'capitalist' culture," and why we wring our hands about where resilience ends and tolerating the intolerable begins: because it seems that the drugs make people less responsive to the irritations and outrages of daily life.

It's particularly hard to avoid noticing that those twenty years have been an age of irrational exuberance, when confidence and flexibility have been deployed to no other ends than more confidence and flexibility, when the only rebellion that assertiveness has served is a rebellion against certain realities: that you can't count on an economy built on debt that can't be paid back, that neither the planet nor the housing markets can sustain growth forever, that optimism, no matter how good it feels, is not always warranted. Of course, we have no way of knowing how many of those forward-looking masters of the universe were taking SSRIs while they confidently invented and implemented increasingly bizarre ways to make money for nothing. We do have some idea of how many of the rest of us had a little chemical help to feel contented while they were robbing us blind— and it's a lot. I have a hard time believing this is a coincidence.

Still, as tempting as it is to say that Kramer is wrong, that Prozac is anything but a neutral technology, that *so what* is an essential, and therefore dangerous, effect of the drug, we have to remember what Norman Zinberg said about drug, set, and setting. Maybe the reason that the SSRIs seem to foster conformity is that they are prescribed by doctors as the cure for a disease. If you think that what was wrong with you was that your brain chemicals were imbalanced, and the drug makes you feel better, then why look elsewhere for the source of your suffering? Why not just take your improvement and go home to consume mental health services for your chronic illness?

So Kramer may well be right to say that SSRIs are neutral—not because they have no real effects on the mind or because they merely help us to overcome our sickness and get back to the "healthy" way of being human, but because those effects are so protean that they can be shaped in virtually any way. Perhaps doctors could replace Frank Ayd's spiel with a rap about how the drugs are an antidote to a toxic society, something you need because consumer capitalism is much better at creating need than satisfaction, that this indeed is how the system works, and when that gap opens up between what you long for and what you can have, and you conclude, as you must, that the fault is yours, or when you reproach yourself for your failure to be clever or strong or smart enough to reap the bounty—material and otherwise—that is surely out there, or when your disappointment and demoralization seem unbearable, in short, when you feel like Job, the drugs may well give you comfort. Perhaps if the story about the drugs were told this way, if people thought that it was their social arrangements, and not their brain chemistry, that had made it necessary to trade in their orgasms for antidepression, they would use their newfound energy and confidence "in service of social reform."

Kramer's concern is misplaced: the dangers of complacency are not in the drugs but in the diagnosis. Kramer is not the first observer of the American landscape to worry about this kind of hazard. Traveling through America in the early nineteenth century, Alexis de Tocqueville thought he had glimpsed a crucial problem:

that the pursuit of happiness alone was not enough to ensure a vibrant and free society. "A nation that asks nothing of its government but the maintenance of order," he wrote, "is already a slave at heart, the slave of its own well-being awaiting nothing but the hand that binds it."

Because that's what calling our suffering a disease provides: a way to ask something of government—which, in a democracy, means one another. And surely there are other demands we can imagine—and other ways to take care of one another—than better treatments and the money to pay for them. What would happen to the depression statistics if people were less worried about paying for health care, for college, for retirement? Or if we weren't left to our own devices to figure out how to work and take care of our families and have a little time left over to actually enjoy ourselves and one another? Or if we weren't constantly being reminded by advertisers of all the ways we fall short of achieving the good life? Or if we felt that we had some influence with our legislators, our presidents, our financiers, or anyone else who exerts power over what we care about? Or if we thought there was something we could actually change other than our neurochemistry?

There's no double-blind study that can answer these questions, and anyway it's not up to doctors to tell us that there's something wrong with the way we're living. But—and here is another unintended benefit of the diseasing of depression—they have managed to point out that people are suffering from recalcitrant unhappiness in epidemic numbers. I don't doubt that this is true, but I also don't think that it is merely because of something that has gone awry in our wiring. The depression doctors would have us think that this is the case, and that we should ask of one another nothing but the drugs and the cognitive-behavioral therapy that will ensure our well-being. They would have us believe that the subject of our unhappiness, the content of our discontents, is just so much electromolecular static.

Writing in 1840 in America, Tocqueville would of course have been worried about slavery, just as writing in 5000 B.C. a Sumerian poet would have worried about the wrath of the gods. But slavery is gone, at least in most of the world, and the gods have been replaced by science. Now that we are responsible for our own destinies, the danger we face is one that the manufactured version of depression can only deepen: not that a hand will bind us or that a god will destroy us, but that we will bind and destroy ourselves, that we will mistake our anguish and our rage and our terror of what we have become for the symptoms of a disease and dismiss them, that we will find our solace in the privacy of our own medicine chests and seek normalcy as our most magnificent aspiration.

Science is not a democracy. We cannot choose whether molecules interact in our brains to give us our experience, including our experience of the suffering we call depression, any more than we can choose whether the planets circle the sun or whether gravity pulls us to the earth. Neither can we determine the particulars of those interactions. But we can choose what we make of these facts, and what, if anything, to do about them. Now that I've told you this story, we both know that we don't have to put our discontents into the hands of the drug companies and their doctors. When it comes to understanding and alleviating the suffering now known as depression, we don't have to give up biography for biochemistry. We don't have to give up the ghost for the machine.

And then we will be free to fashion our own stories about our unhappiness. Maybe yours will follow the disease model. Maybe it will include taking antidepressants (or other drugs) or entering psychotherapy. Or maybe you will find options that involve neither. That happened to me once—when, four years after my Holiday Inn miracle cure, I fell into another depression. There was a reason for this one too. My wife and I were trying to have a baby. We had made the sacrifices to the great gods of medicine required of would-be parents of a certain age—endless doctors' visits and bodily indignities and the transformation of our sex life into a factory job—but without success. Not only that, but my inexhaustible penchant for dithering had led me not to know which side I was on, whether our failure was a blessing or a curse, whether I should be relieved or devastated when the monthly bad news came. In the midst of it, we decided to build a house, on the assumption, I suppose, that if we built it, the child would come. I threw myself into the project, and by the time we were putting up the downstairs walls, I was just as childless, and just as frustrated about that, but no longer depressed. I was, I suppose, resilient.

A few years later, a researcher at Princeton University suggested to me that because large muscle movement is known to increase serotonin metabolism, all that hammering was what had cured me. I didn't disagree entirely. The hammer was surely important to the cure. I had swung it with a redemptive fury, and as the house took shape, I found reason to hope that I could indeed bring something, if not a human life, into being. Maybe the serotonin explanation is correct, but I choose to believe that giving my burning anger a place to go, vanquishing my helplessness, and losing myself in a task as I had once lost myself in my then future wife's eyes are what cured me.

I didn't intend to treat my depression by building a house—which, by the way, is a very expensive and time-consuming cure. But had I considered myself diseased, would I have stumbled on this cure for my unhappiness? Had I taken antidepressants, would my recovery have gone down in my biography as a lesson about the value of losing myself or just as another illness cured by a drug? And would I have noticed that this was the same lesson that I had learned from my MDMA experience: that the redemption of despair lies in involvement in the world and engagement with others—to put it briefly, in love?

I don't think I'm done with being depressed. I don't think it's going to be all that much fun to get older, and as hopeful as the recent regime change has made me, I think we might have fouled our nests irretrievably. My tragic view is getting pretty fixed, and not in the sense of repaired. So I am sure that I'll have all sorts of opportunities to deploy this lesson. But I doubt I would have learned it if

I thought my problem was a chemical imbalance, and if I believed that Princeton researcher—or, for that matter, if I believed Daniel Amen and George Papakostas, both of whom told me that my MDMA cure was the result of unleashing a flood of serotonin, as if the rest of it—Angel and Grace and my wife's bottomless blue eyes—didn't matter.

I suppose I'll never know whose story is the right one. But I know what mine is, and I'm sticking to it for now. The greatest injustice that Eliphaz and his friends inflicted on Job was that they refused to let him have his version of events. That's what the depression doctors want to do to you.

Call your sorrow a disease or don't. Take drugs or don't. See a therapist or don't. But whatever you do, when life drives you to your knees, which it is bound to do, which maybe it is meant to do, don't settle for being sick in the brain. Remember that's just a story. You can tell your own story about your discontents, and my guess is that it will be better than the one that the depression doctors have manufactured.

NOTES

CHAPTER 1

Page

2 *a question first posed:* Smith, "Pavlov and Integrative Physiology," R747.

3 *On Easter night:* Loewi, *From the Workshop of Discoveries,* 30–34. See also Finger, *Minds Behind the Brain,* 268–73.

5 *Twarog's first paper:* Twarog, "Responses of a Molluscan Smooth Muscle."

5 *Her paper with Irvine Page:* Twarog and Page, "Serotonin Content of Some Mammalian Tissues."

6 *27 million Americans:* These figures are notoriously hard to pin down. However, for a good analysis of both the numbers and their meaning, see Barber, "The Medicated Americans." For raw numbers, see Olfson and Marcus, "National Patterns in Antidepressant Medication Treatment," and the Pharmacy Facts and Figures pages on the Drug Topics website, http://drugtopics.modernmedicine.com/drugtopics/article/articleList.jsp?categoryId=7604.

8 *no more effective at treating depression:* Geddes et al., "Selective Serotonin Reuptake Inhibitors."

8 *the drugs fail to outperform placebos:* Kirsch et al., "The Emperor's New Drugs," http://journals.apa.org/prevention/volume5/pre0050023a.html.

8 *"remake the self":* Kramer, *Listening to Prozac,* t.p.

9 *By now, asking about the virtue:* Ibid., 300.

9 *W. H. Auden's elegy:* Auden, *Collected Poems,* 271.

10 *"the common cold of mental illness":* Centers for Disease Control and Prevention, "Understanding Depression," http://www.cdc.gov/nasd/docs/d001201-d001300/d001247/d001247.html.

10 *"the leading cause of disability":* World Health Organization, "Depression," http://www.who.int/mental_health/management/depression/definition/en/.

10 *his drug was the first SSRI:* Wong, Bymaster, and Engleman, "Prozac (Fluoxetine, Lilly 110140), the First Selective Serotonin Reuptake Inhibitor"; Carlsson and Wong, "Correction: A Note on the Discovery of Selective Serotonin Reuptake Inhibitors," 1203.

11 *"seven preeminent medical, advocacy, and civic groups":* Depression Is Real Coalition, "Link to Us," http://www.depressionisreal.org/depression-link.html.

12 *Some say depression is all in your head:* Depression Is Real Coalition, "Right and Wrong," http://www.depressionisreal.org/depression-dr-greengard.html.

14 *the 1980 release of the third edition:* American Psychiatric Association, *Diagnostic and Statistical Manual,* 3rd ed.

19 *manic-depressive illness:* American Psychiatric Association, *Diagnostic and Statistical Manual,* 2nd ed., 36.

19 *involutional psychotic reaction:* American Psychiatric Association, *Diagnostic and Statistical Manual,* 24.

20 *depressive neurosis:* American Psychiatric Association, *Diagnostic and Statistical Manual,* 2nd ed., 40.

20 *"Depression is neither more nor less":* Kramer, *Against Depression,* 41.

20 *"an occupying government":* Ibid., 25.

21 *"the eradication of depression":* Ibid., 111.

22 *depressionisreal.org is quietly funded:* In particular, by Wyeth. See Depression Is Real Coalition, "What Is," http://www.depressionisreal.org/depression-about-coalition.html.

CHAPTER 2

Page

25 *It is customary:* See, for example, Jackson, *Melancholia & Depression,* 7–8; Horwitz and Wakefield, *The Loss of Sadness,* 57–61.

25 *"it appears to me to be no more divine":* Hippocrates, *Hippocratic Writings,* 154.

25 *"Fear and sadness":* Schmidt, *Melancholy and the Care of the Soul,* 32.

26 *He is rumored to have cured:* Roccoatagliata, *History of Ancient Psychiatry,* 163–64.

26 *"it is a deadly symptom":* Hippocrates, *Hippocratic Writings,* 20.

26 *he devoted an entire book:* Ibid., 152–54.

26 *according to one scholar:* Kramer, *History Begins at Sumer,* 105–15.

27 *"mark among all the people":* Job 1:3, Jerusalem Bible.

27 *Job is not God-fearing:* Ibid., 1:9–11.

28 *"malignant ulcers":* Ibid., 2:7.

28 *"Curse God," she says:* Ibid., 2:9–10.

28 *May the day perish:* Ibid., 3:3–4.

28 *I should now:* Ibid., 3:13–17.

29 *"Why give light":* Ibid., 3:20.

29 *"Why make this gift":* Ibid., 3:24.

29 *"only food is sighs":* Ibid.

29 *Can you recall:* Ibid., 4:7.

29 *Was ever any man:* Ibid., 4:17.

29 *And now your turn:* Ibid., 4:3.

30 *"charlatans, physicians":* Ibid., 13:4–5.

30 *Is not man's life:* Ibid., 7:1–4.

31 *Why do the wicked:* Ibid., 21:7–8.

31 *But man?:* Ibid., 14:10–12.

31 *"I mean to remonstrate":* Ibid., 3:13.

31 *"that of a dog":* Richardson, *William James*, 14.

32 *Does a wise man:* Job 15:2–6.

32 *"it is man who breeds":* Ibid., 5:7.

32 *You shall be safe:* Ibid., 5:21–26.

33 *"In its commonest form":* Kramer, *Against Depression*, 44.

33 *"a fixed tragic view":* Ibid.

33 *"toward assertiveness":* Ibid., 4.

34 *"Our aesthetic and intellectual preferences":* Ibid., 106.

35 *"On this medication":* Ibid., 4.

CHAPTER 3

Page

38 *Prozac, or Sarafem:* Daw, "Is PMDD Real?" http://www.apa.org/monitor/oct02/pmdd.html.

39 *the diagnosis ran into stiff opposition:* Caplan, *They Say You're Crazy*, 122–67.

39 *the nine depression criteria:* For diagnostic criteria, see American Psychiatric Association, *Diagnostic and Statistical Manual*, 4th ed., text revision, 775–77.

39 *The normal process of life:* James, *The Varieties of Religious Experience*, 136.

40 *"uncontrollable urge":* GlaxoSmithKline, "New Survey Reveals," www.gsk.com/press_archive/press2003/press_06102003.htm.

41 *a study by David Rosenhan:* Rosenhan, "On Being Sane in Insane Places."

41 *A cottage industry:* See, for example, the "Letters" section of *Science* 180: 356–65. Also Spitzer, "On Pseudoscience in Science."

42 *So when I found out:* Information about most government-funded clinical trials can be found at the U.S. National Institutes of Health website www.ClinicalTrials.gov. The National Institute of Mental Health (NIMH) minor depression study can be found at http://www.clinicaltrials.gov/

ct2/show/NCT00048815?term=minor+depression&rank=1. The Massachusetts General Hospital trial I eventually entered is described at http://www.clinicaltrials.gov/ct2/show/NCT00361374?term=major+depression+Omega&rank=1.

44 M. trunculus *looks like:* For good pictures of these two snails, see Monfils, "Murex Shells," http://www.manandmollusc.net/Shell_photos/dye-murex.html.

44 *according to legend:* Told by the Alexandrian mythographer Julius Pollux. See Garfield, *Mauve: How One Man Invented a Color,* 39; Hazel, *Who's Who in the Greek World,* 197.

44 *"exactly the colour":* Pliny, *Natural History,* 447.

44 *"Mauve Measles":* Garfield, *Mauve,* 60–62.

44 *bat guano and from certain lichens:* Travis, "Perkin's Mauve," 62–64.

45 *By the time William Perkin:* Garfield, *Mauve,* 19–20; Travis, "Perkin's Mauve," 53–54.

45 *recruited from Germany:* Garfield, *Mauve,* 21.

45 *Hofmann had an interest:* Garfield, *Mauve,* 26; Travis, "Perkin's Mauve," 54.

46 *An enterprising Scotsman, Charles Macintosh:* Garfield, *Mauve,* 24.

46 *Hofmann, however, thought he could extract value:* Travis, "Perkin's Mauve," 55; Garfield, *Mauve,* 33.

46 *Malaria was not only a scourge:* Garfield, *Mauve,* 31–34.

46 *Cinchona bark:* Shapiro and Shapiro, *The Powerful Placebo,* 20–22.

46 *a pair of Frenchmen isolated quinine:* Garfield, *Mauve,* 31; Sherman, *Twelve Diseases,* 139.

47 *enough to hinder the business of empire:* Garfield, *Mauve,* 30–33.

47 *"happy experiment":* Ibid., 34.

47 *Perkin soon determined:* Travis, "Perkin's Mauve," 55.

47 *a powder with a reddish tint:* Garfield, *Mauve,* 35–39; Travis, "Perkin's Mauve," 55–56.

47 *"Perfectly black":* Garfield, *Mauve,* 36.

48 *Perkin's invention proved:* Travis, "Perkin's Mauve."

48 *"When I felt . . . miserable and forsaken":* Marquardt, *Paul Ehrlich,* 159.

48 *Ehrlich had been introduced:* Bäumler, *Paul Ehrlich,* 5–6.

49 *"They cared so little":* Marquardt, *Paul Ehrlich,* 73.

49 *"I can see the structural formula":* Bäumler, *Paul Ehrlich,* 6.

49 *"Substances act only when they are linked":* Ibid., 10.

49 *"brilliant eccentric":* Ibid., 13.

50 *"minute creatures that live [there]":* Quoted in Amici, "The History of Italian Parasitology," 4.

51 *Theriac, for instance:* Watson, *Theriac and Mithridatium,* 71–82; Bartisch, *Theriac.*

51 *George Washington:* Shapiro and Shapiro, *The Powerful Placebo,* 25.

51 *"If the whole* materia medica": Holmes, "Currents and Countercurrents," 108.

51 *They even had a theory:* Arikha, *Passions and Tempers,* 3–37.

51 *time-proven, if poorly understood, remedies:* Jouanna, *Hippocrates,* 205–6.

52 *"If we were once to admit":* Quetel, *History of Syphilis,* 79.

52 *Koch announced that he had discovered:* Bäumler, *Paul Ehrlich,* 14–16; Ullmann, "Pasteur-Koch."

53 *Ehrlich had found that methylene violet:* Bäumler, *Paul Ehrlich,* 16.

53 *Pasteur came to focus on:* Ullmann, "Pasteur-Koch."

53 *It should be possible:* Marquardt, *Paul Ehrlich,* 91.

53 *"straight onward":* Ibid., 93.

53 *"learn how to take aim":* Bäumler, *Paul Ehrlich,* 143.

54 *"There must be":* Marquardt, *Paul Ehrlich,* 94.

54 *In 1903, one of Ehrlich's dyes:* Mann, *Elusive Magic Bullet,* 7.

54 *he had discovered something significant:* Marquardt, *Paul Ehrlich,* 142–45.

54 *he determined that the test animals:* Bäumler, *Paul Ehrlich,* 128.

54 *The first three years:* Ibid., 129–30.

55 *a colleague of Ehrlich's:* Quetel, *History of Syphilis,* 139–40.

55 *"putrid liquid":* Sherman, *Twelve Diseases,* 83.

55 *he injected some of that liquid:* Nuland, *Doctors,* 183–86.

55 *"gonorrhea and the chancre":* Quetel, *History of Syphilis,* 82.

55 *when his aorta burst:* Sherman, *Twelve Diseases,* 90.

55 *Hunter treated himself:* Ibid., 184.

56 *Mercury had been the treatment of choice:* Quetel, *History of Syphilis,* 86–87.

56 *"A night with Venus":* Marquardt, *Paul Ehrlich,* 48.

56 *doctors' ability to publicize:* Quetel, *History of Syphilis,* 114–20.

56 *Public health measures:* Ibid., 120–23.

56 *Nineteenth century man:* Ibid., 119.

56 *a French scientist gave even more reason:* Ibid., 162–64.

57 *word about 606 got out:* Marquardt, *Paul Ehrlich,* 163–66, 175.

57 *"remarkable effect":* Mann, *Elusive Magic Bullet,* 10.

57 *Patients showed up:* Ibid., 156–60.

57 *Hoechst had distributed 65,500 free doses:* Mann, *Elusive Magic Bullet,* 10–12.

58 *375,000 doses:* Bäumler, *Paul Ehrlich,* 167–69.

58 *There's hardly a child:* Ibid., 167.

58 *"long, slow, painful and expensive grind":* Benedek and Erlen, "The Scientific Environment of the Tuskegee Study." See also Berdin and Flavin, "The Least of My Brothers," http://web.archive.org/web/20080124141121/http://wisdomtools.com/poynter/syphilis.html.

CHAPTER 4

Page

62 *Diagnostic trends varied:* See, for example, Sandifer et al., "Psychiatric Diagnosis," and Beck et al., "Reliability of Psychiatric Diagnoses."

64 *"The name hysteria":* Richardson, *William James*, 336.

64 *His doctoral dissertation:* Shorter, *History of Psychiatry*, 101.

65 *"we cannot afford to pay":* Kraepelin, "Manifestations of Insanity," 512.

65 *an analysis of his dreams:* Ibid.

65 *"from the medical point of view":* Kraepelin, *Lectures on Clinical Psychiatry*, 1.

66 *Julien Offray de La Mettrie:* For biographical information on La Mettrie, see Frederick the Great, "Eulogy on La Mettrie," and Wellman, *La Mettrie: Medicine, Philosophy and Enlightenment.*

66 *"it possesses muscles":* La Mettrie, *Machine Man*, 28.

66 *a "machine that winds itself up":* Ibid., 7.

66 *a "vain term":* Ibid., 26.

66 *"everything can be explained":* Ibid., 28.

66 *the brain was divided:* Davies, *Phrenology*, 3–12.

66 *"This doctrine concerning the head":* Van Wyhe, "The Authority of Human Nature," 25.

67 *"There is always unending applause":* Ibid., 29.

67 *Spurzheim claimed:* Davies, *Phrenology*, 7.

67 *nearly half the mental patients:* Steinach, "Etiology of General Paresis," 877.

67 *whose eyes were weak:* Shorter, *History of Psychiatry*, 101.

68 *"masturbatory insanity":* Shorter, *History of Psychiatry*, 103.

68 *"Pathological anatomy":* Kraepelin, *Lectures*, 27.

69 *"Deus creavit":* Blunt, *Linnaeus*, 184.

70 *"cut nature at its joints":* Shorter, *History of Psychiatry*, 105.

70 *His method was straightforward:* Ibid.

71 *dementia praecox:* Kraepelin, *Lectures*, 25.

71 *involution psychosis:* Ibid., 15.

72 *"All the insane are dangerous":* Ibid., 2–3.

72 *to prevent the marriage of the insane:* Ibid., 3.

72 *"the growing degeneration":* Ibid., 4. See also Zilboorg and Henry, *A History of Medical Psychology*, 453–54.

73 *"In the course of years":* Quoted in Jackson, *Melancholia*, 190.

73 *"psychomotor excitement":* Kraepelin and Diefendorf, *Clinical Psychiatry*, 381.

74 *Gentlemen, the patient:* Kraepelin, *Lectures*, 12.

75 *Here is a case:* Ibid., 14–15.

75 *"numberless . . . cases of maniacal-depressive insanity":* Ibid., 19.

76 *The mildest form:* Kraepelin and Diefendorf, *Clinical Psychiatry,* 400–401.

76 *Georges Dreyfus:* Dreyfus, *Die Melancholie.* I have not been able to find an English version of this monograph. For a discussion of its particulars, see Hoch and MacCurdy, "The Prognosis of Involution Melancholia," and Shorter, *History of Psychiatry,* 356.

77 *It includes all the morbidly anxious states:* Kraepelin and Diefendorf, *Clinical Psychiatry,* 348–49.

78 *"one of the most frequent forms":* Hoch and MacCurdy, "The Prognosis of Involution Melancholia," 1.

78 *"zeal outran his judgment":* Ibid., 2–3.

78 *Variations of the emotional status:* Ibid., 3.

78 *"individual taste":* Ibid., 16.

CHAPTER 5

Page

82 *"She had been one of the sanest":* Meyer, *Commonsense Psychiatry,* 22–23. Also, Meyer, "Presidential Address," 21.

82 *he had wanted to stay on in Forel's lab:* Meyer, *Commonsense Psychiatry,* 21–22.

83 *an "old humbug":* Ibid., 24.

83 *His office was upstairs:* Ibid., 43–46.

84 *"hopelessly sunk into routine":* Ibid., 47.

84 *their reasoning for diagnosis:* Ibid., 48.

85 *"the existence of a pathologist":* Ibid.

85 *"Now, doctor, show us":* Ibid.

85 *"mind cannot be diseased":* Meyer, *Commonsense Psychiatry,* 5; see also Lidz, "Adolf Meyer and the Development of American Psychiatry," 321.

85 *"lasting wish":* Meyer, *Commonsense Psychiatry,* 51.

86 *Meyer met Jane Addams:* Ibid.

86 *"accept the disposition":* Ibid., 71.

86 *"early prevention of danger":* Ibid., 71–75.

87 *"The human organism":* Meyer, "Presidential Address," 3.

87 *"Steering clear of useless puzzles":* Lidz, "Adolf Meyer," 326.

87 *his newfound pragmatism:* Meyer, *Commonsense Psychiatry,* 546–47. Also Meyer, "Presidential Address."

87 *the job of psychiatrists:* Meyer, "Presidential Address."

87 *There he found his mother:* Meyer, *Commonsense Psychiatry,* 83.

88 *"the supposed disease":* Ibid., 174.

88 *"neurologizing tautologies":* Ibid., 381.

88 *can we not use general principles:* Ibid., 156.

89 *"There is no advantage":* Meyer, "The 'Complaint' as the Center of Genetic-Dynamic and Nosological Teaching," 366.

89 *to view the abnormal:* Meyer, *Commonsense Psychiatry,* 136.

89 *"The public here":* Ibid., 57.

90 *"give us a clue for progress":* Meyer, "A Few Demonstrations of the Pathology of the Brain," 242.

90 *a rearguard action:* Meyer, *Commonsense Psychiatry,* 51.

90 *neurologists had already cornered:* Shorter, *History of Psychiatry,* 114–19.

91 *"the list":* Lutz, *American Nervousness,* 19.

91 *"insomnia, flushing, drowsiness":* Beard, *American Nervousness,* 7–8.

91 *"modern civilization":* Ibid., 96.

92 *"agnostic philosophy":* Ibid., 123–25.

92 *"Of our fifty millions":* Ibid., 97.

92 *Sooner or later:* Ibid., 99.

93 *If a physician:* Gilman, "The Yellow Wallpaper," 1.

93 *"I have no confidence":* Richardson, *William James,* 400–401.

94 *it had "become marginal":* Shorter, *History of Psychiatry,* 144.

94 *"days when real science":* Meyer, "Presidential Address," 3.

95 *Psychiatry became real:* Ibid., 20–21.

95 *"The great mistake":* Meyer, *Commonsense Psychiatry,* 4.

95 *"commonsense psychiatry":* Lidz, "Adolf Meyer," 323.

96 *Kraepelin's manic-depressive insanity:* Meyer, *Commonsense Psychiatry,* 163.

96 *"constitutional depression":* Jackson, *Melancholia and Depression,* 196–97.

97 *There are conditions:* Meyer, *Commonsense Psychiatry,* 175.

97 *"normal depression":* Jackson, *Melancholia and Depression,* 200.

98 *"the person himself":* Ibid.

98 *"brain mythology":* Meyer, *Commonsense Psychiatry,* 134.

98 *"the dominant figure":* Zilboorg and Henry, *History of Medical Psychology,* 502–3.

98 *The physician can offer:* Quoted in Jackson, *Melancholia and Depression,* 200.

99 *"a second-rate thinker":* Shorter, *History of Psychiatry,* 111–12.

99 *"acquiring a Main Street beachhead":* Ibid., 161.

99 *Give me a dozen healthy infants:* Watson, *Behaviorism,* 82.

100 *And Bernays:* Lears, "From Salvation to Self-Realization," 20.

100 *"therapeutic ethos":* Ibid., 23.

100 *the mental hygiene movement:* Beers, *The Mind That Found Itself.* For Meyer's view, see Meyer, *Commonsense Psychiatry,* 312. Also Shorter, *History of Psychiatry,* 161.

CHAPTER 6

Page

103 *rejection sensitive:* See Kramer, *Listening to Prozac,* 67–77, 87–107.

103 *The HAM-D:* Hamilton, "A Rating Scale for Depression."

106 *"This Be the Verse"*: Larkin, *High Windows*, 30.

110 *Robert Louis Stevenson's "Requiem"*: Stevenson, "Requiem," http://www .poetry-archive.com/s/requiem.html.

110 *"Life is either mostly adventure"*: Elkin, *The MacGuffin*, 233.

111 *the change in human character*: Woolf, "Mr. Bennett and Mrs. Brown," 194.

111 *Mourning, with its "painful mood"*: Freud, "Mourning and Melancholia," 204.

111 *"consider interfering with it"*: Ibid.

112 *Not so with melancholia*: By *melancholia* Freud has in mind something similar to what Kraepelin at first kept in and then removed from his *Lehrbuch*: a state characterized by despondency, fear, delusions, and hypochondria and without obvious connection to the circumstances of a patient's life. But Freud wasn't following Kraepelin's nosology here; indeed, he considered Kraepelin a professional enemy. Instead, he was drawing on a common understanding of melancholia as the condition first described by Hippocrates.

112 *"In mourning"*: Ibid., 205–6.

112 *It would be fruitless*: Ibid., 206.

113 *"related to the child's"*: Freud, *Beyond the Pleasure Principle*, 9.

113 *"carried along with it"*: Ibid., 9–10.

114 *Whatever insult set off*: Ibid., 211.

114 *Patients manage to avenge themselves*: Ibid.

114 *"Prince Hamlet has ready"*: Freud, "Mourning and Melancholia," 206.

115 *"The loss of the love-object"*: Ibid., 210.

115 *Karl Popper*: Popper, *The Poverty of Historicism*, 131–35.

116 learned helplessness: Seligman, "Learned Helplessness in the Rat."

116 *a series of studies*: Alloy and Abramson, "Judgment of Contingency in Depressed and Nondepressed Students."

117 *"depressed people are 'sadder but wiser'"*: Ibid., 479–80.

117 *a "crucial question"*: Ibid., 480.

118 *"brought our instincts"*: Freud, "Transience," 199.

118 *Shell shock*: Zaretsky, *Secrets of the Soul*, 121–26.

119 *"What lives, wants to die again"*: Quoted in Gay, *Freud*, 395.

119 Todestrieb *("death instinct")*: Gay, *Freud*, 391–92; Freud, *Civilization and Its Discontents*, 68–69.

119 *Once mourning is overcome*: Freud, "Transience," 200.

119 *"far more a matter"*: Gay, *Freud*, 553.

119 *The fateful question*: Freud, *Civilization and Its Discontents*, 81–82.

120 *"The life imposed on us"*: Ibid., 13.

120 *"the threat of external unhappiness"*: Ibid., 64.

122 *"psychology for winners"*: Kovel, *The Age of Desire*, 267.

122 *"Freudianism helped construct"*: Zaretsky, *Secrets of the Soul*, 142–43.

122 *"life in America"*: New York Times, "Warns of Danger in American Life," June 5, 1927.

123 *"I believe that I have not given"*: Freud, *Civilization and Its Discontents*, 81.

123 *"My pessimism appears"*: Gay, *Freud*, 552–53.

124 *"declared that his health"*: New York Times, "American Loses Suit Against Freud," May 25, 1927.

125 *"It burdens [a doctor]"*: Freud, *Question of Lay Analysis*, 95.

125 *"suggest somatic rather than psychogenic affections"*: Freud, "Mourning and Melancholia," 203.

125 *"include elements from the mental sciences"*: Freud, *Question of Lay Analysis*, 88.

125 *"self knowledge," he wrote*: Ibid., 90.

125 *"specialized branch of medicine"*: Ibid., 91.

125 *"As long as I live"*: Gay, *Freud*, 491.

125 *"an attempt at repression"*: Freud, *Question of Lay Analysis*, 96.

126 *"expect nervous disorders"*: Ibid., 5.

CHAPTER 7

Page

133 *"Come on, you motherfuckers"*: Rosenhan, "On Being Sane," 256.

133 *"the bereaved widow"*: Dufresne, *Against Freud*, 13.

134 *"literary and scientific hero"*: Wortis, *Fragments of an Analysis with Freud*, 1.

134 *"very unusually talented"*: Ibid., 3.

134 *"I had no wish"*: Ibid.

134 *"the views of the widow"*: Dufresne, *Against Freud*, 15.

134 *"Though I am myself skeptical"*: Wortis, *Fragments*, 5.

135 *"He [Freud] would have thrown me out"*: Dufresne, *Against Freud*, 16.

135 *"I was taunting Freud"*: Ibid., 11.

135 *"It is true you have no palpable symptoms"*: Wortis, *Fragments*, 154.

135 *"You know shit about psychoanalysis"*: Dufresne, *Against Freud*, 13.

135 *"He seemed to be a bit hard of hearing"*: Wortis, *Fragments*, 24.

135 *"Dreaming is nothing but"*: Wortis, *Fragments*, 88.

135 *"No man could tell the truth"*: Ibid., 121.

136 *"You have not yet completed"*: Ibid., 61.

136 *"A person who professes to believe"*: Ibid., 44.

136 *"I feel sure"*: Ibid., 154.

136 The New York Times *was satisfied*: New York Times, "Dr. Sakel Is Dead; Psychiatrist, 57," December 3, 1957.

137 *"vagotropic nervous system"*: Sakel, *Schizophrenia*, 189.

137 *"deliberately abandoned the normal scientific procedure"*: Ibid., 188.

137 *"accidents"*: Sakel, *Schizophrenia*, 188.

137 *"courageously persisted in his experiments"*: Frostig, "Clinical Observations

in Insulin Treatment of Schizophrenia"; Jessner and Ryan, *Shock Treatment in Psychiatry*, 4.

138 *"Beyond this point"*: Jessner and Ryan, *Shock Treatment*, 11.

138 *"One is frequently surprised"*: Ibid., 27–28.

139 *One of our patients*: Ibid., 44.

139 *"the result of research"*: von Meduna, "The Significance of the Convulsive Reaction," 140.

140 *Paracelsus, the sixteenth-century Swiss physician*: Pearce, "Leopold Auenbrugger," 105.

140 *Although one doctor went so far*: Jessner and Ryan, *Shock Treatment*, 65; see also Kennedy, "Critical Review: The Treatment of Mental Disorders by Induced Convulsions."

141 *the patient got out*: Fink, "Meduna and the Origins of Convulsive Therapy," 1036.

141 *"a swindler, a humbug, a cheat"*: Ibid.

142 *The Hungarian missed no opportunity*: von Meduna, "The Significance of the Convulsive Reaction."

142 *a vast anti-Semitic conspiracy*: Shorter and Healy, *Shock Therapy*, 20.

142 *His technique was simple*: Shorter, *History of Psychiatry*, 218.

142 *"not using the much simpler method"*: Quoted in Shorter and Healy, *Shock Therapy*, 35.

142 *no one raised an objection*: Shorter, *History of Psychiatry*, 219.

143 *a pair of electrified forceps*: Impastatao, "Story of the First Electroshock Treatment," 1113.

143 *the seizure-inducing dose*: Shorter and Healy, *Shock Therapy*, 36.

143 *"calm, well oriented"*: Ibid., 37–42.

143 *A new word entered the Italian language*: Ibid., 44.

144 *even Meduna hailed*: Jessner and Ryan, *Shock Treatment*, xiv.

144 *"Our patients seem"*: Shorter and Healy, *Shock Therapy*, 73.

144 *"driv[ing] the Devil out of our patients"*: Ibid., 30.

144 *"bedevil the psychotic"*: Jessner and Ryan, *Shock Treatment*, xi.

144 *"Shock therapy has thrust"*: Ibid., xiii.

144 *One . . . patient, when coming out of coma*: Ibid., 42.

145 *Walter Freeman*: El-Haj, *The Lobotomist*.

145 *doctors who used drugs*: Kalinowsky and Hoch, *Shock Treatments*, 200–216.

145 *The various reflexes disappear*: Sakel, "The Methodical Use of Hypoglycemia in the Treatment of Psychoses," 117.

146 *"the therapeutic effect"*: Jessner and Ryan, *Shock Treatment*, 47; see also Cobb, "Review of Neuropsychiatry for 1938."

146 *the improvement "due to the patient's experience"*: Jessner and Ryan, *Shock Treatment*, 47.

146 *"the mistakes in theory"*: Quoted in Kalinowsky and Hoch, *Shock Treatments*, 229.

146 *when Sakel noticed:* Sakel, *Schizophrenia*, 235.

146 *when Cerletti concluded that he was getting better results:* Shorter and Healy, *Shock Therapy*, 79–82, 94–96.

147 *"the results [with neurotics]":* Kalinowsky and Hoch, *Shock Treatments*, 184.

148 *"As treating physicians":* Freeman, "Advantages Noted in Shock Therapy."

148 *"Indiscriminate use":* Kalinowsky and Hoch, *Shock Treatments*, 185–86.

148 *I am pleased to hear:* Wortis, *Fragments*, 113.

149 *I said incidentally:* Ibid., 110.

149 *Psychoanalysis [Freud said]:* Ibid.

149 *"Analysis never claimed a prerogative":* Ibid.

149 *"I said that in New York":* Ibid., 111.

150 *"what [Wortis] really wished":* Ibid., 112.

150 *"As Charcot [Freud's early mentor]":* Ibid., 138.

150 *"secret love":* Shorter and Healy, *Shock Therapy*, 84–92.

151 *He made his own contribution to the method:* Kalinowsky and Hoch, *Shock Treatments*, 37.

151 *the program was terminated:* Shorter and Healy, *Shock Therapy*, 55.

151 *"Pasteur of psychiatry":* Shorter and Healy, *Shock Therapy*, 55–56; Laurence, "Psychiatrist Hits Misuse of Shocks."

CHAPTER 8

Page

154 *so is Prozac:* Kalia et al., "Comparative Study of Fluoxetine."

158 *MDMA has a very powerful effect:* For a comprehensive and readable guide to the neurochemistry of MDMA, see Malberg and Bronson, "How MDMA Works in the Brain."

160 *"expedites lingering parturition":* Valenstein, *The War of the Soups and the Sparks*, 40.

161 *he experienced "an uninterrupted stream":* Hofmann, *LSD: My Problem Child*, 47.

161 *"I decided to conduct some experiments":* Grof, "Stanislav Grof Interviews Dr. Albert Hofmann," 22; http://www.maps.org/news-letters/v11n2/11222gro.html.

161 *"I was open to the fact":* Ibid., 24.

162 *"Beginning dizziness":* Hofmann, *LSD*, 48.

162 *Every exertion of my will:* Ibid., 49.

163 *"The effects were still":* Ibid., 52.

163 *"model psychoses":* Stevens, *Storming Heaven*, 11–12.

163 *"spectacular, and almost unbelievable"*: Schmiege, "The Current State of LSD as a Therapeutic Tool," 203.

164 *"subjects appeared more interested"*: Rinkel et al., "Experimental Psychiatry II," 884.

164 *the CIA, which, upon hearing of LSD*: For this early history, see Stevens, *Storming Heaven*, 80–87.

164 *"happy and dreamy feeling"*: Rinkel, DeShon, and Solomon, "Experimental Schizophrenia-like Symptoms," 574.

164 *"subjects became hostile"*: Rinkel et al., "Experimental Psychiatry II," 883.

164 *"By taking Delysid himself"*: Stevens, *Storming Heaven*, 12.

165 *To make the trivial world sublime*: Ibid., 57.

165 *bringing scientific research*: It's not strictly illegal to conduct research with LSD or other illegal drugs, but the bar is set high. Researchers must pass muster with the Drug Enforcement Agency, which tends to be suspicious of people who want to use illegal drugs for any purpose. In the last five years, however, some scientists have succeeded at launching some small studies in which people take LSD.

166 *Gaddum had gotten samples of serotonin*: Gaddum and Hameed, "Drugs Which Antagonize 5-hydroxytryptamine."

166 *"The cats became for a time"*: Green, "Gaddum and LSD," 9.

166 *Lysergic acid diethylamide*: Gaddum, "Antagonism between Lysergic Acid Diethylamide and 5-hydroxytryptamine," 15P.

167 *he finally "experienced some of the known effects"*: Green, "Gaddum and LSD," 9.

167 *"that the mental effects"*: Amin, Crawford, and Gaddum, "The Distribution of the Substance P," 616.

167 *that the mental changes*: Woolley and Shaw, "A Biochemical and Pharmacological Suggestion."

168 *"queer like a monstrous picture"*: Green, "Gaddum and LSD," 10.

168 *according to his daughter*: Healy, *The Creation of Psychopharmacology*, 204.

168 *We have all had wonderful dreams*: Thuillier, *Ten Years That Changed the Face of Mental Illness*, 70–71.

CHAPTER 9

Page

172 *"recurrent substance use"*: American Psychiatric Association, *Diagnostic and Statistical Manual*, 4th ed., text revision, 199.

173 *"It is precisely those communities"*: Freud, *Civilization and Its Discontents*, 50–51.

174 *Carlos Zarate*: Zarate et al., "A Randomized Trial of an n-methyl-D-aspartate Antagonist."

175 *"serendipitous discovery"*: Ibid., 856.

175 *"a single dose [of ketamine]"*: Ibid., 857.

175 *"play an important role"*: Ibid.

175 *anesthesiologists and pain doctors have long noted:* See, for instance, Correll and Futter, "Two Case Studies of Patients with Major Depressive Disorder."

176 *they've been using the drug therapeutically:* For this history see "Ketamine," Erowid website, www.erowid.org/ketamine/.

176 *"perceptual disturbances, confusion"*: Zarate et al., "A Randomized Trial," 861.

178 *"I felt the hammer striking"*: Thuillier, *Ten Years,* 106.

179 *Ehrlich had tried a dye:* Bäumler, *Paul Ehrlich,* 39–41.

179 *Pietro Bodoni, started to give methylene blue:* Healy, *Creation of Psychopharmacology,* 44–45.

179 *"Patents had been obtained"*: Ibid., 45.

179 *they had an overall effect that was interesting:* Ibid., 80.

180 *One day, so the story goes:* Thuillier, *Ten Years,* 112.

180 *In the corridors:* Ibid., 113.

180 *"pharmacological lobotomy"*: Ibid., 109.

181 *Chlorpromazine was slow to catch on:* Healy, *The Creation of Psychopharmacology,* 83–85, 96–99.

182 *the first test subject:* Broadhurst, interview.

182 *Fortunately for them:* Kuhn, "The Imipramine Story," 210.

182 *"rode, in his nightshirt"*: Tansey and Christie, "Drugs in Psychiatric Practice," 141.

183 *he repeated the performance:* Kuhn, "The Treatment of Depressive States."

183 *psychiatrists compared him to Ichabod Crane:* Healy, *The Antidepressant Era,* 57.

183 *"bring a complete change"*: Kuhn, "The Treatment of Depressive States," 462–63.

184 *"a general retardation"*: Ibid., 459.

184 *"Almost any neurotic symptom"*: Ibid., 462.

184 *"had not been so well"*: Kuhn, "The Imipramine Story," 216.

184 *a "world-wide ignorance"*: Ibid., 212.

185 *depression was too small a market:* Healy, *Antidepressant Era,* 56–59.

185 *a story appeared:* Laurence, "Wide New Fields Seen for TB Drug," 1, 3.

186 *"an international wheeler-dealer"*: Tansey and Christie, "Drugs in Psychiatric Practice," 145.

186 *When an observant brat:* Kline, "Monoamine Oxidase Inhibitors," 194.

186 *Rauwolfia serpentina:* Goenka, "Rustom Jal Vakil and the Saga of *Rauwolfia serpentina,*" 196.

187 *the glaziers at his hospital:* Kline, "Use of *Rauwolfia serpentina* Benth. in Neuropsychiatric Conditions," 121.

187 *would relieve simple depression:* Kline, "Monoamine Oxidase Inhibitors," 197.

188 *"immediately led me to speculate":* Ibid., 198.

188 *"Here indeed was a fairly unique situation!":* Ibid., 200.

188 *Kline arranged to testify:* Loomer, Saunders, and Kline, "Iproniazid: An Amine-Oxidase Inhibitor."

188 *"a side effect of an anti-tuberculosis drug":* Harrison, "TB Drug Is Tried in Mental Cases."

189 *"a disease is what the medical profession recognizes":* Jellinek, *The Disease Concept of Alcoholism,* 12.

189 *"On the antidepressant drugs":* Rusk, "Drugs and Depression."

189 *"eudaemonia":* Laurence, "Drug Called a Psychic Energizer Found Useful in Treating Mental Illnesses."

189 *"moral and social implications":* Kuhn, "Treatment of Depressive States," 464.

189 *"could improve ordinary performance":* Laurence, "Drug Called a Psychic Energizer."

190 *"persistent moderate depression":* New York Times, "Death of Woman Laid to 'Pep Pills,'" April 11, 1958.

190 *dose intended for "severe depression":* New York Times, "City Agents Hunt Supplies of Drug," April 12, 1958, 1.

190 *"CITY RESTRICTS SALE OF ENERGIZING DRUGS":* Ibid.

190 *"DRUG INVESTIGATED IN 2 DEATHS IN CITY":* New York Times, April 15, 1958, 1.

190 *reports of jaundice:* Kline, "Monoamine Oxidase Inhibitors," 202.

191 *"headline hunting":* New York Times, "Physician Warns on Wonder Drugs," April 13, 1960.

191 *the roster of "wonder drugs":* Ibid.

192 *Even the Puritans:* Levine, "The Discovery of Addiction," 144–46.

192 *$14 billion a year:* That's the drug czar's estimate for 2008. It doesn't include the costs of incarceration. See Office of National Drug Control Policy, "National Drug Control Strategy," http://www.whitehousedrugpolicy.gov/publications/policy/09budget/exec_summ.pdf.

192 *$14 billion is only a little more:* You can add it up yourself. See "Pharmacy Facts and Figures," Drug Topics website, http://drugtopics.modernmedicine.com/drugtopics/article/articleList.jsp?categoryId=7604.

193 *"If a drug makes you feel good":* Klerman, "Psychotropic Hedonism vs. Pharmacological Calvinism," 3.

193 *From Sad to Glad:* Kline, *From Sad to Glad: Kline on Depression.*

194 *the "inaugural article" of the antidepressant era:* Healy, *Antidepressant Era,* 148.

195 *They ran a controlled clinical trial:* Davies and Shepherd, "Reserpine in the Treatment of Anxious Patients."

195 *an excellent account:* Healy, *Antidepressant Era,* 152–55.

196 *"There is good evidence"*: Schildkraut, "The Catecholamine Hypothesis," 517.

197 *"at best, a reductionistic oversimplification"*: Ibid.

197 *"considerable heuristic value"*: Ibid., 518.

198 *the "soups" and the "sparks"*: Valenstein, *War of the Soups and the Sparks.*

198 *a British team*: Curtis and Eccles, "Excitation of Renshaw Cells."

198 *John Gaddum figured out*: Gaddum, "Push-Pull Cannulae."

198 *brilliant, beautiful photos*: Dahlstrom and Carlsson, "Making Visible the Invisible."

198 *one of the most-cited papers in the medical literature*: Healy, *Antidepressant Era,* 156.

CHAPTER 10

Page

203 *"it was discovered"*: Thomas, *The Medusa and the Snail,* 159.

203 *A total of seventy-four trials*: Turner et al., "Selective Publication of Antidepressant Trials."

204 *another analysis of clinical trials*: Kirsch et al., "The Emperor's New Drugs."

204 *the FDA's own director of clinical research*: Leber, "Approvable Action on Forrest Laboratories, Inc."

204 *according to a team of reviewers*: Turner et al., "Selective Publication of Antidepressant Trials."

204 *severe depressions blunt the placebo effect*: Kirsch et al., "Initial Severity and Antidepressant Benefits."

205 *the American Medical Association, in its first code of ethics*: Brody, *Hooked,* 139–41.

205 *there were two kinds in the world*: See Liebnau, *Medical Science and Medical Industry,* 4–10.

205 *Cuforhedake Brane-Fude*: Young, *The Medical Messiahs,* 3–12.

206 *the U.S. government, just then on the cusp of its Progressive Era*: Temin, *Taking Your Medicine,* 18–37.

206 *a highly publicized stunt*: Ibid., 28.

206 *"the greatest instrument"*: Liebnau, *Medical Science,* 91.

207 *"any statement . . . which shall be false"*: Temin, *Taking Your Medicine,* 30.

207 *the Harrison Anti-Narcotics Law*: Liebnau, *Medical Science,* 94.

207 *Dr. Johnson's Mild Combination Treatment for Cancer*: Young, *Medical Messiahs,* second photo after 204.

207 *"with reference to plain matter of fact"*: Temin, *Taking Your Medicine,* 33.

208 *outlaw "any statement"*: Ibid.

208 *Franklin Roosevelt, the president's son*: "Prontosil," *Time,* December 28, 1936.

209 *"just throw drugs together"*: Brody, *Hooked,* 248.

209 *the FDA found that its only recourse:* Temin, *Taking Your Medicine*, 42–43; Brody, *Hooked*, 250–51.

209 *Had Massengill named its drug:* Marks, *Progress of Experiment*, 82.

209 *doctors would dispense the information necessary:* Marks, *Progress of Experiment*, 83–89; "Prontosil," *Time*, December 28, 1936.

210 *"there is no scientific evidence":* Marks, *Progress of Experiment*, 96.

211 *"the wives of my Congressional group":* Ibid., 96–97.

211 *This was the regulatory environment:* Lasagna, "Congress, the FDA, and New Drug Development," 323.

212 *The doctor of today:* Lasagna, "The Controlled Clinical Trial," 353.

212 *the number of pages of JAMA:* Temin, *Taking Your Medicine*, 85–86.

213 *"He who orders":* Kefauver, *In the Hands of a Few*, 8.

213 *"safe and efficacious in use":* Temin, *Taking Your Medicine*, 122.

213 *The company hawked the drug:* Knightley et al., *Suffer the Children*, 26.

213 *She was concerned about reports:* Ibid., 73–81.

214 *Merrell's medical director:* Ibid., 73–81.

215 *The story was reported on the front page:* Mintz, "'Heroine' of FDA Keeps Bad Drug off Market," A1.

215 *"thalidomide was already barred":* Lasagna, "Congress, the FDA, and New Drug Development," 328.

215 *For the first time:* Temin, *Taking Your Medicine*, 121–24.

215 *The answer was that "substantial evidence":* Ibid., 127.

215 *"even though there may be preponderant evidence":* Lasagna, "Congress, the FDA, and New Drug Development." See also Temin, *Taking Your Medicine*, 124.

216 *Louis Lasagna cited a momentous event:* Lasagna, "The Controlled Clinical Trial," 367.

216 *"Their cases were as similar":* Lind, *A Treatise of the Scurvy*, 192–93. Key passages from this 1753 work can be found at the James Lind Library, http://www.jameslindlibrary.org/trial_records/17th_18th_Century/lind/lind_kp.html.

216 *Franklin and Mesmer had different ideas:* Kaptchuk, "Intentional Ignorance," 394.

217 *"at hazard, and in parts very distant":* Ibid., 396.

219 *an RCT is much more suited:* See Healy, *Mania*, 128–34.

219 *"taken the blindfold test":* "Morgan's Old Gold," *Time*, http://www.time.com/time/magazine/article/0,9171,786005,00.html.

219 *A decade later:* Gold, Kwit, and Otto, "The Xanthines (Theobromine and Aminophylline) in the Treatment of Cardiac Pain."

220 *He cited the Old Gold campaign:* Shapiro and Shapiro, *The Powerful Placebo*, 154.

220 *Fisher was trying to sort out fact from opinion:* Marks, *Progress of Experiment*, 141–48.

221 *to equalize the chance:* Ibid., 144.

222 *"free a researcher":* Ibid., 146.

222 *I have an intense prejudice:* "Conference on Therapy," *American Journal of Medicine*, 727.

223 *Let the experimenter who is driven:* Marks, *Progress of Experiment*, 139–40.

CHAPTER 11

Page

227 *"a psychological response to an identifiable stressor":* American Psychiatric Association, Diagnostic and Statistical Manual of Mental Disorders, 4th ed., 679–80.

227 *"excessive anxiety and worry":* Ibid., 476.

228 *Erving Goffman and Michel Foucault:* See Goffman, *Stigma: Notes on the Management of Spoiled Identity,* and Foucault, *Madness and Civilization.*

229 *socially constructed:* See Berger and Luckmann, *The Social Construction of Reality,* also Gergen, "The Social Constructionist Movement in Modern Psychology."

232 *"fevered polemical discussions":* Spitzer and Wilson, "Nosology and the Official Psychiatric Nomenclature," 837.

233 *"what [was] behind the symptom":* Menninger, *The Vital Balance,* 325.

233 *"We must attempt to explain":* Ibid.

233 *"the force of factors":* Menninger, "Psychiatric Experience in the War," 580.

233 *This synthesis was enshrined:* Grob, "Origins of DSM-I."

234 *"Instead of putting so much emphasis":* Menninger, *The Vital Balance,* 325.

234 *The Group for the Advancement of Psychiatry:* Shorter, *History of Psychiatry,* 173–77.

234 *"an objective critical attitude":* Wilson, "DSM-III and the Transformation of American Psychiatry," 401.

234 *psychologist Philip Ash showed:* Ash, "The Reliability of Psychiatric Diagnosis."

234 *a dismal 42 percent of cases:* Beck, "Reliability of Psychiatric Diagnoses": 213.

235 *a team led by Martin Katz:* Katz, Cole, and Lowery, "Studies of the Diagnostic Process."

235 *manic depression was much more common:* See, for example, Sandifer et al., "Psychiatric Diagnosis."

236 *"There is a terrible sense of shame":* Wilson, "DSM-III," 405.

236 *they disrupted APA meetings:* Bayer, *Homosexuality and American Psychiatry,* 102–3.

236 *the protestors got what they wanted:* Ibid., 112–50.

236 *"Referenda on matters of science":* Ibid., 153.

236 *"If groups of people march":* Ibid., 141.

237 *Laing focused on schizophrenia:* Laing, *The Divided Self.*

237 *"problems of living":* Szasz, *The Myth of Mental Illness.*

238 *"documenting the total number of people":* Wilson, "DSM-III," 403.

238 *"carrying psychiatrists on a mission":* Ibid.

239 *"Some individuals may interpret":* American Psychiatric Association, *Diagnostic and Statistical Manual,* 2nd ed., 122.

239 *some argued, and continue to argue:* See, for example, Klerman et al., "A Debate on DSM-III." More recently, see Kirk and Kutchins, *The Selling of DSM,* and Caplan, *They Say You're Crazy.*

239 *"I was uncomfortable with not knowing":* Spiegel, "The Dictionary of Disorder," http://www.newyorker.com/archive/2005/01/03/050103fa_fact.

240 *depressive neurosis:* American Psychiatric Association, *Diagnostic and Statistical Manual,* 2nd ed., 40.

241 *overall attempt "to avoid terms":* Ibid., viii.

241 *its professional discussions relegated:* Wilson, "DSM-III," 403.

242 *"assumed . . . an underlying process":* Bayer and Spitzer, "Neurosis, Psychodynamics and DSM-III," 189.

242 *neurosis had been first described:* Knoff, "A History of the Concept of Neurosis."

242 *"DSM-III gets rid of the castles of neurosis":* Bayer and Spitzer, "Neurosis, Psychodynamics and DSM-III," 189.

242 *"many patients":* Wilson, "DSM-III," 405.

242 *"a straitjacket":* Bayer and Spitzer, "Neurosis, Psychodynamics and DSM-III," 190.

243 *"wish [neurosis] reinserted":* Ibid., 192.

243 *"scientists attempting to advance":* Ibid., 190.

243 *"pro-neurosis forces":* Ibid., 193.

243 *a large role in "Project Flower":* Wilson, "DSM-III," 407.

243 *"extremely embarrassing":* Bayer and Spitzer, "Neurosis, Psychodynamics and DSM-III," 193–94.

244 *The DSM criteria for MDD:* American Psychiatric Association, *Diagnostic and Statistical Manual,* 3rd ed., 213–15.

244 *dysthymic disorder:* Ibid., 220–22.

244 *adjustment disorder:* Ibid., 300.

244 *"Clerks rather than experts":* Wilson, "DSM-III," 406.

244 *the Feighner criteria:* Feighner et al., "Diagnostic Criteria for Use in Psychiatric Research."

245 *the single most commonly cited article:* Spitzer, Endicott, and Robins, "The Development of Diagnostic Criteria in Psychiatry," 21.

245 *"This communication will present":* Feighner et al., "Diagnostic Criteria," 57.

245 *"criteria for establishing diagnostic validity"*: Ibid.

245 *Allan Horwitz and Jerome Wakefield examined*: Horwitz and Wakefield, *Loss of Sadness*, 93–94.

246 *many people who had recently been bereaved*: Clayton, Halikas, and Maurice, "The Bereavement of the Widowed."

246 *the DSM-III committee were reminded*: Horwitz and Wakefield, *Loss of Sadness*, 100–102.

246 *"A full depressive syndrome"*: American Psychiatric Association, *Diagnostic and Statistical Manual*, 3rd ed., 333.

247 *grief begins to wane*: Clayton and Darvish, "Course of Depressive Symptoms Following the Stress of Bereavement."

248 *the profession would have been back to the bad old days*: As indeed, they turned out to be in Marwit, "Reliability of Diagnosing Complicated Grief."

251 *"depression in the Western world"*: Andrews and Skoog, "Lifetime Prevalence of Depression," 495.

CHAPTER 12

Page

255 *"not typically used for your condition"*: Tilburt et al., "Prescribing 'Placebo Treatments,'" http://www.bmj.com/cgi/content/full/337/oct23_2/a1938.

256 *"As a result of accumulated knowledge"*: Zinberg, *Drug, Set, and Setting*, 187.

257 *secondary anxiety*: Becker, "History, Culture, and Subjective Experience."

257 *Zinberg wrote about subcultures*: Zinberg et al., "Patterns of Heroin Use"; Zinberg, Harding, and Winkeller, "A Study of Social Regulatory Mechanisms"; and Zinberg, *Drug, Set, and Setting*, 135–71.

258 *Harvard's lawyers objected*: Zinberg, *Drug, Set, and Setting*, 199–200.

259 *An ad that ran in a 1945 issue of the* American Journal of Psychiatry: "If the individual is depressed . . . ," xiii.

260 *an indication that was worth $1.5 billion*: "Pharmacy Facts and Figures," Drug Topics website, http://drugtopics.modernmedicine.com/drug topics/article/articleList.jsp?categoryId=7604.

260 *"outstanding effectiveness . . . with which Miltown relieves"*: "Relieves Anxiety and Anxious Depression," Miltown advertisement, 20.

260 *people were "miltowning"*: Callahan and Berrios, *Reinventing Depression*, 106. See also Shorter, *Before Prozac*, 45.

260 *"No Miltown today"*: Shorter, *Before Prozac*, 45.

260 *The industry pushed the minor tranquilizers*: Callahan and Berrios, *Reinventing Depression*, 110.

260 *"when the patient rambles"*: "When the patient rambles . . . ," advertisement, 420.

260 *Valium eventually took up more:* Callahan and Berrios, *Reinventing Depression*, 110.

261 *The minor tranquilizers' success:* Wheatley, "A Comparative Trial of Imipramine"; Hollister, "Mental Disorders"; Blackwell, "Psychotropic Drugs in Use Today."

261 *the first "product of the pharmaceutical industry":* Callahan and Berrios, *Reinventing Depression*, 109.

262 *14 million prescriptions:* Ibid., 112.

262 *"penicillin for the blues":* Ibid., 106.

262 *"It's hard to appreciate the difficulty":* Martin, "Pharmaceutical Virtue," 161.

262 *"chemical revolution in psychiatry":* Ayd, *Recognizing the Depressed Patient,* iii.

262 *"depressions are among the most common illnesses":* Ibid., 5.

263 *"Not all depressed people require the services":* Ibid., 117.

263 *"Many melancholics can be cared for":* Ibid., 119.

263 *"treatment can be just as effective":* Ibid., 119–20.

263 *"to explain to the patient":* Ibid., 117.

263 *"Depressed people are very suggestible":* Ibid., 119.

263 *You have an illness called a depression:* Ibid., 117.

264 *"I found a musicologist":* Martin, "Pharmaceutical Virtue," 161.

264 *We gave this to doctors:* Ibid.

265 *one sales rep for every eight doctors:* Callahan and Berrios, *Reinventing Depression*, 110.

266 *the dangers of "mood drugs":* Shorter, *Before Prozac,* 117.

266 *"other products which also affect the mind":* Callahan and Berrios, *Reinventing Depression*, 112.

266 *one of "the most frequently abused drugs":* Shorter, *Before Prozac,* 118.

266 *"these drugs have produced":* Ibid., 117.

266 *"the greatest commercial success":* Shorter, *Before Prozac,* 99.

267 *their sales continued to languish:* Callahan and Berrios, *Reinventing Depression*, 111–12.

268 *a team at Eli Lilly:* Wong et al., "A Selective Inhibitor of Serotonin Uptake."

268 *this wasn't in the company's game plan:* This story can also be found in Shorter, *Before Prozac,* 177, and Healy, *Antidepressant Era,* 167–68.

268 *zimelidine syndrome:* Shorter, *Before Prozac,* 174.

269 *the company was finally interested:* Bremner, "Fluoxetine in Depressed Patients"; see also Wong, Bymaster, and Engleman, "Prozac (Fluoxetine, Lilly 110140) the First Selective Serotonin Uptake Inhibitor."

270 *they suddenly can't reach orgasm:* Nurnberg et al., "Sildenafil Treatment of Women," 395.

270 *nearly 70 percent:* Lin et al., "The Role of the Primary Care Physician in

Patients' Adherence to Antidepressant Therapy," 70; see also Nurnberg et al., "Sildenafil Treatment of Women."

270 *"similar findings"*: Leber, "Approvable action."

270 *antidepressants had become*: Spielmans et al., "The Accuracy of Psychiatric Medication Advertisements in Medical Journals," 268.

271 *"Unfortunately, our internal policies"*: Ibid., 271.

271 *80 percent of "high prescribers"*: Neslin, "Executive Summary: ROI Analysis of Pharmaceutical Promotion," www.rxpromoroi.org/rapp/exec_sum .html. See also Hunt, "Interaction of Detailing and Journal Advertising," www.rxpromoroi.org/pdf/interaction_whitepaper.pdf.

271 *after reading in the* Journal of the American Medical Association: See, for instance, Hirschfeld et al., "The National Depressive and Manic-Depressive Association Consensus Statement on the Undertreatment of Depression," and Kessler, "The Epidemiology of Major Depressive Disorder."

272 *four times more likely*: Olfson and Marcus, "National Patterns in Antidepressant Medication Treatment."

272 *the drug's $1.5 billion in sales*: "Prozac Making Lilly a Little Edgy," *Business-Week*, June 22, 1992.

273 *"the king of antidepressants"*: Langreth, "Mending the Mind," B1.

273 *its sales were still up by 18 percent*: Ibid.

273 *still only 10 percent*: Hirschfeld et al., "National Depressive and Manic-Depressive Association Consensus Statement."

273 *"patients [become] informed consumers"*: Ibid., 338.

273 *Eli Lilly hired the Leo Burnett Company*: Elliott, "A New Campaign by Leo Burnett Will Try to Promote Prozac Directly to Consumers."

274 *"one of the most serious assignments"*: Ibid.

274 *depression "isn't just feeling down"*: Stanfel, "Prozac Print Campaign," 507.

274 *"assisting people"*: Ibid., 506.

274 *The first ad*: You can find this ad in, among other outlets, the May 1998 issue of *Reader's Digest*, 182–84.

274 *Lilly and Burnett took Prozac*: Hume, "Prozac Getting a New Prescription."

275 *Have you stopped doing the things you enjoy*: I am grateful to Joseph Dumit, who provided me with DVD copies of the television ads discussed here.

275 *companies were spending*: Block, "Costs and Benefits of Direct-to-Consumer Advertising," 513.

275 *Pfizer introduced a cartoon character*: Aurthur, "Little Blob, Don't Be Sad (or Anxious or Phobic)."

276 *"the science of arresting the human intelligence"*: Gilbody, Wilson, and Watts, "Direct-to-Consumer Advertising of Psychotropics."

276 *Hey you*: Ad copy can be found in Williams, "Effexor XR Warning Letter," http://www.fda.gov/cder/warn/2004/Effexor.pdf.

276 *"by failing to draw a clear distinction"*: Ibid.

277 *"there is no clear"*: Lacasse and Leo, "Serotonin and Depression," e392.

277 *Ad industry research*: Neslin, "RAPP Study," http://www.rxpromoroi.org/rapp/index.html.

277 *"six percent of the increase"*: Block, "Costs and Benefits," 514.

277 *"treating everyone"*: Ibid., 519.

278 *The team, led by Richard Kravitz*: Kravitz et al., "Influence of Patients' Requests for Direct-to-Consumer Advertised Antidepressants."

278 *"Some things about the ad"*: Ibid., 1997.

281 *"Depression doesn't mean you have something wrong with your character"*: Pfizer, "Myths and Facts about Depression," http://www.zoloft.com/depr_myths_facts.aspx.

281 *"Like other illnesses such as diabetes"*: Eli Lilly and Company, "Prozac Makes History," http://www.prozac.com/disease_information/treatment_depression.jsp?reqNavId=1.1.4.

CHAPTER 13

Page

288 *"caught up in the contagion"*: Beck, "Evolution of the Cognitive Model of Depression," 969.

288 *He dabbled*: Beck, "Reliability of Psychiatric Diagnoses."

288 *"that the dreams"*: Beck and Ward, "Dreams of Depressed Patients."

289 *therapist and patient work together*: Rush et al., "Comparative Efficacy of Cognitive Therapy and Pharmacotherapy in the Treatment of Depressed Outpatients," 17.

292 *Beck is going by the book*: Beck, *Cognitive Therapy*.

294 *Therapeutic outcomes are dependent*: Luborsky et al., "The Researcher's Own Therapy Allegiances," 65.

298 *Rafael Osheroff*: This version of the story follows Klerman, "The Patient's Right to Effective Treatment"; see also Shorter, *History of Psychiatry*, 309–10, and Healy, *The Antidepressant Era*, 245–50.

299 *"The case left the strong impression"*: Shorter, *History of Psychiatry*, 310.

299 *If a pharmaceutical firm makes a claim*: Klerman, "The Patient's Right," 416.

300 *"Everyone Has Won"*: Rosenzweig, "Some Implicit Common Factors in Diverse Methods of Psychotherapy."

300 *Luborsky subjected the dodo bird*: Luborsky, Luborsky, and Singer, "Comparative Studies of Psychotherapies."

300 *"The different forms of psychotherapy"*: Ibid., 1006.

300 *Luborsky's work got updated*: Smith and Glass, "Meta-analysis of Psychotherapy Outcome Studies," and Smith, Glass, and Miller, *The Benefits of Psychotherapy*.

300 *"convincing evidence that therapy"*: Consumers Union, "Mental Health: Does Therapy Help?" 734. The report is available at www.consumerreports .org, but you have to sign up and pay for a membership to see it.

301 *"psychotherapies are not doing"*: Klein, "Preventing Hung Juries about Therapy Studies."

301 *"If clinical psychology is to survive"*: American Psychological Association, "Training in and Dissemination of Empirically-Validated Psychological Treatment," 21.

302 *Beck got a chance:* Rush et al., "Comparative Efficacy."

303 *"a profound effect"*: DeRubeis and Beck, "Cognitive Therapy," 293.

303 *Beck could then plausibly claim:* Rush et al., "Comparative Efficacy," 25.

303 *researchers replicated:* See, for instance, Blackburn, "The Efficacy of Cognitive Therapy in Depression"; Blackburn, Eunson, and Bishop, "A Two-Year Naturalistic Follow-up"; Dobson, "A Meta-analysis"; Robinson, Berman, and Neimeyer, "Psychotherapy for the Treatment of Depression"; D. Shapiro et al., "Effects of Treatment Duration"; and D. Shapiro, "Meta-analysis of Comparative Therapy Outcome Studies."

303 *"the most scientifically tested"*: Spiegel, "More and More, Favored Psychotherapy Lets Bygones Be Bygones."

304 *"the best-documented effectiveness"*: American Psychiatric Association, "Practice Guidelines for the Treatment of Patients with Major Depressive Disorder," 11.

304 *"the most widely practiced approach"*: Spiegel, "More and More."

304 *Critics complain:* Westen and Morrison, "A Multidimensional Meta-analysis of Treatments for Depression."

305 *"Psychotherapy is essentially"*: Holmes, "All You Need Is Cognitive Behaviour Therapy?" 290.

305 *they also assert:* Westen and Morrison, "A Multidimensional Meta-analysis," 878.

305 *"empirically supported therapies"*: Hollon, Thase, and Markowitz, "Treatment and Prevention of Depression," 39.

305 *what happens when researchers try:* Wampold, "Methodological Problems in Identifying Efficacious Psychotherapies," 27.

306 *"unconditional support"*: Foa et al., "Treatment of Post-traumatic Stress Disorder in Rape Victims," 721.

306 *"that something intended"*: Westen and Bradley, "Empirically Supported Complexity," 267; see also Parker and Fletcher, "Treating Depression with the Evidence-Based Psychotherapies," 354.

306 *Allegiances do matter:* Robinson, Berman, and Neimeyer, "Psychotherapy for the Treatment of Depression"; see also Wampold, *The Great Psychotherapy Debate,* and Smith et al., *The Benefits of Psychotherapy.*

307 *two independent groups:* Wampold et al., "A Meta-analysis of Outcome Studies Comparing Bona Fide Psychotherapies," and Depression Guideline Panel, *Treatment of Major Depression.*

308 *Westen and Morrison acknowledge:* Westen and Morrison, "A Multidimensional Meta-analysis," 886–87.

308 *all the critics:* See, for instance, Parker and Fletcher, "Treating Depression with the Evidence-Based Psychotherapies"; Wampold et al., "A Meta-analysis"; and Seligman, "The Effectiveness of Psychotherapy."

308 *a group of loyal cognitive therapists:* Jacobson et al., "A Component Analysis of Cognitive-Behavior Treatment for Depression."

309 *according to the meta-analysts:* See Parker and Fletcher, "Treating Depression with the Evidence-Based Psychotherapies," for a review of this research.

309 *"How therapy is conducted":* Wampold, "Establishing Specificity in Psychotherapy Scientifically," 197.

313 *"A person fails to live":* Peale, *The Power of Positive Thinking,* 173.

313 *"The world in which you live":* Ibid.; quotes, in order, from 166, 166, 173, 167.

CHAPTER 14

Page

316 *"The DSM-IV . . . has 100 percent reliability":* Amen, *Healing the Hardware of the Soul,* xx-xxi. Amen says he transcribed this passage from a recording of Insel's speech. The recording is not available. Insel told me that he did not keep a copy of this speech, but did not dispute the content of the quote.

317 *Brain imaging in clinical practice:* Ibid., 20.

317 *"the basic pathophysiology":* Insel and Quirion, "Psychiatry as a Clinical Neuroscience Discipline," 2223.

318 *"I was suspicious":* Amen, *Healing the Hardware,* 8–9.

322 *"One is often":* Ibid., 13.

322 *Johann Spurzheim, who had redrawn:* Davies, *Phrenology,* 5–9.

322 *Orson Squire Fowler:* Martin, "Saints, Sinners, and Reformers," www .crookedlakereview.com /books /saints_sinners/martin12.html.

323 *"what they are":* Davies, *Phrenology,* 38.

323 *"Would you become great mentally":* Fowler and Fowler, *The Self Instructor in Phrenology and Physiology,"* 11.

323 *a therapeutic empire:* Martin, "Saints, Sinners."

323 *President James Garfield:* Martin, "Saints, Sinners." Davies has his own list in *Phrenology,* 38.

323 *Horace Greeley published:* Davies, *Phrenology,* 50.

323 *"the greatest discovery":* Martin, "Saints, Sinners."

323 *"Phrenology . . . has assumed"*: Davies, *Phrenology*, 121.

324 *"Phrenology, it must be confessed"*: Mackey, "Phrenological Whitman," www.conjunctions.com/archives/c29-nm.htm.

324 *The relationship between Whitman and the Fowlers*: Davies, *Phrenology*, 123–25.

324 *I made a small test*: Quoted in Martin, "Saints, Sinners."

325 *"The equilibrium or balance"*: Quoted in Finger, *Minds behind the Brain*, 166–67.

325 *he didn't report the fact*: Barker, "Phineas among the Phrenologists."

325 *Broca's area*: Finger, *Minds*, 137–54.

325 *Wernicke's area*: Finger, *Minds*, 150; Shorter, *History of Psychiatry*, 80–81.

325 *David Ferrier*: Finger, *Minds*, 155–75.

326 *At least that's the story*: See for example Finger, *Minds*, 135–36; Barker, "Phineas among the Phrenologists"; and Simpson, "Phrenology and the Neurosciences."

326 *empathy*: Preston and de Waal, "Empathy."

326 *racial prejudice*: Richeson et al., "An fMRI Investigation of the Impact of Interracial Contact on Executive Function."

326 *sexual orientation*: Fitzgerald, "A Neurotransmitter System Theory of Sexual Orientation"; LeVay, *Queer Science*.

326 *the October 16, 2008, issue of* Nature: Krishnan and Nestler, "The Molecular Neurobiology of Depression."

327 *a team of French doctors reported*: Bejjani et al., "Transient Acute Depression Induced by High-Frequency Brain Stimulation."

330 *subjectivity is a forbidden topic*: For a discussion of the religious proportions of the banishment of subjectivity from talk about consciousness, see Wallace, *The Taboo of Subjectivity*.

330 *"the mind is a set of operations"*: Kandel, "The New Science of Mind," 69.

331 *Empirical reductionism*: Woese, "A New Biology for a New Century," 174.

331 *With the progress*: Popper and Eccles, *The Mind and Its Brain*, 97.

CHAPTER 15

Page

342 *Various critics of the current diagnostic system*: Regier, "State-of-the-Art Psychiatric Diagnosis," 26.

343 *"clinical significance"*: American Psychiatric Association, *Diagnostic and Statistical Manual*, 4th ed., text revision, xxxi.

343 *Despite the prominence of clinical significance*: Narrow et al., "Revised Prevalence Estimates of Mental Disorders in the United States," 118.

344 *the prevalence of depression is cut nearly in half*: Narrow et al., "Revised Prevalence Estimates," 120.

344 *the current categorical approach:* See, for example, Goldberg, "A Dimensional Model for Common Mental Disorders"; also Kessler et al., "Prevalence, Correlates, and Course of Minor Depression and Major Depression in the National Comorbidity Survey."

344 *These doctors have already struck back:* Kessler et al., "Mild Disorders Should Not Be Eliminated From the DSM-V."

344 *"should be addressed":* Ibid., 1121.

345 *The major problem for mental disorders:* Kupfer, First, and Regier, *A Research Agenda for* DSM-V, 208.

345 *the pathophysiologically based classification system:* See Kupfer, First, and Regier, *Research Agenda*, chapter 1.

346 *Fink thinks:* Taylor and Fink, "Restoring Melancholia in the Classification of Mood Disorders," and Fink, "The Medical Evidence-Based Model for Psychiatric Syndromes."

347 *"that what is now considered":* Taylor and Fink, "Restoring Melancholia," 4.

347 *a highly effective treatment for melancholia:* See Shorter and Healy, *Shock Therapy*, especially 83–102 and 164–80.

348 *"on, or shortly after, penetration":* American Psychiatric Association, *Diagnostic and Statistical Manual*, 4th ed., text revision, 552.

348 *repeated studies:* For a review, see Waldinger, "Premature Ejaculation."

349 *"focusing on the ejaculation-delaying effects":* Ibid., 553.

349 *the package insert for Paxil:* GlaxoSmithKline, "Prescribing Information," http://us.gsk.com/products/assets/us_paxil.pdf.

349 *researchers gave Viagra:* Nurnberg, "Sildenafil Treatment of Women."

350 *female sexual dysfunction:* American Psychiatric Association, *Diagnostic and Statistical Manual*, 4th ed., text revision, 543.

351 *"for the vast majority of . . . the walking wounded":* Shorter, *Before Prozac*, 201.

352 *"proves to be a sort of instant super-Prozac":* Kramer, *Listening to Prozac*, 290.

352 *a "neutral technology":* Parens, "Kramer's Anxiety," 26–27.

352 *"low self-worth":* Kramer, *Listening to Prozac*, 8.

352 *socially capable:* Ibid., 11.

355 *"as many as 50 percent of the patients":* For a review, see Fava, "Prospective Studies."

356 *Prozac poop-out:* Glenmullen, *Prozac Backlash*, 91–94.

356 *tachyphylaxis:* Solomon et al., "Tachyphylaxis in Unipolar Major Depressive Disorder."

357 *a good working definition of disease:* For a brilliant discussion of this idea, see Sedgwick, "Illness—Mental and Otherwise."

360 *decreasingly sympathetic support systems:* For a devastating critique of this response, see Wallace, "The Depressed Person."

361 *"modern opiate":* Ibid, 272.

361 *"induces conformity"*: Kramer, "The Valorization of Sadness," 52.

361 *"on balance a progressive force"*: Kramer, *Listening to Prozac*, 272.

362 *"so-what drugs"*: Shorter, *Before Prozac*, 199.

362 *"facilitat[ing] better performance"*: Parens, "Kramer's Anxiety," 415.

364 *"A nation that asks"*: Tocqueville, *Democracy in America, vol. 2*, 142.

BIBLIOGRAPHY

Alloy, L. B., and L. Y. Abramson. "Judgment of Contingency in Depressed and Nondepressed Students: Sadder but Wiser?" *Journal of Experimental Psychology* 108, no. 4 (1979): 441–85.

Amen, D. G. *Healing the Hardware of the Soul.* New York: Free Press, 2002.

American Psychiatric Association. *Diagnostic and Statistical Manual: Mental Disorders.* Washington, DC: American Psychiatric Association, 1954.

———. *Diagnostic and Statistical Manual of Mental Disorders.* 2nd. ed. Washington, DC: American Psychiatric Association, 1968.

———. *Diagnostic and Statistical Manual of Mental Disorders,* 3rd. ed. Washington, DC: American Psychiatric Association, 1980.

———. *Diagnostic and Statistical Manual of Mental Disorders,* 4th ed., text revision, Washington, DC: American Psychiatric Association, 2000.

———. "Practice Guidelines for the Treatment of Patients with Major Depressive Disorder." *American Journal of Psychiatry* 157, suppl. (2000): 1–45.

American Psychological Association. "Training in and Dissemination of Empirically-Validated Psychological Treatment: Report and Recommendations." *The Clinical Psychologist* 48 (1995): 3–23.

Amici, R. R. "The History of Italian Parasitology." *Veterinary Parasitology* 98 (2001): 3–30.

Amin, A. H., T. B. B. Crawford, and J. H. Gaddum. "The Distribution of the Substance P and 5-hydrotoxytryptamine in the Central Nervous System of the Dog." *Journal of Physiology* 126 (1954): 596–618.

Andrews, G., R. Poulton, and I. Skoog. "Lifetime Prevalence of Depression: Restricted to a Minority or Waiting for Most?" *British Journal of Psychiatry* 187 (2005): 495–96.

Arikha, N. *Passions and Tempers: A History of the Humours.* New York: Little, Brown, 2007.

Ash, P. "The Reliability of Psychiatric Diagnosis." *Journal of Abnormal and Social Psychology* 44 (1949): 272–77.

Auden, W. H. *Collected Poems.* New York: Modern Library, 2007.

Aurthur, Kate. "Little Blob, Don't Be Sad (or Anxious or Phobic)." *New York Times,* January 2, 2005.

Ayd, F. *Recognizing the Depressed Patient: With Essentials of Management and Treatment.* New York: Grune & Stratton, 1961.

Barber, C. "The Medicated Americans: Antidepressant Prescriptions on the Rise." *Scientific American,* February 2008, pp. 37–48.

Barker, F. G. "Phineas among the Phrenologists: The American Crowbar Case and Nineteenth-Century Theories of Cerebral Localization." *Journal of Neurosurgery* 82 (1995): 673–82.

Bartisch, G. *Theriac.* Translated by D. L. Blanchard. Portland, OR: Blanchard's Books, 2000.

Bäumler, Ernst. *Paul Ehrlich: Scientist for Life.* New York: Holmes & Meier, 1984.

Bayer, R. *Homosexuality and American Psychiatry.* Princeton, NJ: Princeton University Press, 1987.

Bayer, R., and R. L. Spitzer. "Neurosis, Psychodynamics and DSM-III: A History of the Controversy." *Archives of General Psychiatry* 42 (1985): 187–96.

Beard, G. M. *American Nervousness; its Causes and Consequences, a Supplement to Nervous Exhaustion.* New York: G. P. Putnam's Sons, 1881.

Beck A. T. "Evolution of the Cognitive Model of Depression and Its Neurobiological Correlates." *American Journal of Psychiatry* 165, no. 8 (2008): 969–78.

———. "Reliability of Psychiatric Diagnoses: 1. A Critique of Systematic Studies." *American Journal of Psychiatry* 119 (1962): 210–16.

Beck, A. T., and C. H. Ward. "Dreams of Depressed Patients. Characteristic Themes in Manifest Content." *Archives of General Psychiatry* 5, no. 5 (1961): 462–67.

Beck, A. T., C. H. Ward, M. Mendelson, J. E. Mock, and J. K. Erbaugh. "Reliability of Psychiatric Diagnoses: 2. A Study of Consistency of Clinical Judgments and Ratings." *American Journal of Psychiatry* 119 (1962): 351–57.

Beck, J. S. *Cognitive Therapy: Basics and Beyond.* New York: Guilford Press, 1995.

Becker, H. S. "History, Culture, and Subjective Experience: An Exploration of the Social Bases of Drug-Induced Experiences." *Journal of Health and Social Behavior* 8 (1967): 162–76.

Beers, C. *Mind That Found Itself: A Memoir of Madness and Recovery.* West Valley City, UT: Waking Lion Press, 2007 (originally published 1908).

Bejjani, B. P., P. Damier, I. Arnulf, L. Thivard, A. Bonnet, D. Dormont, P. Cornu, B. Pidoux, Y. Samson, and Y. Agid. "Transient Acute Depression Induced by High-Frequency Brain Stimulation." *New England Journal of Medicine* 340, no. 19 (1999): 1476–80.

Benedek, Thomas G., and Jonathon Erlen. "The Scientific Environment of the Tuskegee Study of Syphilis, 1920–1960." *Perspectives in Biology and Medicine* 43, no. 1 (Autumn 1999): 1–30.

Berdin, Victoria, and Jennifer Flavin. "The Least of My Brothers: Syphilis in History." Internet Archive Wayback Machine website. Poynter Center for the Study of Ethics and American Institutions, Indiana University Bloomington, and the National Institutes of Health (Grant Number 1 T15 AI07601). Available online at http://web.archive.org/web/20080124141121/http://wisdomtools.com/poynter/syphilis.html.

Berger, P. L., and T. Luckmann. *The Social Construction of Reality: A Treatise in the Sociology of Knowledge.* Garden City, NJ: Anchor Books, 1966.

Blackburn, I. M., K. M. Eunson, and S. Bishop. "A Two-Year Naturalistic Follow-up of Depressed Patients Treated with Cognitive Therapy, Pharmacotherapy and a Combination of Both." *Journal of Affective Disorders* 10, no. 1 (Jan.–Feb. 1986): 67–75.

Blackburn, I. M., S. Bishop, A. I. Glen, L. J. Whalley, and J. E. Christie. "The Efficacy of Cognitive Therapy in Depression: A Treatment Trial Using Cognitive Therapy and Pharmacotherapy, Each Alone and in Combination." *British Journal of Psychiatry* 139 (1981): 181–89.

Blackwell, B. "Psychotropic Drugs in Use Today: The Role of Diazepam in Medical Practice." *Journal of the American Medical Association* 225 (1973): 1637–41.

Block, A. E. "Costs and Benefits of Direct-to-Consumer Advertising." *Pharmacoeconomics* 25, no. 6 (2007): 511–21.

Blunt, W. *Linnaeus: The Compleat Naturalist.* Princeton, NJ: Princeton University Press, 2002.

Book of Job. Jerusalem Bible. Garden City, NY: Doubleday, 1966.

Bourquin, Avril. Man and Mollusc. www.manandmollusc.net.

Bremner, J. D. "Fluoxetine in Depressed Patients: A Comparison with Imipramine." *Journal of Clinical Psychiatry* 45, no. 10 (1984): 414–19.

Brody, H. *Hooked: Ethics, the Medical Profession, and the Pharmaceutical Industry.* Lanham, MD: Rowman & Littlefield, 2007.

Callahan, C. M., and G. E. Berrios. *Reinventing Depression: A History of the Treatment of Depression in Primary Care, 1940–2004.* New York: Oxford University Press, 2005.

Caplan, P. *They Say You're Crazy: How the World's Most Powerful Psychiatrists Decide Who's Normal.* Reading, MA: Addison-Wesley, 1995.

Carlsson, A., and D. T. Wong. "Correction: A Note on the Discovery of Selective Serotonin Reuptake Inhibitors." *Life Sciences* 61, no. 12 (1997): 1203.

Centers for Disease Control and Prevention. "Understanding Depression— Yours and Theirs." NASD National Ag Safety Database website, edited by Brenda J. Thames and Deborah J. Thomason, Clemson University Cooperative Extension Service. Available online at http://www.cdc.gov/NASD/docs/d001201-d001300/d001247/d001247.html.

Clayton, P. J., and H. S. Darvish. "Course of Depressive Symptoms Following the Stress of Bereavement." In *Stress and Mental Disorder*, edited by J. E. Barrett, R. M. Rose, and G. Klerman. New York: Raven Press, 1979, pp. 121–36.

Clayton, P. J., J. A. Halikas, and W. L. Maurice. "The Bereavement of the Widowed." *Diseases of the Nervous System* 32 (1971): 597–604.

Cobb, S. "Review of Neuropsychiatry for 1938." *Archives of Internal Medicine* 62 (1938): 883–99.

"Conference on Therapy: How to Evaluate a New Drug." *American Journal of Medicine* 17, no. 4 (1954): 722–27.

Conjunctions: The Web Forum of Innovative Writing. Available online at www.conjunctions.com.

Consumers Union. *Consumer Reports*. Available online at www.consumerreports.org.

———. "Mental Health: Does Therapy Help?" *Consumer Reports* 60, no. 11 (November 1995): 734–39.

Correll, G. E., and G. E. Futter. "Two Case Studies of Patients with Major Depressive Disorder Given Low-Dose (Subanaesthetic) Ketamine Infusions." *Pain Medicine* 7, no. 1 (2006): 92–95.

Crooked Lake Review. Available online at www.crookedlakereview.com.

Curtis, D. R., and R. M. Eccles. "The Excitation of Renshaw Cells by Pharmacological Agents Applied Electrophoretically." *Journal of Physiology* 141 (1958): 434–45.

Dahlstrom, A., and A. Carlsson. "Making Visible the Invisible: Recollections of the First Experiences with the Histochemical Fluorescence Method for Visualization of Tissue Monoamines." In *Discoveries in Pharmacology, vol. 3: Pharmacological Methods, Receptors, and Chemotherapy*, edited by M. J. Parnham and J. Bruinvels. Amsterdam: Elsevier, 1983, pp. 97–125.

Davies, D. L., and M. Shepherd. "Reserpine in the Treatment of Anxious Patients." *Lancet* 2 (1955): 117–20.

Davies, J. D. *Phrenology: Fad and Science*. Hamden, CT: Archon Books, 1971 (originally published 1955).

Daw, J. "Is PMDD Real?" *APA Monitor* 33, no. 9 (October 2002). Available online at http://www.apa.org/monitor/oct02/pmdd.html.

Depression Guideline Panel, Agency for Health Care Policy and Research, US Department of Health and Human Services. *Treatment of Major Depression: Clinical Practice Guideline Number 5*. Rockville, MD: Public Health Service, 1993.

Depression Is Real Coalition. "Link to Us." DepressionIsReal.org website. Available at http://www.depressionisreal.org/depression-link.html.

———. "'Right and Wrong' Radio Advertisement Featuring Dr. Greengard." DepressionIsReal.org website, The Depression Is Real Campaign. Available online at http://www.depressionisreal.org/depression-dr-green gard.html.

———. "What Is." DepressionIsReal.org website. Available at http://depression isreal.org/depression-about-coalition.html.

DeRubeis, R. J., and A. T. Beck. "Cognitive Therapy." In *Handbook of Cognitive-Behavioral Therapies*, edited by K. S. Dobson. New York: Guilford Press, 1988.

Dobson, K. S. "A Meta-analysis of the Efficacy of Cognitive Therapy for Depression." *Journal of Consulting and Clinical Psychology* 57 (1989): 414–19.

Dreyfus, G. L. *Die Melancholie: Ein Zustandsbild des Manisch-Depressiven Irreseins.* Jena: Gustave Fischer, 1907.

Dufresne, T. *Against Freud: Critics Talk Back.* Stanford, CA: Stanford University Press, 2007.

El-Haj, J. *The Lobotomist: A Maverick Medical Genius and His Tragic Quest to Rid the World of Mental Illness.* Hoboken, NJ: John Wiley & Sons, 2005.

Eli Lilly and Company. "Prozac Makes History." Prozac, Fluoxetine Hydrochloride website, Lilly USA LLC, 2009. Available online at http://www.prozac.com/disease_information/treatment_depression. jsp?reqNavId=1.1.4.

Elkin, S. *The MacGuffin.* Champaign, IL: Dalkey Archive Press, 1999.

Elliott, S. "A New Campaign by Leo Burnett Will Try to Promote Prozac Directly to Consumers." *New York Times*, July 1, 1997.

Erowid. Website. Available online at www.erowid.org.

Fava, M. "Prospective Studies of Adverse Events Related to Antidepressant Discontinuation." *Journal of Clinical Psychiatry*, 67, suppl. 4 (2007): 14–21.

Feighner, J. P., R. Munoz, G. Winokur, R. A. Woodruff, Jr., S. B. Guze, and E. Robins. "Diagnostic Criteria for Use in Psychiatric Research." *Archives of General Psychiatry* 26 (1972): 57–63.

Finger, S. *Minds behind the Brain: A History of the Pioneers and Their Discoveries.* Oxford: Oxford University Press, 2000.

Fink, M. "The Medical Evidence-Based Model for Psychiatric Syndromes: Return to a Classical Paradigm." *Acta Psychiatrica Scandinavica* 116 (2008): 81–84.

———. "Meduna and the Origins of Convulsive Therapy." *American Journal of Psychiatry* 141 (September 1984): 1034–41.

Fitzgerald, P. J. "A Neurotransmitter System Theory of Sexual Orientation." *Journal of Sexual Behavior* 5, no. 3 (2008): 746–48.

Flynn, J. "Prozac Is Making Lilly a Little Edgy." *BusinessWeek,* June 22, 1992.

Foa, E. B., B. O. Rothbaum, D. S. Riggs, and T. B. Murdock. "Treatment of Post-traumatic Stress Disorder in Rape Victims: A Comparison Between Cognitive-Behavioral Procedures and Counseling." *Journal of Consulting and Clinical Psychology* 59 (1991): 715–23.

Foucault, M. *Madness and Civilization.* New York: Random House, 1965.

Fowler, O. S., and L. N. Fowler. *The Self Instructor in Phrenology and Physiology: With Over One Hundred New Illustrations, Including a Chart for the Use of Practical Phrenologists.* New York: Fowler & Wells, 1859.

Frederick the Great. "Eulogy on La Mettrie." In J. O. de La Mettrie, *Man a Machine (Homme-machine),* edited by Gertrude Carman Bussey, 3–9. La Salle, IL: Open Court, 1912.

Freeman, Lucy. "Advantages Noted in Shock Therapy." *New York Times,* December 18, 1949.

Freud, Sigmund. *Beyond the Pleasure Principle.* New York: W. W. Norton, 1961.

———. *Civilization and Its Discontents.* London: Penguin Books, 2002.

———. "Mourning and Melancholia." In *On Murder, Mourning, and Melancholia,* 201–19. London: Penguin Books, 2005.

———. *The Question of Lay Analysis: Conversations with an Impartial Person.* New York: W. W. Norton, 1989.

———. "Transience." In *On Murder, Mourning, and Melancholia,* 195–200. London: Penguin Books, 2005.

Frostig, J. P. "Clinical Observations in Insulin Treatment of Schizophrenia: Preliminary Report." *American Journal of Psychiatry* 96 (1940): 1167–90.

Gaddum, J. H. "Antagonism between Lysergic Acid Diethylamide and 5-hydroxytryptamine." *Journal of Physiology* 121, suppl., no. 1 (1953): 15P.

———. "Push-Pull Cannulae," *Journal of Physiology* 155, suppl. no. 1 (1961): 1P–2P.

Gaddum, J. H., and K. A. Hameed. "Drugs Which Antagonize 5-hydroxytryptamine." *British Journal of Pharmacology* 9 (1954): 240–48.

Gaddum, J. H., and M. Vogt. "Some Central Actions of 5-hydroxytryptamine and Various Antagonists." *British Journal of Pharmacology* 11 (1956): 175–79.

Garfield, S. *Mauve: How One Man Invented a Color That Changed the World.* New York: W. W. Norton, 2000.

Gay, P. *Freud: A Life for Our Time.* New York: W. W. Norton, 1988.

Geddes, J. R., N. Freemantle, J. Mason, M. P. Eccles, and J. Boynton. SSRIs versus other antidepressants for depressive disorder. *Cochrane Database of Systematic Reviews,* 2 (2000): CD001851. Available online at http://www.cochrane.org/reviews/en/info_427099072210033858.html.

Gergen, K. J. "The Social Constructionist Movement in Modern Psychology." *American Psychologist* 40 (1985): 266–75.

Gilbody, S., P. Wilson, and I. Watts. "Direct-to-Consumer Advertising of Psychotropics: An Emerging and Evolving Form of Pharmaceutical Company Influence." *British Journal of Psychiatry* 185 (2004): 1–2.

Gilman, C. P. "The Yellow Wallpaper." In *The Yellow Wallpaper and Other Writings*, 1–18. New York: Bantam, 2006 (originally published 1892).

GlaxoSmithKline. "New Survey Reveals Common yet Underrecognized Disorder—Restless Legs Syndrome—Is Keeping America Awake at Night." Press release, June 10, 2003. Available online at www.gsk.com/press_archive/press2003/press_06102003.htm.

———. "Prescribing Information: Paxil (Paroxetine Hydrochloride)." Available online at http://us.gsk.com/products/assets/us_paxil.pdf.

Glenmullen, J. *Prozac Backlash: Overcoming the Dangers of Prozac, Zoloft, Paxil, and other Antidepressants with Safe, Effective Alternatives.* New York: Simon & Schuster, 2000.

Goenka, A. H. "Rustom Jal Vakil and the Saga of *Rauwolfia serpentina*." *Journal of Medical Biography* 15 (2007): 195–200.

Goffman, E. *Stigma: Notes on the Management of Spoiled Identity.* New York: Touchstone, 1986.

Gold, H., N. T. Kwit, and H. Otto. "The Xanthines (Theobromine and Aminophylline) in the Treatment of Cardiac Pain." *Journal of the American Medical Association* 108 (1937): 2173–79.

Goldberg, D. "A Dimensional Model for Common Mental Disorders." *British Journal of Psychiatry* 1996, suppl. 30: 44–49.

Green, A. R. "Gaddum and LSD: the Birth and Growth of Experimental and Clinical Neuropharmacology Research on 5-HT in the UK." *British Journal of Pharmacology* 154, no. 8 (2008): 1–17.

Greenberg, G. "Is It Prozac? Or Placebo?" *Mother Jones*, Nov.–Dec. 2003, pp. 76–80.

Grob, G. "Origins of DSM-I: A Study in Appearance and Reality." *American Journal of Psychiatry* 128 (1991): 421–31.

Grof, S. "Stanislav Grof Interviews Dr. Albert Hofmann." *MAPS* (Multidisciplinary Association for Psychedelic Studies) 11 (2002): 22–35. Available online at http://www.maps.org/news-letters/v11n2/11222gro.html.

Hamilton, M. "A Rating Scale for Depression." *Journal of Neurology, Neurosurgery, and Psychiatry* 23 (1960): 56–62.

Harrison, E. "TB Drug Is Tried in Mental Cases." *New York Times*, April 7, 1957.

Hazel, J. *Who's Who in the Greek World.* New York: Routledge, 2001.

Healy, D. *The Antidepressant Era.* Cambridge, MA: Harvard University Press, 1997.

———. *The Creation of Psychopharmacology.* Cambridge, MA: Harvard University Press, 2002.

———. *Mania: A Short History of Bipolar Disorder.* Baltimore, MD: Johns Hopkins University Press, 2008.

Hippocrates. *Hippocratic Writings,* translated by Francis Adams. Chicago: Encyclopaedia Britannica, 1952.

———. *On the Nature of Man.* Vol. 4 in *Hippocrates,* translated by W. H. S. Jones. Cambridge, MA: Loeb Classical Library, 1931.

Hirschfeld, R. M., M. Keller, S. Panico, B. Arons, D. Barlow, F. Davidoff, J. Endicott, et al. "The National Depressive and Manic-Depressive Association Consensus Statement on the Undertreatment of Depression." *JAMA* 277, no. 4 (1997): 333–40.

Hoch, A., and J. T. MacCurdy. "The Prognosis of Involution Melancholia." *Archives of Neurology and Psychiatry* 7 (January 1922): 1–18.

Hofmann, Albert. *LSD: My Problem Child.* Sarasota, FL: Multidisciplinary Association for Psychedelic Studies (MAPS), 2005 (originally published 1979).

Hollister, L. E. "Mental Disorders: Antianxiety and Antidepressant Drugs." *New England Journal of Medicine* 286 (1972): 1195–98.

Hollon, S. D., M. E. Thase, and J. C. Markowitz. "Treatment and Prevention of Depression." *Psychological Science in the Public Interest* 3, no. 2 (2002): 39–77.

Holmes, J. "All You Need Is Cognitive Behaviour Therapy?" *British Medical Journal* 324 (2002): 290.

Holmes, O. W. "Currents and Countercurrents in Medical Science." In *Medical Essays 1842–1882,* 93–110. N.p.: Plain Label Books, n.d.

Horwitz, A. V., and J. C. Wakefield. *The Loss of Sadness.* Oxford: Oxford University Press, 2007.

Hume, S. "Prozac Getting a New Prescription: Burnett Tries Varied Approaches in Bringing Antidepressants to TV." *Adweek,* Sept. 21, 1998.

Hunt, Charles. "Interaction of Detailing and Journal Advertising: How Detailing and Journal Advertising Impact New Prescriptions." Rx PromoROI website, 2005. Available online at www.rxpromoroi.org/pdf/interaction_whitepaper.pdf.

"If the individual is depressed . . ." Advertisement. *American Journal of Psychiatry* 101 (1945): xiii.

Impastatao, D. "The Story of the First Electroshock Treatment." *American Journal of Psychiatry* 116 (1960): 1113–14.

Insel, T. R., and R. Quirion. "Psychiatry as a Clinical Neuroscience Discipline." *Journal of the American Medical Association* 294, no. 17 (2005): 2221–24.

Jackson, S. *Melancholia & Depression: From Hippocratic Times to Modern Times.* New Haven: Yale University Press, 1990.

Jacobson, N. S., K. S. Dobson, P. A. Truax, M. E. Addis, K. Koerner, J. K. Gollan, E. Gortner, and S. E. Prince. "A Component Analysis of Cognitive-Behavior Treatment for Depression." *Journal of Consulting and Clinical Psychology* 64 (1996): 295–304.

James Lind Library website, The Royal College of Physicians of Edinburgh. Available online at www.jameslindlibrary.org.

James, William. *The Varieties of Religious Experience.* Cambridge, MA: Harvard University Press, 1985.

Jellinek, E. *The Disease Concept of Alcoholism.* Piscataway, NJ: Alcohol Research Documentation, 1960.

Jessner, L., and V. G. Ryan. *Shock Treatment in Psychiatry.* New York: Grune & Stratton, 1941.

Jouanna, J. *Hippocrates.* Baltimore, MD: Johns Hopkins University Press, 1999.

Kalia, M., J. P. O'Callaghan, D. B. Miller, and M. Kramer, "Comparative Study of Fluoxetine, Sibutramine, Sertraline and Dexfenfluramine on the Morphology of Serotonergic Nerve Terminals Using Serotonin Immunohistochemistry." *Brain Research* 859, no. 1 (2000), 92–105.

Kalinowsky, L., and P. Hoch. *Shock Treatments and Other Somatic Procedures in Psychiatry.* New York: Grune & Stratton, 1946.

Kandel, E. R. "The New Science of Mind." In *Best of the Brain from Scientific American,* edited by F. E. Bloom, 66–75. New York: Dana Press, 2007.

Kaptchuk, T. "Intentional Ignorance: A History of Blind Assessment and Placebo Controls in Medicine." *Bulletin of the History of Medicine* 72, no. 3 (1998): 389–433.

Katz, M. M., J. O. Cole, and H. A. Lowery. "Studies of the Diagnostic Process: The Influence of Symptom Perception, Past Experience and Ethnic Background on Diagnostic Decisions." *American Journal of Psychiatry* 125, no. 7 (1969): 937–47.

Kefauver, E. *In the Hands of a Few: Monopoly Power in America.* New York: Pantheon, 1965.

Kennedy, A. "Critical Review: The Treatment of Mental Disorders by Induced Convulsions." *Journal of Neurology and Psychiatry* 3 (1950): 49–82.

Kessler, R. C., S. Z. Hao, D. G. Blazer, and M. Swartz. "Prevalence, Correlates, and Course of Minor Depression and Major Depression in the National Comorbidity Survey." *Journal of Affective Disorders* 45 (1997): 19–30.

Kessler, R. C., K. R. Merikangas, P. Berglund, W. W. Eaton, D. S. Koretz, and E. E. Walters. "Mild Disorders Should Not Be Eliminated from the DSM-V." *Archives of General Psychiatry* 60 (2003): 1117–22.

Kessler, R. C., P. Berglund, O. Demler, R. Jin, D. Koretz, K. R. Merikangas, A. J. Rush, E. E. Walters, and P. S. Wang. "The Epidemiology of Major Depressive Disorder: Results from the National Comorbidity Survey Replication (NCS-R)." *JAMA* 289, no. 23 (2003): 3095–3105.

"Ketamine." *Erowid* ("earth wisdom") website, Erowid.org. Available online at www.erowid.org/ketamine/.

Kirk, S. A., and H. Kutchins. *The Selling of DSM: The Rhetoric of Science in Psychiatry.* New Brunswick, NJ: Aldine Transactions, 2008 (originally published 1992).

Kirsch, Irving, Thomas J. Moore, Alan Scoboria, and Sarah S. Nicholls. "The Emperor's New Drugs: An Analysis of Antidepressant Medication Data Submitted to the U.S. Food and Drug Administration." *Prevention & Treatment*, July 2002. Available online at http://journals.apa.org/prevention/volume5/pre0050023a.html.

Kirsch, I., B. J. Deacon, T. B. Huedo-Medina, A. Scoboria, T. J. Moore, and B. T. Johnson. "Initial Severity and Antidepressant Benefits: A Meta-analysis of Data Submitted to the Food and Drug Administration." *PloS Medicine* 5, no. 2 (2008): 260–68.

Klein, D. F. "Preventing Hung Juries about Therapy Studies." *Journal of Clinical and Consulting Psychology* 64 (1996): 81–87.

Klerman, G. L. "The Patient's Right to Effective Treatment: Implications of *Osheroff v. Chestnut Lodge.*" *American Journal of Psychiatry* 147, no. 4 (1990): 410–11.

———. "Psychotropic Hedonism vs. Pharmacological Calvinism." *The Hastings Center Report* 2, no. 4 (1972): 1–3.

Klerman, G. L., G. E. Vaillant, R. L. Spitzer, and R. Michels. "A Debate on DSM-III." *American Journal of Psychiatry* 141 (1984): 539–55.

Kline, N. *From Sad to Glad: Kline on Depression.* New York: Ballantine, 1974.

———. "Monoamine Oxidase Inhibitors." In *Discoveries in Biological Psychology*, edited by F. J. Ayd and B. Blackwell, 194–204. Philadelphia: J. B. Lippincott, 1970.

———. "Use of *Rauwolfia serpentina* Benth. in Neuropsychiatric Conditions." *Annals of the New York Academy of Sciences* 59 (1954): 107–32.

Knightley, P., H. Evans, E. Potter, et al. *Suffer the Children: The Story of Thalidomide.* New York: Viking, 1979.

Knoff, W. F. "A History of the Concept of Neurosis, with a Memoir of William Cullen." *American Journal of Psychiatry* 127 (1970): 80–84.

Kovel, J. *The Age of Desire: Case Histories of a Radical Psychoanalyst.* Boston: Beacon Books, 1981.

Kraepelin, E. *Lectures on Clinical Psychiatry.* New York: Hafner, 1968.

———. "Manifestations of Insanity." *History of Psychiatry* 3, no. 12 (1992): 509–29.

Kraepelin, E., and A. R. Diefendorf. *Clinical Psychiatry: A Text-Book for Students and Physicians.* London: Macmillan, 1912.

Kramer, P. D. *Against Depression.* New York: Viking, 2005.

————. *Listening to Prozac: A Psychiatrist Explores Antidepressant Drugs and the Remaking of the Self.* New York: Penguin, 1994.

————. "The Valorization of Sadness: Alienation and the Melancholic Temperament." In *Prozac as a Way of Life*, edited by C. Elliott and T. Chambers, 52. Chapel Hill, NC: University of North Carolina Press, 2004.

Kramer, S. *History Begins at Sumer: Thirty-Nine Firsts in Recorded History.* Philadelphia: University of Pennsylvania, 1988.

Kravitz, R. L., R. M. Epstein, M. D. Feldman, C. E. Franz, R. Azari, M. S. Wilkes, L. Hinton, and P. Franks. "Influence of Patients' Requests for Direct-to-Consumer Advertised Antidepressants: A Randomized Controlled Trial." *JAMA* 293, no. 16 (2005): 1995–2002.

Krishnan, V., and E. J. Nestler. "The Molecular Neurobiology of Depression." *Nature* 455 (2008): 894–902.

Kuhn, R. "The Imipramine Story." In *Discoveries in Biological Psychiatry*, edited by F. J. Ayd and B. Blackwell, 205–17. Philadelphia: J. B. Lippincott, 1970.

————. "The Treatment of Depressive States with G22355 (Imipramine Hydrochloride)." *American Journal of Psychiatry* 115 (1958): 459–64.

Kupfer, D. J., M. B. First, and D. A. Regier. *A Research Agenda for DSM-V.* Washington, DC: American Psychiatric Association, 2002.

Lacasse, J. R., and J. Leo. "Serotonin and Depression: A Disconnect between the Advertisements and the Scientific Literature." *PloS Medicine* 2, no. 12 (2005): e392.

Laing, R. D. *The Divided Self: An Existential Study in Sanity and Madness.* Harmondsworth: Penguin, 1960.

La Mettrie, J. O. de. "Machine Man." In *La Mettrie: Machine Man and Other Writings*, edited by A. Thomson. Cambridge: Cambridge University Press, 1996.

————. "Man a Machine (*Homme-machine*)." In *Man a Machine (Homme-machine)*, edited by Gertrude Carman Bussey. La Salle, IL: Open Court, 1912.

————. "Treatise on the Soul." In *La Mettrie: Machine Man and Other Writings*, edited by A. Thomson. Cambridge: Cambridge University Press, 1996.

Langreth, R. "Mending the Mind. High Anxiety: Rivals Threaten Prozac's Reign." *Wall Street Journal*, May 9, 1996, p. B1.

Larkin, Philip. *High Windows.* New York: Farrar, Straus and Giroux, 1984.

Lasagna, L. "Congress, the FDA, and New Drug Development: before and after 1962." *Perspectives in Biology and Medicine* 32, no. 3 (1989): 322–43.

————. "The Controlled Clinical Trial: Theory and Practice." *Journal of Chronic Diseases* 1, no. 4 (1955): 353–67.

Laurence, W. L. "Drug Called a Psychic Energizer Found Useful in Treating Mental Illnesses." *New York Times*, Science in review, December 22, 1957.

————. "Psychiatrist Hits Misuse of Shocks." *New York Times*, August 21, 1953.

————. "Wide New Fields Seen for TB Drug Including Aid to Narcotics Addicts." *New York Times*, July 5, 1952, pp. 1, 3.

Lears, T. J. J. "From Salvation to Self-Realization: Advertising and the Therapeutic Roots of the Consumer Culture." In *The Culture of Consumption: Critical Essays in American History, 1880–1980*, edited by R. Fox and T. J. J. Lears, 1–30. New York: Pantheon, 1983.

Leber, P. "Approvable Action on Forrest Laboratories, Inc., NDA 20-822 Celexa (Citalopram HBr) for the Management of Depression." Memorandum, Department of Health and Human Services, Public Health Service, Food and Drug Administration, Center for Drug Evaluation and Research, May 4, 1998.

LeVay, S. *Queer Science: The Use and Abuse of Research into Homosexuality.* Cambridge, MA: MIT Press, 1996.

Levine, H. G., . "The Discovery of Addiction: Changing Conceptions of Habitual Drunkenness in America." *Journal of Studies on Alcohol* 39, no. 1 (1978): 144–51.

Lidz, T. "Adolf Meyer and the Development of American Psychiatry." *American Journal of Psychiatry* 123, no. 3 (Sept. 1966): 320–32.

Liebnau, J. *Medical Science and Medical Industry: The Formation of the American Pharmaceutical Industry.* Baltimore: Johns Hopkins University Press, 1987.

Lin, E. H. B., M. Von Korff, W. Katon, T. Bush, G. E. Simon, E. Walker, and P. Robinson. "The Role of the Primary Care Physician in Patients' Adherence to Antidepressant Therapy." *Medical Care* 33, no. 1 (1995): 67–74.

Lind, James. *A Treatise of the Scurvy.* Edinburgh: A. Kincaid & A. Donaldson, 1753.

————. "A Treatise of the Scurvy." James Lind Library website. The Royal College of Physicians of Edinburgh. Available online at http://www.james lindlibrary.org/trial_records/17th_18th_Century/lind/lind_kp.html.

Loewi, O. *From the Workshop of Discoveries.* Lawrence, KS: University of Kansas Press, 1953.

Loomer, H. P., J. C. Saunders, and N. S. Kline. "Iproniazid: an Amine-Oxidase Inhibitor, as an Example of a Psychic Energizer." *Congressional Record*, 1957, pp. 1382–90.

Luborsky, Lester, Lisa Luborsky, and B. Singer. "Comparative Studies of Psychotherapies: Is It True That 'Everyone Has Won and All Must Have Prizes'?" *Archives of General Psychiatry* 32 (1975): 995–1008.

Luborsky, L., L. Diguer, D. A. Seligman, R. Rosenthal, E. D. Krause, S. Johnson, G. Halperin, M. Bishop, J. S. Berman, and E. Schweizer. "The Researcher's Own Therapy Allegiances: A 'Wild Card' in Comparisons of Treatment Efficacy." *Clinical Psychology: Science and Practice* 6, no. 1 (1999): 95–106.

Lutz, T. *American Nervousness, 1903: An Anecdotal History.* Ithaca, NY: Cornell University Press, 1991.

Mackey, N. "Phrenological Whitman." *Conjunctions,* Fall 1997. Available online at www.conjunctions.com/archives/c29-nm.htm.

Malber, J., and K. R. Bronson. "How MDMA Works in the Brain." In *Ecstasy: The Complete Guide,* edited by J. Holland, 29–38. Rochester, VT: Park Street Press, 2001.

Mann, J. *Elusive Magic Bullet: The Search for the Perfect Drug.* Oxford: Oxford University Press, 1999.

Marks, H. M. *The Progress of Experiment: Science and Therapeutic Reform in the United States, 1900–1990.* Cambridge: Cambridge University Press, 2000.

Marquardt, M. *Paul Ehrlich.* New York: Schuman, 1951.

Martin, E. "Pharmaceutical Virtue." *Culture, Medicine and Psychiatry* 30 (2006): 157–74.

Martin, J. H. "Saints, Sinners, and Reformers: The Burned-Over District Revisited." *Crooked Lake Review,* 2005. Available online at www.crooked lakereview.com/books/saints_sinners/martin12.html.

Marwit, S. J. "Reliability of Diagnosing Complicated Grief: A Preliminary Investigation." *Journal of Clinical and Consulting Psychiatry* 64, no. 3 (1996): 563–68.

Massachusetts General Hospital. "Safety and Effectiveness of Omega-3 Fatty Acids, EPA Versus DHA, for the Treatment of Major Depression." ClinicalTrials.gov website, U.S. National Institutes of Health. 2006–2009. Available online at http://www.clinicaltrials.gov/ct2/show/NCT00361 374?term=major+depression+Omega&rank=1.

Menninger, K. *The Vital Balance.* New York: Viking, 1963.

Menninger, W. "Psychiatric Experience in the War, 1941–1946." *American Journal of Psychiatry* 103, no. 5 (March 1947): 577–86.

Meyer, A. *The Commonsense Psychiatry of Dr. Adolf Meyer,* edited by Alfred Lief. New York: McGraw Hill, 1948.

——. "The 'Complaint' as the Center of Genetic-Dynamic and Nosological Teaching in Psychiatry." *New England Journal of Medicine* 199, no. 8 (1928): 360–70.

——. "A Few Demonstrations of the Pathology of the Brain." *American Journal of Insanity* 52 (1895): 238–45.

——. "Presidential Address: Twenty-Five Years of Psychiatry in the United States and Our Present Outlook." *American Journal of Psychiatry* 8 (July 1928): 1–31.

Mintz, M. "'Heroine' of FDA Keeps Bad Drug off Market." *Washington Post,* July 15, 1962: p. A1.

Monfils, Paul. "Murex Shells Used to Produce Dye." Man and Mollusc website, edited by Avril Bourquin. Available online at http://www.manand mollusc.net/Shell_photos/dye-murex.html.

"Morgan's Old Gold." *Time*, May 28, 1928. Available online at http://www .time.com/time/magazine/article/0,9171,786005,00.html.

Multidisciplinary Association for Psychedelic Studies website. Available online at www.maps.org.

Narrow, W. E., D. S. Rae, L. N. Robins, and D. A. Regier. "Revised Prevalence Estimates of Mental Disorders in the United States: Using a Clinical Significance Criterion to Reconcile 2 Surveys' Estimates." *Archives of General Psychiatry* 59, no. 2 (2002): 115–30.

National Institute of Mental Health. "Drug Therapy to Treat Minor Depression." Clinicaltrials.gov website, U.S. National Institutes of Health, 2002– 2009. Available online at http://www.clinicaltrials.gov/ct2/show/NCT0 0048815?term=minor+depression&rank=1.

National Institutes of Health. www.ClinicalTrials.gov.

Neslin, Scott. "Executive Summary: ROI Analysis of Pharmaceutical Promotion." RxPromoROI website. Available online at www.rxpromoroi.org/ rapp/exec_sum.html.

———. "RAPP Study." RxPromorROI website, 2002. Available online at http://www.rxpromoroi.org/rapp/index.html.

New York Times. "American Loses Suit against Freud." May 25, 1927.

———. "City Agents Hunt Supplies of Drug." April 12, 1958, p. 1.

———. "Death of Woman Laid to 'Pep Pills.'" April 11, 1958.

———. "Dr. Sakel Is Dead; Psychiatrist, 57." Dec. 3, 1957.

———. "Physician Warns on Wonder Drugs." April 13, 1960.

———. "Warns of Danger in American Life." June 5, 1927.

Nuland, S. *Doctors: The Biography of Medicine*. New York: Knopf, 1988.

Nurnberg, H. G., P. L. Hensley, J. R. Heiman, H. A. Croft, C. DeBattista, and S. Paine. "Sildenafil Treatment of Women with Antidepressant-Associated Sexual Dysfunction: A Randomized Controlled Trial." *JAMA* 300, no. 4 (2008): 395–404.

Office of National Drug Control Policy. "National Drug Control Strategy, FY 2009 Budget Summary." Executive Summary. Available online at http:// www.whitehousedrugpolicy.gov/publications/policy/09budget/exec _summ.pdf.

Olfsun, M., and S. Marcus. "National Patterns in Antidepressant Medication Treatment." *Archives of General Psychiatry* 66, No. 8 (2009): 848–56.

Parens, E. "Kramer's Anxiety." In *Prozac as a Way of Life*, edited by C. Elliott and T. Chambers, 21–32. Chapel Hill, NC: University of North Carolina Press, 2004.

Parker, G., and K. Fletcher. "Treating Depression with the Evidence-Based Psychotherapies: A Critique of the Evidence." *Acta Psychiatrica Scandinavica* 115 (2007): 352–59.

Peale, Norman Vincent. *The Power of Positive Thinking.* New York: Fawcett Columbine, 1966.

Pearce, J. M. S. "Leopold Auenbrugger: Camphor-Induced Epilepsy—Remedy for Manic Psychosis." *European Neurology* 59 (2008): 105–7.

Pfizer. "Myths and Facts about Depression." Zoloft (Sertraline HCI) website, 2008. Available online at http://www.zoloft.com/depr_myths_facts.aspx.

"Pharmacy Facts and Figures." Drug Topics website. Available online at http://drugtopics.modernmedicine.com/drugtopics/article/articleList.jsp?categoryId=7604.

Pliny. [Gaius Plinius Secundus], *Natural History: A Selection.* London: Penguin Classics, 1991.

Popper, K. *The Poverty of Historicism.* New York: Harper & Row, 1957.

Popper, K. R., and J. C. Eccles. *The Mind and Its Brain: An Argument for Interactionism.* London: Routledge & Kegan Paul, 1977.

Preston, S. D., and F. B. de Waal. "Empathy: Its Ultimate and Proximate Bases." *The Behavioral and Brain Sciences* 25 (2002): 1–20.

"Prontosil." *Time,* December 28, 1936.

Quetel, C. *History of Syphilis,* translated by J. Braddock and B. Pike. Baltimore, MD: Johns Hopkins University Press, 1992.

Reader's Digest. "Depression Hurts: Prozac Can Help." Advertisement. May 1998, pp. 182–84.

Regier, D. A. "State-of-the-Art Psychiatric Diagnosis." *World Psychiatry* 3, no. 1 (2004): 25–26.

"Relieves Anxiety and Anxious Depression." Miltown advertisement. *Archives of General Psychiatry* 8, no. 2 (1963), following p. 135.

Richardson, R. D. *William James: In the Maelstrom of American Modernism.* Boston: Houghton Mifflin, 2006.

Richeson, J. S., A. A. Baird, H. L. Gordon, T. F. Heatherton, C. L. Wyland, S. Trawalter, and J. N. Shelton. "An fMRI Investigation of the Impact of Interracial Contact on Executive Function." *Nature Neuroscience* 6, no. 12 (2003): 1323–28.

Rinkel, M., H. J. DeShon, and H. C. Solomon. "Experimental Schizophrenia-like Symptoms." *American Journal of Psychiatry* 108 (1952): 572–78.

Rinkel, M., R. W. Hyde, H. C. Solomon, and H. Hoagland. "Experimental Psychiatry II: Clinical and Physio-chemical Observations in Experimental Psychoses." *American Journal of Psychiatry* 111, no. 12 (1955): 881–95.

Robinson, L. A., J. S. Berman, and R. A. Neimeyer. "Psychotherapy for the Treatment of Depression: A Comprehensive Review of Controlled Outcome Research." *Psychological Bulletin* 108 (1990): 30–49.

Roccatagliata, G. *A History of Ancient Psychiatry.* Santa Barbara, CA: Greenwood Press, 1986.

Rosenhan, D. "On Being Sane in Insane Places." *Science* 179, no. 70 (Jan. 1973): 250–58.

Rosenzweig, S. "Some Implicit Common Factors in Diverse Methods of Psychotherapy." *American Journal of Orthopsychiatry* 6 (1936): 412–15.

Rush, A. J., A. T. Beck, M. Koracs, and S. Hollon. "Comparative Efficacy of Cognitive Therapy and Pharmacotherapy in the Treatment of Depressed Outpatients." *Cognitive Therapy and Research* 1, no. 1 (1977): 17–37.

Rusk, H. A. "Drugs and Depression." *New York Times*, September 6, 1959.

RxPromoROI: A Resource for Pharmaceutical Promotion ROI Results website. Available at www.rxpromoroi.org.

Sakel, M. "The Methodical Use of Hypoglycemia in the Treatment of Psychoses." *American Journal of Psychiatry* 94 (1937): 111–29.

———. *Schizophrenia.* New York: Philosophical Library, 1958.

Sandifer, M. G., A. Hordern, G. C. Timbury, and L. M. Green. "Psychiatric Diagnosis: A Comparative Study in North Carolina, London, and Glasgow." *British Journal of Psychiatry* 114 (1968): 1–9.

Schildkraut, J. "The Catecholamine Hypothesis of Affective Disorders." *American Journal of Psychiatry* 122 (1965): 509–22.

Schmidt, J. *Melancholy and the Care of the Soul: Religion, Moral Philosophy and Madness in Early Modern England.* London: Ashgate Publishing, 2007.

Schmiege, G. R. "The Current State of LSD as a Therapeutic Tool: A Summary of the Clinical Literature." *Journal of the Medical Society of New Jersey* 60 (1963): 203–7.

Sedgwick, P. "Illness—Mental and Otherwise." *Hastings Center Reports* 1, no. 3 (1973): 19–40.

Seligman, M. "The Effectiveness of Psychotherapy: The *Consumer Reports* Study." *American Psychologist* 50 (1996): 965–74.

———. "Learned Helplessness in the Rat." *Journal of Comparative and Physiological Psychology* 88 (1975): 534–41.

Shapiro, A. K., and E. Shapiro. *The Powerful Placebo: From Ancient Priest to Modern Physician.* Baltimore, MD: Johns Hopkins University Press, 1997.

Shapiro, D. A. "Meta-analysis of Comparative Therapy Outcome Studies: A Replication and Refinement." *Psychological Bulletin* 92 (1982): 581–604.

Shapiro, D. A., M. Barkham, A. Rees, G. E. Hardy, S. Reynolds, and M. Startup. "Effects of Treatment Duration and Severity of Depression on the Effectiveness of Cognitive-Behavioral Therapy and Psychodynamic-Interpersonal Psychotherapy." *Journal of Consulting and Clinical Psychology* 62, no. 3 (1994): 522–34.

Sherman, I. W. *Twelve Diseases That Changed Our World.* Washington, DC: AMS Press, 2007.

Shorter, E. *Before Prozac: The Troubled History of Mood Disorders in Psychiatry.* Oxford: Oxford University Press, 2008.

———. "A History of Psychiatry: From the Era of the Asylum to the Age of Prozac." New York: John Wiley & Sons, 1997.

Shorter, E., and D. Healy. *Shock Therapy: A History of Electroconvulsive Treatment in Mental Illness.* New Brunswick, NJ: Rutgers University Press, 2007.

Simpson, D. "Phrenology and the Neurosciences: Contributions of FJ Gall and JG Spurzheim." *ANZ Journal of Surgery* 75 (2005): 475–82.

Smith, G. P. "Pavlov and Integrative Physiology." *American Journal of Physiology—Regulatory, Integrative, and Comparative Physiology* 279 (2000): R743–55.

Smith, M. L., and G. V. Glass. "Meta-analysis of Psychotherapy Outcome Studies." *American Psychologist* 32 (1977): 752–60.

Smith, M. L., G. V. Glass, and T. I. Miller. *The Benefits of Psychotherapy.* Baltimore: Johns Hopkins University, 1980.

Solomon, D. A., A. C. Leon, T. I. Mueller, W. Coryell, J. J. Teres, M. A. Posternak, L. L. Judd, J. Endicott, and M. B. Keller. Tachyphylaxis in Unipolar Major Depressive Disorder. *Journal of Clinical Psychiatry* 66, no. 3 (March 2005): 283–90.

Spiegel, A. "The Dictionary of Disorder: How One Man Revolutionized Psychiatry." *The New Yorker,* January 3, 2005. Available online at http://www.newyorker.com/archive/2005/01/03/050103fa_fact.

———. "More and More, Favored Psychotherapy Lets Bygones Be Bygones." *New York Times,* Feb. 14, 2006.

Spielmans, G. I., S. A. Thielges, A. L. Dent, and R. P. Greenberg. "The Accuracy of Psychiatric Medication Advertisements in Medical Journals." *Journal of Nervous and Mental Disease* 196, no. 4 (2008): 267–73.

Spitzer, R. L. "On Pseudoscience in Science, Logic in Remission, and Psychiatric Diagnosis: A Critique of Rosenhan's 'On Being Sane in Insane Places'" and accompanying commentary. *Journal of Abnormal Psychology* 84 (1975): 433–74.

Spitzer, R. L., and P. T. Wilson. "Nosology and the Official Psychiatric Nomenclature." In *Comprehensive Textbook of Psychiatry,* edited by A. Freedman, H. Kaplan, and B. Sadock, 826–45. Baltimore: Williams & Wilkins, 1975.

Spitzer, R. L., J. Endicott, and E. Robins. "The Development of Diagnostic Criteria in Psychiatry." *Current Contents* 19 (January 1989).

Stanfel, R. "Prozac Print Campaign." In *Encyclopedia of Major Marketing Campaigns,* vol. 2, 506–7. Florence, KY: Gale Thompson, 2007.

Steinach, I. "Etiology of General Paresis." *Medical Record* 54 (1898): 877–78.

Stevens, Jay. *Storming Heaven: LSD and the American Dream.* London: Heinemann, 1987.

Stevenson, Robert Louis. "Requiem." Poetry-Archive.com website. Available online at http://www.poetry-archive.com/s/requiem.html. Accessed March 29, 2009.

Szasz, T. *The Myth of Mental Illness.* New York: Harper & Row, 1961.

Tansey, E. M., and D. A. Christie. "Drugs in Psychiatric Practice." In *Transcript of a Witness Seminar Held at the Wellcome Institute for the History of Medicine, London, on 11 March 1997.* London: Wellcome Institute, 1997.

Taylor, M. A., and M. Fink. "Restoring Melancholia in the Classification of Mood Disorders." *Journal of Affective Disorders* 105 (2008): 1–14.

Temin, P. *Taking Your Medicine: Drug Regulation in the United States.* Cambridge, MA: Harvard University Press, 1980.

Thomas, L. *The Medusa and the Snail: More Reflections of a Biology Watcher.* New York: Viking, 1970.

Thuillier, J. *Ten Years That Changed the Face of Mental Illness.* London: Martin Dunitz, 1999 (originally published 1981).

Tilburt, J. C., E. J. Emanuel, T. J. Kaptchuk, F. A. Curlin, and F. G. Miller. "Prescribing 'Placebo Treatments': Results of National Survey of US Internists and Rheumatologists." *British Medical Journal* 337 (2008). Available online at http://www.bmj.com/cgi/content/full/337/oct23_2/a1938.

Tocqueville, Alexis de. *Democracy in America,* vol. 2. New York: Modern Library, 1990.

Travis, A. S. "Perkin's Mauve: Ancestor of the Organic Chemical Industry." *Technology and Culture* 31, no. 1 (1990): 51–82.

Turner, E. H., A. M. Matthews, E. Linardatos, R. A. Tell, and R. Rosenthal. "Selective Publication of Antidepressant Trials and Its Influence on Apparent Efficacy." *New England Journal of Medicine* 358, no. 3 (2008): 252–60.

Twarog, B. M. "Responses of a Molluscan Smooth Muscle to Acetylcholine and 5-hydroxytryptamine." *Journal of Cellular and Comparative Physiology* 44, no. 1 (1954): 141–63.

Twarog, B. M., and I. H. Page. "Serotonin Content of Some Mammalian Tissues and Urine and a Method for its Determination." *American Journal of Physiology* 175, no. 1 (1953): 157–61.

Ullmann, A. "Pasteur-Koch: Distinctive Ways of Thinking about Infectious Diseases." *Microbe* 2, no. 8 (2007): 383–87.

Valenstein, E. S. *The War of the Soups and the Sparks: The Discovery of Neurotransmitters and the Dispute over How Nerves Communicate.* New York: Columbia University Press, 2005.

Van Wyhe, J. "The Authority of Human Nature: The *Schädellehre* of Franz Joseph Gall." *British Journal of the History of Science* 35 (2002).

von Meduna, L. "The Significance of the Convulsive Reaction during the Insulin and the Cardiazol Therapy of Schizophrenia." *Journal of Nervous and Mental Disease* 87, no. 2 (Feb. 1938): 133–40.

Waldinger, M. D. "Premature Ejaculation: Definition and Drug Treatment." *Drugs* 67, no. 4 (2007): 547–68.

Wallace, B. A. *The Taboo of Subjectivity: Toward a New Science of Consciousness.* Oxford: Oxford University Press, 2000.

Wallace, D. F. "The Depressed Person." In *Brief Interviews with Hideous Men,* 37–69. New York: Little, Brown, 2000.

Wampold, B. E. "Establishing Specificity in Psychotherapy Scientifically: Design and Evidence Issues." *Clinical Psychology Science and Practice* 12 (2005): 197.

———. *The Great Psychotherapy Debate: Models, Methods, and Findings.* Mahwah, NJ: Erlbaum, 2001.

———. "Methodological Problems in Identifying Efficacious Psychotherapies." *Psychotherapy Research* 7, no. 1 (1997): 21–43.

Wampold, B. E., G. W. Mondin, M. Moody, F. Stich, K. Benson, and H. Ahn. "A Meta-analysis of Outcome Studies Comparing Bona Fide Psychotherapies: Empirically, 'All Must Have Prizes.'" *Psychological Bulletin* 122, no. 3 (1997): 203–15.

Watson, G. *Theriac and Mithridatium: A Study in Therapeutics.* London: Wellcome Historical Medical Library, 1966.

Watson, J. B. *Behaviorism,* rev. ed. Chicago: University of Chicago Press, 1930.

Wellman, K. *La Mettrie: Medicine, Philosophy and Enlightenment.* Durham, NC: Duke University Press, 1992.

Westen, D., and R. Bradley. "Empirically Supported Complexity: Re-thinking Evidence-Based Practice in Psychotherapy." *Current Directions in Psychological Science* 14 (2005): 266–71.

Westen, D., and S. Morrison. "A Multidimensional Meta-analysis of Treatments for Depression, Panic, and Generalized Anxiety Disorder: An Empirical Examination of the Status of Empirically Supported Therapies." *Journal of Consulting and Clinical Psychology* 69, no. 6 (2001): 875–913.

Wheatley, D. "A Comparative Trial of Imipramine and Phenobarbital in Depressed Patients Seen in General Practice." *Journal of Nervous and Mental Disorders* 148 (1969): 542–49.

"When the patient rambles . . ." Advertisement. *Archives of General Psychiatry* 119, no. 2 (1962): 420.

Williams, Rebecca. "Effexor XR Warning Letter." U.S. Food and Drug Administration (FDA) website, Division of Drug Marketing, Advertising, and Communications, March 18, 2004. Available online at http://www.fda.gov/cder/warn/2004/Effexor.pdf.

Wilson, M. "DSM-III and the Transformation of American Psychiatry." *American Journal of Psychiatry* 150, no. 3 (1993): 399–410.

Woese, C. R. "A New Biology for a New Century." *Microbiology and Molecular Biology Reviews* 68, no. 2 (2004): 173–86.

Wong, D. T., F. P. Bymaster, and E. A. Engleman. "Prozac (Fluoxetine, Lilly 110140), the First Selective Serotonin Reuptake Inhibitor and an Antidepressant Drug: Twenty Years Since Its First Publication." *Life Sciences* 57, no. 5 (1995): 411–41.

Wong, D. T., J. S. Horng, F. P. Bymaster, K. L. Hauser, and B. B. Molloy. "A Selective Inhibitor of Serotonin Uptake: Lilly 110140, 3-(p-trifluoromethyl-phenoxy)-N-methyl-3-phenylpropylamine." *Life Sciences* 15 (1974): 471–79.

Woolf, Virginia. "Mr. Bennett and Mrs. Brown." In *Virginia Woolf Reader*, edited by M. E. Leaska, 192–211. Orlando, FL: Harcourt, 1984.

Woolley, D., and E. Shaw. "A Biochemical and Pharmacological Suggestion about Certain Mental Disorders." *Science* 119 (1954): 587–88.

World Health Organization. "Depression." World Health Organization website, United Nations. Available online at http://www.who.int/mental_health/management/depression/definition/en/.

Wortis, J. *Fragments of an Analysis with Freud.* New York: McGraw Hill, 1975 (originally published 1954).

Young, J. H. *The Medical Messiahs: A Social History of Health Quackery in Twentieth Century America.* Princeton, NJ: Princeton University Press, 1967.

Zarate, C. A., Jr., J. B. Singh, P. J. Carlson, N. E. Brutsche, R. Ameli, D. A. Luckenbaugh, D. S. Charney, and H. K. Manji. "A Randomized Trial of an n-methyl-D-aspartate Antagonist in Treatment-Resistant Major Depression." *Archives of General Psychiatry* 63 (2006): 856–64.

Zaretsky, E. *Secrets of the Soul: A Social and Cultural History of Psychoanalysis.* New York: Knopf, 2004.

Zilboorg and Henry, G. *A History of Medical Psychology.* New York: W. W. Norton, 1941.

Zinberg, N. E. *Drug, Set, and Setting: The Basis for Controlled Intoxicant Use.* New Haven: Yale University, 1984.

Zinberg, N. E., W. M. Harding, and M. Winkeller. "A Study of Social Regulatory Mechanisms in Controlled Illicit Drug Users." In *Classic Contributions to the Addictions*, edited by H. Shaffer and M. E. Burglass, 277–300. New York: Brunner/Mazel, 1981.

Zinberg, N. E., W. M. Harding, S. M. Stelmack, and R. A. Marblestone. "Patterns of Heroin Use." *Annals of the New York Academy of Sciences* 311 (1978): 10–24.

ACKNOWLEDGMENTS

At Simon & Schuster: Dedi Felman, David Rosenthal, Michele Bové, Gypsy da Silva, and, above all, Sarah Hochman.

At *Harper's Magazine,* in which I first told the story of my clinical trial: Jennifer Szalai, Roger Hodge, and Ted Ross.

For research materials: David Healy, who generously provided transcripts of his interviews with the early psychopharmacologists, and Joseph Dumit, whose recordings of television ads were indispensable.

For granting me interviews, all of which were lengthier than they bargained on: Per Bech, Alan Broadhurst, Arvid Carlsson, Donald Klein, Janis Schoenfeld, and Betty Twarog.

For research and editorial assistance: Beth Card, Lucy Green, Richard Green, Wally Izbicki, Richard Kokoska, Rosalba Onofrio, Sian Oram, and Amy Ryan.

For close reading and commentary: Bill Musgrave, Jennifer Percy-Dowd, Stuart Vyse, and Rand Cooper (who also provided German lessons).

For general aiding and abetting: Carrie Billy, Henry Bowers, Glenn Cheney, Bill Finnegan, Ruth Grahn, Karen Greenberg, Richard Greenberg, Julie Holland, Garret Keizer, Maureen Kelleher, Lindsey Lane, Bart Laws, Karyn Marcus, Errol Morris, Jim Rutman, Michael Silverstone, Jeff Singer, and John Vaillant.

For sharing me with my book: Joel Powers.

For putting up with me, and for her blue eyes: Susan Powers.

INDEX

About the Author

Gary Greenberg is a practicing psychotherapist and author of *The Noble Lie* and *The Self on the Shelf.* He has written about the intersection of science, politics, and ethics for many publications, including *Harper's, The New Yorker, Wired, Discover, Rolling Stone,* and *Mother Jones,* where he's a contributing writer. He lives in Connecticut.